THE HANDBOOK OF MANUFACTURING ENGINEERING

Second Edition

Factory Operations

Planning and Instructional Methods

THE HANDBOOK OF MANUFACTURING ENGINEERING

Second Edition

Factory Operations

Planning and Instructional Methods

EDITED BY
Richard Crowson

CRC Taylor & Francis
Taylor & Francis Group
Boca Raton London New York

A CRC title, part of the Taylor & Francis imprint, a member of the Taylor & Francis Group, the academic division of T&F Informa plc.

Published in 2006 by
CRC Press
Taylor & Francis Group
6000 Broken Sound Parkway NW, Suite 300
Boca Raton, FL 33487-2742

Printed in the United States of America on acid-free paper
10 9 8 7 6 5 4 3 2 1

International Standard Book Number-10: 0-8493-5550-8 (Hardcover)
International Standard Book Number-13: 978-0-8493-5550-9 (Hardcover)
Library of Congress Card Number 2005020352

Library of Congress Cataloging-in-Publication Data

Crowson, Richard.
Factory operations : planning and instructional methods / Richard Crowson and Jack Walker.
p. cm.
Includes bibliographical references and index.
ISBN-13: 978-0-8493-5550-9 (0849355508 : alk. paper)
1. Manufacturing processes. 2. Production management. I. Walker, Jack M., 1924- II. Title.

TS183.C76 2005
658.5--dc22 2005020352

T&F informa

Taylor & Francis Group
is the Academic Division of T&F Informa plc.

Visit the Taylor & Francis Web site at
http://www.taylorandfrancis.com

and the CRC Press Web site at
http://www.crcpress.com

Preface

Handbooks are generally considered to be concise references for specific subjects. Today's fast-paced manufacturing culture demands that such reference books provide the reader with how-to information with no frills. Some use handbooks to impart buzzwords on a particular technical subject that will allow the uninitiated to gain credibility when discussing a technical situation with more experienced practitioners.

The second edition of *The Manufacturing Engineering Handbook* was written to equip executives, manufacturing professionals, and shop personnel with enough information to function at a certain level on a variety of subjects. This level is determined by the reader.

The second edition of this handbook is divided into four main sections on issues that face the mechanical engineer as he or she attempts to learn the process of manufacturing. The progression from product and factory development, factory operations, parts fabrication, and assembly processes is a natural progression of information for one learning how a product flows through a manufacturing facility.

A manufacturing engineer is expected to be a problem solver and a person who is capable of working closely with all involved departments to resolve issues and improve designs on a daily basis. The manufacturing engineer is also challenged with the task of improving products and facilities to make the entire process more efficient.

As a manufacturing engineer uses this handbook to study history and apply principles to an existing manufacturing firm, new ideas will be spawned that will allow improvements in process flow and product flow. The successful efforts of many years' experience are captured in these chapters and can be used profitably by any reader willing to think out of the box when facing challenges on a daily basis.

Volume II of this book focuses on the role of the manufacturing engineer as a key component of the operation of the factory. Planning and instruction in the factory fall to the manufacturing engineer. This is the reason that detailed descriptions of successful methods are presented in this section.

As many manufacturing engineers develop firsthand knowledge of engineering principles, some will accept positions as design engineers or managers of design engineering.

This book and the knowledge gained as a manufacturing engineer will serve as a reminder that designing something that is not properly communicated to the fabricators and assemblers will never achieve the design goals desired. The manufacturing engineer may change titles and blend responsibilities, but will always be a manufacturing engineer at heart, if the goals of design and manufacturing are merged.

RICHARD D. CROWSON,
SET, CMfgT, CMfgE

Editor

Richard D. Crowson

Richard Crowson is currently a mechanical engineer at Controlled Semiconductor, Inc., in Orlando, Florida. He has worked in the field of engineering, especially in the area of lasers and in the development of semiconductor manufacturing equipment, for over 25 years. He has experience leading multidisciplinary engineering product development groups for several Fortune 500 companies as well as small and start-up companies specializing in laser integration and semiconductor equipment manufacture.

Crowson's formal engineering training includes academic undergraduate and graduate studies at major universities including the University of Alabama at Birmingham, University of Alabama in Huntsville, and Florida Institute of Technology. He presented and published technical papers at Display Works and SemiCon in San Jose, California.

He has served on numerous SEMI task forces and committees as a voting member. His past achievements include participating in writing the SEMI S2 specification, consulting for the 9th Circuit Court as an expert in laser welding, and sitting on the ANSI Z136 main committee that regulates laser safety in the United States.

Contributors

Greg Chandler
Manufacturing Engineering
Hubbell Premise Wiring, Inc.
Wilmington, North Carolina

Stephen C. Cimorelli
Learjet, Inc.
Wichita, Kansas

Richard D. Crowson
Melbourne, Florida

Alexander Houtzeel
Houtzeel Manufacturing Systems
Software, Inc.
Waltham, Massachusetts

Robert L. Lints
Quality Assurance Systems
St. Louis, Missouri

Clyde S. Mutter
Titusville, Florida

John P. Tanner
Tanner and Associates
Orlando, Florida

Kjell Zandin
H. P. Maynard & Company, Inc.
Pittsburgh, Pennsylvania

Contents

1 Practical Cost Estimating for Manufacturing

John P. Tanner

Editor's note: This chapter was condensed from a book manuscript prepared by John P. Tanner. The editor wishes to thank Mr. Tanner for generously allowing use of this material.

1.0 INTRODUCTION TO COST ESTIMATING

The manufacturing cost estimate is the key in new or follow-on business acquisition and the continued growth of the company. It must be built around a sound, well-thought-out manufacturing plan. It must address the cost of facilities, equipment, tooling, materials, and support labor, as well as direct labor to actually fabricate and assemble the product. The estimate must be responsive to customer delivery requirements and production rates, and reflect manufacturing a product of the desired quality and reliability. The cost estimate is a prediction of what the product will cost to manufacture at some future point in time. Estimating is not an exact science, yet a good cost estimate will come very close to actual costs incurred. The accuracy of any cost estimate depends on:

1. The time allocated for the preparation of the estimate
2. The knowledge and skill of the estimator
3. The depth and completeness of preproduction planning
4. The amount of product description information available
5. The accuracy of the material estimate

The higher the percentage of labor cost in the estimate, the greater is the need for precise labor standards. If engineered standards or estimated standards are not available, the estimator must use historical data and his or her own judgment, experience, and knowledge to develop the labor estimate (see also chapter 4 of this volume, "Work Measurement").

In many products, material and subcontract costs can be as much as 70% of product cost. This means that these costs must be examined very carefully at the time the manufacturing cost estimate is prepared. For best results, direct quotes should be obtained from suppliers and subcontractors. Time constraints in completing the cost estimate may not allow sufficient time to solicit these quotes, forcing the use of historical cost data for the same or similar parts and materials, factored for inflation and anticipated cost growth.

Once the basic estimates of labor hours and material dollars have been put together, the judgment exercised in determining the initial costs and the rate of improvement to achieve the eventual cost will have a major effect on the final accuracy of the estimate. An important part of the estimator's job is to prevent ill-advised management decisions by making certain that the methodology, assumptions, and ground rules are understood. Management must look beyond the estimate and consider a bigger picture than the cost estimates at hand, but should understand the risks associated with arbitrary changes to a well-prepared manufacturing cost estimate.

Figure 1.1 shows the basic structure of a manufacturing cost estimate. The sum of direct labor and direct material is known as prime cost. When prime cost is added to manufacturing overhead, we obtain factory cost, or cost of goods manufactured. Total product cost, then, is the sum of selling expense, general and administrative expense, development cost, and factory cost contingencies. The addition of profit yields the price to the customer.

The problems encountered in developing a sound manufacturing cost estimate can be many; however, they can usually be categorized into the following seven categories:

1. Inadequate data on which to develop the cost estimate
2. Inadequate staff and time to prepare the estimate
3. Poor estimator selection
4. Careless estimating
5. Optimistic estimating
6. Inadequate preproduction planning
7. Management inertia

The seven problem categories are not listed in order of importance. Any one or several can be critical to the development of a sound manufacturing cost estimate, depending on the circumstances and the situation that prevails at the time.

The cost estimator may be a manufacturing engineer, an industrial engineer, or a manufacturing technical specialist with heavy experience in the manufacturing technology in which he or she is preparing cost estimates. Cost estimating is highly demanding work, often requiring extended overtime and short deadlines. It requires the ability to quickly formulate a preproduction plan, and to visualize work flows, equipment, and tooling. A labor estimate that comes close to actually incurred costs must then be accepted by management. Not only must this be done under considerable pressure, but it must handle last-minute changes to the requirements the estimate was built on, as well as management-directed changes. In many large companies, an independent cost estimate may be developed by the fiscal or marketing groups, and is used as a check against the more detailed analysis described in this chapter.

1.0 PROGRAM REQUIREMENTS

Program strategy and objectives. What are the issues? What needs to happen? What is the critical path? What are the assumptions and ground rules? Who is the customer? Is co-production involved? If so, what is the split?

1.1 PRODUCT/HARDWARE

Product definition. How will it change from concept, through development to production? How will configuration be controlled? Will a technical documentation package be provided? When? To what level? What quantities of deliverable hardware will be provided in development? Pilot production? Production? What about spare parts? Will there be any GFE or CFE?

1.2 PROGRAM TIME PHASING

When will the program start? Development? Pilot production? Production? What are the key program milestones? What are anticipated peak production rates? When will they occur?

1.3 PROGRAM CONSTRAINTS

Potential problem areas and risks. New or advanced manufacturing processes and technologies. Unusual inspection, testing, or acceptance requirements.

2.0 DEVELOPMENT PLAN

What will be accomplished in the development program? What deliverable hardware will result? Engineering built? Manufacturing built? How will producibility and DTUPC be addressed? How will the hardware change as it evolves through the development cycle?

3.0 MANUFACTURING PLAN

Fabrication and assembly sequences and flows. Estimated times. Equipment requirements. Tooling requirements. Overall block layouts. Space required. Manufacturing flows. Processes.

3.1 MAKE OR BUY PLAN

What is the rationale for the make or buy plan? Identify major items to be subcontracted.

3.2 TOOLING PLAN

Tooling philosophy. Tooling requirements. Is interchangeability a requirement? Will tooling masters and gages be required? How will tooling differ from development to pilot production to production? Will tooling be design tooling, shop aid tooling? Will NC be used? How will tools and tapes be controlled?

3.3 MANPOWER PLAN

Skills requirements. Availability of manpower. Projected manpower needs of program. Anticipated training requirements. Support personnel needs.

4.0 FACILITIES PLAN

Identify new or additional equipment needed and estimated cost. Additional or existing building floor space requirements and estimated cost. Would include engineering, lab, manufacturing, test, storage, bunker, and special process area. Identify any requirements for special or unusual facilities such as clean rooms, dark rooms, specially reinforced floors, ESD protection, etc. Provide estimated cost and time when facilities would be needed.

FIGURE 1.1 Structure of a cost estimate for a manufactured product.

5.0 MATERIAL ACQUISITION PLAN

Identify long lead items, source selection plans. Who are major subcontractors and how will they be controlled and managed? What are plans for stores and kitting of material? What are the plans for dual sourcing? What is the material handling plan for receiving, stores, staging for production? How will vendor and subcontractor follow-up and expediting be accomplished to ensure on-schedule delivery of material? How will engineering changes and shop overloads be handled?

6.0 SCHEDULES AND TIME-PHASING

6.1 PROGRAM MILESTONE SCHEDULE

Overview schedule of entire program showing all phases with key milestones and events, including follow-on work. Should show development, procurement, facilities, production, etc.

6.2 DEVELOPMENT SCHEDULE

Key events in engineering and development, in sufficient detail to clearly portray development time-phasing.

6.3 MANUFACTURING SCHEDULES

Initial low rate production and full production. Manufacturing buildlines, block release plan.

6.4 PROCUREMENT SCHEDULE

Long lead and subcontract deliveries. Should show entire procurement cycle from requisition through order placement to material kitting and issue from production.

6.5 FACILITIES AND EQUIPMENT SCHEDULE

Must show all key milestones for new equipment and facilities acquisition, including order placement, ship date if equipment, ground breaking if new construction, through to available for production use date.

7.0 QUALITY ASSURANCE PLAN

What is the inspection and test plan for the product? What specifications and standards will apply? How will vendor and material quality levels be maintained? What is the plan for the rework and/or disposition of discrepant hardware?

8.0 NEW MANUFACTURING TECHNOLOGIES

Describe new manufacturing technologies associated with the program and plans for training and qualification in production, including equipment and process shakedown and debug. How and when this will be accomplished ahead of the production phase?

FIGURE 1.1 (Continued)

The cost of preparing the estimate must be borne by the company whether it results in new business or not. Most new contracts for manufactured products are awarded based on lowest cost, best delivery, and quality of the product, not necessarily in that order. Management must decide the "win probability" on any estimate for new business, and from that decide the effort to be expended in preparing the cost estimate. It may even decide not to submit a bid. Management should prepare a bid strategy and plan that includes reviews at critical stages in the preparation of the cost estimate. There are several good reasons for this:

1. Basic errors or omissions can be found and corrected before the estimate goes beyond the preliminary stage, when changes would be costly and time consuming.
2. The preliminary estimate may be sufficient to satisfy the requirements, and early management review ensures that no further cost estimating effort will be authorized beyond the preliminary stage.
3. Management review brings the best minds and talent available in the company to bear on the manufacturing cost estimate, serving as a check on the estimate and its assumptions.

Constraints of time and cost often leave no opportunity to explore and verify many of the premises and the assumptions used in preparing the estimate. In spite of this, cost estimates for manufactured products are prepared every day that accurately reflect manufacturing costs for the product and are truly competitive in bringing in new business for the company. In the following subchapters, the steps required to prepare a manufacturing cost estimate that is both accurate and competitive are explained. Examples and cost estimating data in practical form are also provided to help in preparing estimates.

1.0.1 Bibliography

Tanner, J. P., *Manufacturing Engineering: An Introduction to the Basic Functions*, 2nd ed., Marcel Dekker, New York, 1991.

1.1 UNDERSTANDING THE ESTIMATE REQUIREMENTS

1.1.1 Determination of Cost Estimate Requirements

In any engineering work, the solution is usually readily apparent once the problem is fully defined. The same holds true in cost estimating for manufacturing. If the cost estimate requirements are known and fully understood, preparation of the estimate is usually routine. To do this requires answers to the following questions:

1. Who is the prospective customer?
2. What is the bid strategy?

3. What are the requirements?
4. What are the assumptions and ground rules?
5. Is the product to be manufactured fully defined?
6. What are the potential problem areas and risks?
7. What are the key milestones, delivery requirements, and production rates?

Type of Solicitation

The type of solicitation is important in the formulation of a bid strategy, and in deciding the resources that will be committed to preparing the manufacturing cost estimate. A quotation for an additional quantity of parts to a current customer is one example. A quotation for a newly designed item, which has never been built in production before, is quite a different matter if your company is one of several submitting bids. Firms in the defense industry, whose primary customer is the government, can survive only by winning contracts awarded to the lowest bidder among several submitting bids. This winning bid must be supported by a detailed and valid cost estimate. Figure 1.2 shows the outline for a production program plan used by an aerospace company as the basis for its cost estimate and proposal to the government. Such an outline forces consideration of all the requirements.

The majority of manufacturing cost estimating involves job-order production in the metalworking and allied industries in the United States and the rest of the Western world. Castings, forgings, formed and machined parts, and assemblies are produced by thousands of small and medium-sized firms to exacting specifications within the limits of the estimated cost. Solicitations are usually "firm fixed price," which means that if the cost estimate is in error on the low side, the difference is absorbed by the company.

A high estimate may mean extra profit for the company, or it may mean that the contract is awarded to another firm whose price was more in line with what the item should cost. On bids for an additional quantity of the same or a similar item ordered previously, customers will often expect a price reduction because of the learning

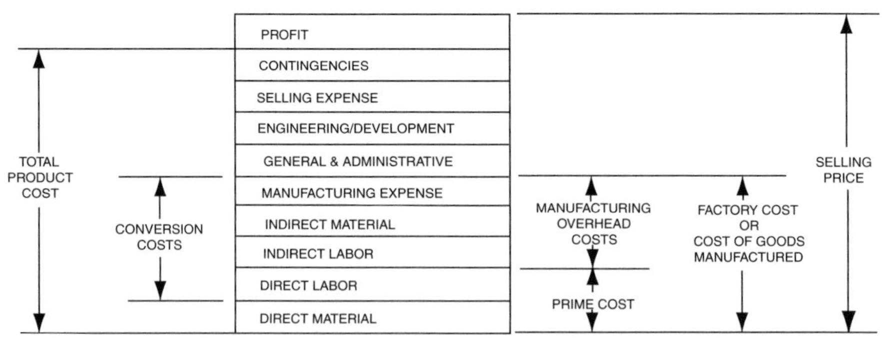

FIGURE 1.2 Production program plan checklist.

curve effect, which implies continuous cost reduction the longer the product remains in production (see Subchapter 1.4).

Product or Hardware to Be Delivered

Adequate product definition is critical to developing a meaningful manufacturing cost estimate. This may include a set of engineering drawings that fully describes the product or parts to be produced. In many instances, shop process or routing sheets and sample parts are available. The product may currently be in production. Should this be the case, it is a relatively easy matter to determine the sequences of manufacture.

If the drawing package describes a product that has never been built in your plant before, a different approach is required to develop a sound manufacturing cost estimate. The drawing package must be broken down into piece parts, subassemblies and assemblies, and a parts list constructed. This will show (in the case of an assembled product) how the assembly goes together and will form the basis for the bill of material and the fabrication and assembly processes.

There are differing levels of detail and description provided by product technical-data packages. Drawing packages may show the lowest level of technical detail down to the smallest piece part and assembly, each on a separate drawing. Other engineering drawings provide a minimum of detail, showing assembly and detail parts on the same drawing. Engineering drawings and other technical documentation cost money to prepare and to update with the latest changes. Technical documentation packages provided to bidders for cost estimates and quotations do not always fully and correctly represent the product. There may be errors in dimensioning and tolerancing that were noted the last time the product was manufactured, but these changes were never picked up and documented by formal engineering change orders to the drawings.

Many times in bidding the product as depicted in the engineering drawing package, Defense Department contractors, with no other product data or knowledge to go by, have seriously underbid production contracts on hardware and systems designed by other contractors simply because all the information needed to fabricate and assemble the product was not shown on the drawings. In preparing a manufacturing cost estimate, every effort should be made to make certain that the drawings are current and that all engineering change information is included.

Engineering drawings must often be supplemented by manufacturing engineering documentation such as process routings, methods sheets, visual aids, tool drawings, test procedures, and process specifications to determine what is really required to manufacture the product. If possible, this documentation should be provided by the prospective customer as part of the bid package. If your company is submitting a cost estimate for the manufacture of a build-to-print product that you have never produced before, it is imperative that whatever shop-process documentation that exists be obtained to aid in developing the cost estimate. If not available, the bidder may have to prepare preliminary documents in order to form the basis of a bid.

Definition or Concept

The estimator must work with many kinds of drawings, specifications, and shop documents. Included will be specification control drawings similar to Figure 1.3, which clearly spell out all critical parameters of the jacketing material for a cable assembly, and may list approved or qualified sources for this material. Figure 1.4 shows a typical detail or fabricated part drawing for a 0.020-in. thick gasket which would be stamped with a die. A typical assembly drawing is shown in Figure 1.5, for a voltage regulator assembly. Notes on such a drawing might include:

1. Prime and seal threads using Loctite
2. Torque fasteners to 3 to 5 in.-lb.
3. Ink stamp assembly number as shown

Column A in the parts list shows the number of items per assembly, column B is the part number, column C is the item name, and column D is the find number shown in the leader arrows on the drawing.

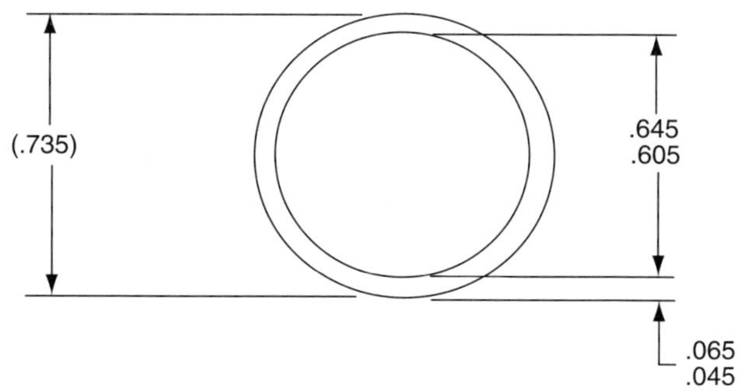

NOTES- UNLESS OTHERWISE SPECIFIED

1. MATERIAL: SILICONE RUBBER CONFORMING TO ZZ-R-765, CLASS 2A OR 2B, GRADE 50, WHITE.
2. IDENTIFY PER MIL-STD-130 WITH MANUFACTURER'S PART NUMBER AND FSCM NUMBER; CONTROL NUMBER ENCLOSED IN PARENTHESIS; BAG AND TAG.
3. SHAPE AND CONCENTRICITY: ID AND OD SHALL BE NOMINALLY CONCENTRIC WITH A UNIFORM WALL THICKNESS. JACKET MAY BE ELLIPTICAL IN FREE FORM EXCEPT OPPOSING SURFACES SHALL NOT ADHERE TO EACH OTHER.
4. LENGTH SHALL BE FURNISHED CONTINUOUS AS SPECIFIED (10 FEET MINIMUM).

FIGURE 1.3 Specification control drawing (SCD) for a hig-reflectance silicone cable jacket.

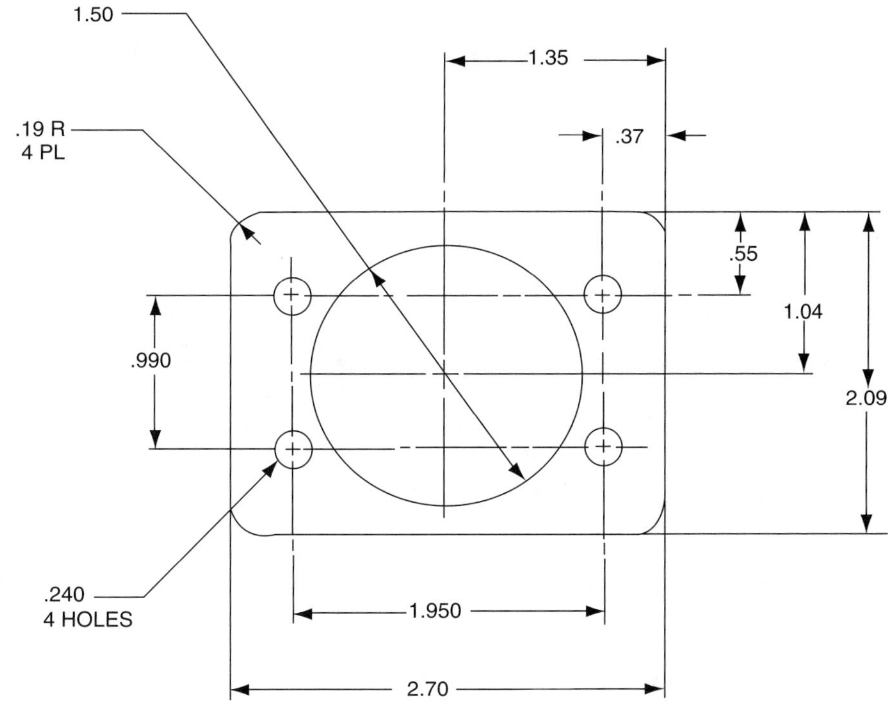

FIGURE 1.4 Gasket made from 0.020-in. material.

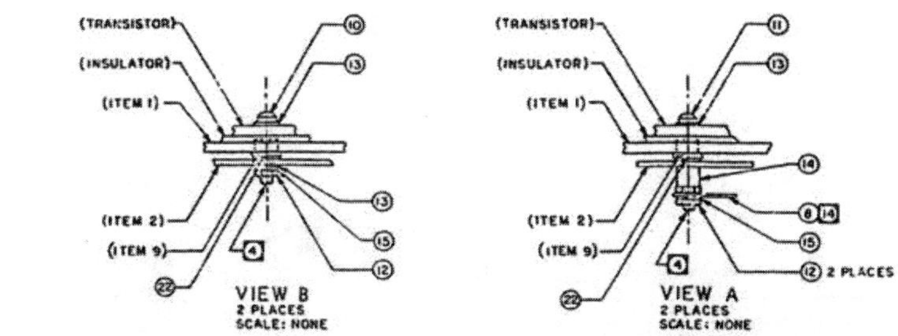

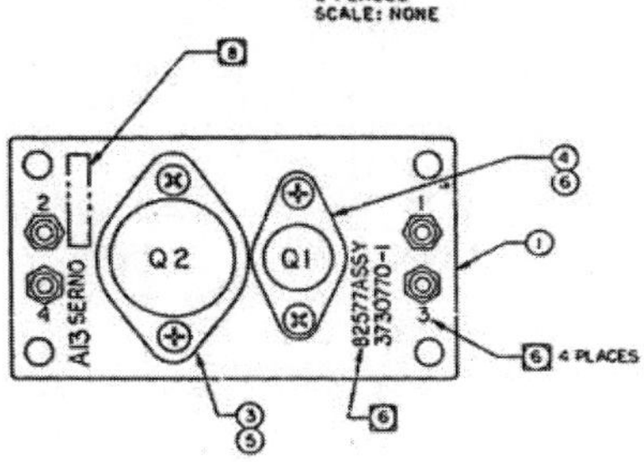

A		B	C			D
4		MS35338-40	WASHER, LOCK			15
2		NAS43000-10	SPACER			14
6		NAS620-4	WASHER			13
6		NAS671-4	NUT			12
2		NAS1100-04-14	SCREW			11
2		NAS1100-04-12	SCREW			10
4		M38527/9-025	WASHER, INSULATING, MIL-M-38527/9			9
2		MS35431-1	TERMINAL			8
4		FT020A01	TERMINAL MIL-T-55155/2			7
1		653206-2	INSULATOR, TRANSISTOR	9		6
1		653206-1	INSULATOR, TRANSISTOR	9		5
1		PA60309-1	TRANSISTOR	13		4
1		JANTX2N5038	TRANSISTOR MIL-S-19500/439			3

FIGURE 1.5 Voltage regulator assembly.

Often it becomes necessary to estimate the cost to produce an item that is not yet fully designed. Such estimates are made from sketches, concept drawings, or design layouts. Preparation of these estimates requires the estimator to fully understand the design concept envisioned by the designer. It is not uncommon for such preliminary designs and preliminary bills of material to grow by as much as 40% in complexity and parts count by the time the final manufacturing drawings are released. Figure 1.6 shows a design layout for a mortar round that, once fully designed, developed, and qualified, would be produced in high volume for the army.

Delivery Requirements

The estimator must know if the customer's delivery schedule can be met, considering the lead time required for planning, tooling, and obtaining the necessary parts and material, plus the number of days to actually manufacture the product. Analysis of the delivery requirements determines peak production rates, and whether one or two shift operations are needed. Perhaps the required delivery rate can be met only by extended overtime. All of these elements affect cost.

Analysis of shop flow times, material lead times, capacity of shop machines, and shop and machine loads in the time period the proposed work would be performed can be crucial in developing any manufacturing cost estimate. A firm may manufacture to stock or inventory, based on a sales forecast. This offers the advantage of smoothing the shop workload and being able to plan well in advance for manufacturing operations. Delivery requirements requiring higher rates of production may require more units in work in a given time. This allows better labor utilization and production efficiency. This has to traded off against the cost of carrying inventory, as

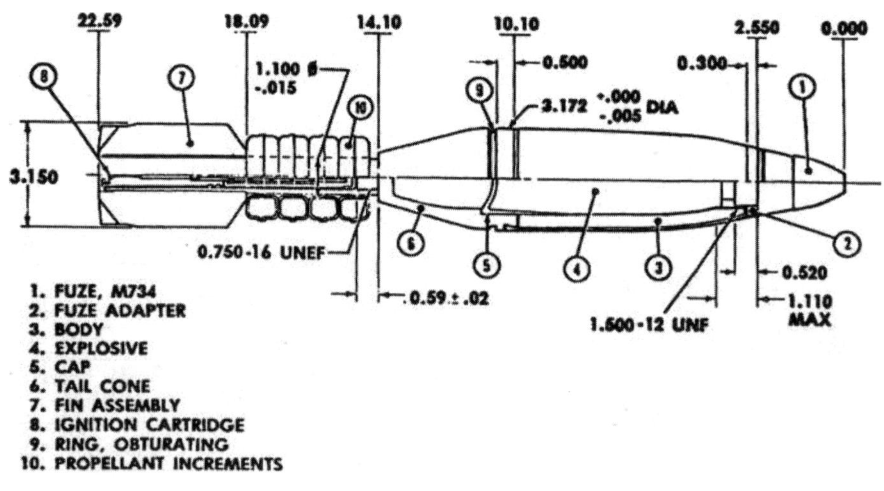

FIGURE 1.6 Design layout for a new mortar round.

opposed to delivering for payment upon product completion. Lower production rates result in smaller lot sizes, increasing the number of setups for the same number of units. This lowering of production rates can have a profoundly adverse effect on the attainment of projected improvement curves. Setup and teardown caused by small production lot sizes should be avoided as much as possible.

Special Provisions

A contingency factor should be applied to any manufacturing cost estimate in which the product is not fully defined or is still undergoing change. The amount of contingency is a judgment call which would:

1. Vary with the stage of product development
2. Depend on the newness of the program, the market application area, the technology, the industrial processes, and the organization
3. Depend on the time allowed for development
4. Consider the degree of development required
5. Vary with the general economic conditions

Figure 1.7 shows a contingency policy used by a large manufacturer of computers and point-of-sale terminals.

In addition to the importance of product definition, a complete bill of material or engineering parts list is vital to estimating manufacturing costs. This must be carefully reviewed, and the make-or-buy decisions for pricing purposes made. The make-or-buy plan determines the form the estimate will take and establishes the basis for material pricing, labor content estimate, and the long-range impact on facilities and capital equipment.

In government contracts there may be special provisions for placing work with small or minority businesses in economically depressed areas, or clauses that parts or materials can be obtained only from government-qualified sources. Careful attention should be given to such provisions specified in the request for quote when formulating a bid strategy.

"Understanding of estimate requirements" implies that management and the estimator understand all new processes and manufacturing technologies that may be introduced if the job is won. The costs that are associated with training, technical support, safety, toxic waste handling and disposal, etc. can be substantial and should be recognized. Such costs may force companies with limited resources to go outside for these products and services.

1.1.2 Estimate or Bid Proposal Strategy

Formulation of a bid or pricing strategy follows a thorough understanding of the requirements. Such a strategy would consider the probability of winning the job when competition is involved or when no competition is involved. Marketing intelligence concerning the prospective customer and the competition will be a major

PHASES OF COST ESTIMATE	CONTINGENCY GUIDELINES AS % OF MLB
I. ENGINEERING DESIGN GOAL	30 – 50%
PRELIMINARY DESIGN "A" MODEL AVAILABLE IF REQUIRED	
II. ENGINEERING	20 – 30%
"B" MODEL IN FABRICATION PRELIMINARY B/M AVAILABLE PARTIAL DRAWINGS AVAILABLE	
III. PRELIMINARY PRODUCTION	10 – 20%
"B" MODEL TEST COMPLETED AND ACCEPTED MANUFACTURING B/M AVAILABLE MAJORITY OF DRAWINGS RELEASED	
IV. PRODUCTION	0 – 10%
B/M COMPLETE COMPLETE DRAWING PACKAGE AVAILABLE SPECIFICATIONS RELEASED	

Note: MLB=Material, Labor and Burden

FIGURE 1.7 Contingency policy.

factor in developing the bid or estimate strategy. "Win probability" determines the number of company resources that will be devoted to preparing the cost estimate. A low win probability could only justify a minimal investment of resources, or a no-bid decision.

The bid strategy defines the rules and guidelines to follow in formulating a tooling philosophy, spelling out ground rules and assumptions, assigning personnel, and determining how close to shave the final price quotation. The bid strategy should then be made known to the estimators and all key people involved in preparing, reviewing, and approving the cost estimate. It will be the baseline for preparing and issuing a proposal authorization, or directive, specifically spelling out who is going to do what and when. Such a proposal authorization or directive should contain the following information as a minimum.

1. What is to be bid, and who is to do what in preparing the cost estimate
2. A time-phased plan or schedule for preparing the estimate
3. A list of ground rules and assumptions

4. A make-or-buy plan
5. Specific description of any special provisions contained in the request for quote from the customer
6. The specifications that apply to the estimate, and in case of conflict, which takes precedence, the drawings or the specifications
7. Peak production rates that processes, equipment, tools, and facilities must be able to support

The cost estimators now have a well-thought-out and well-researched plan to follow, and management is providing the necessary leadership and direction to reach the goals determined to be advantageous for the manufacturing cost estimate at hand.

1.1.3 Estimate or Proposal Plan, and Time Phasing

It seems that there is rarely sufficient time to prepare the manufacturing cost estimate. The customer wants a response within days or even hours after requesting price and delivery, and management must have time to review and approve the cost estimate. This often creates a situation requiring long hours of overtime and much pressure on the estimating team. Temptation is strong to provide expedient answers and estimates that cannot be supported when analyzed in depth by management and supervision.

There should be a proposal or cost estimate schedule plan that highlights the critical milestones in the estimate preparation and review, and maximizes the available (limited) time that is allocated for this purpose. Figure 1.8 shows such a plan for an aerospace product. This represents a minimum plan. Where time is very limited, such as a few days to a week, the plan must be much more detailed, covering the actual steps in estimate preparation and review.

Resources to Be Allocated

A key decision of how much time and money to invest in preparation of the manufacturing cost estimate is required, since this investment may not result in winning the new business. A cost estimate for a manufactured product can be used for a number of reasons other than to establish the bid price of a product for quotation. These include:

1. To verify quotations submitted by vendors
2. To determine whether a proposed product or item can be manufactured and marketed profitably
3. To provide the basis for a make-or-buy decision
4. To determine the most economical methods, processes, tools, or materials for a manufactured product
5. As an aid in evaluating design alternatives
6. To determine whether to bid

Preparation of the cost estimate will also aid in the determination of resources to be finally expended in its preparation. One criterion used to determine the resources

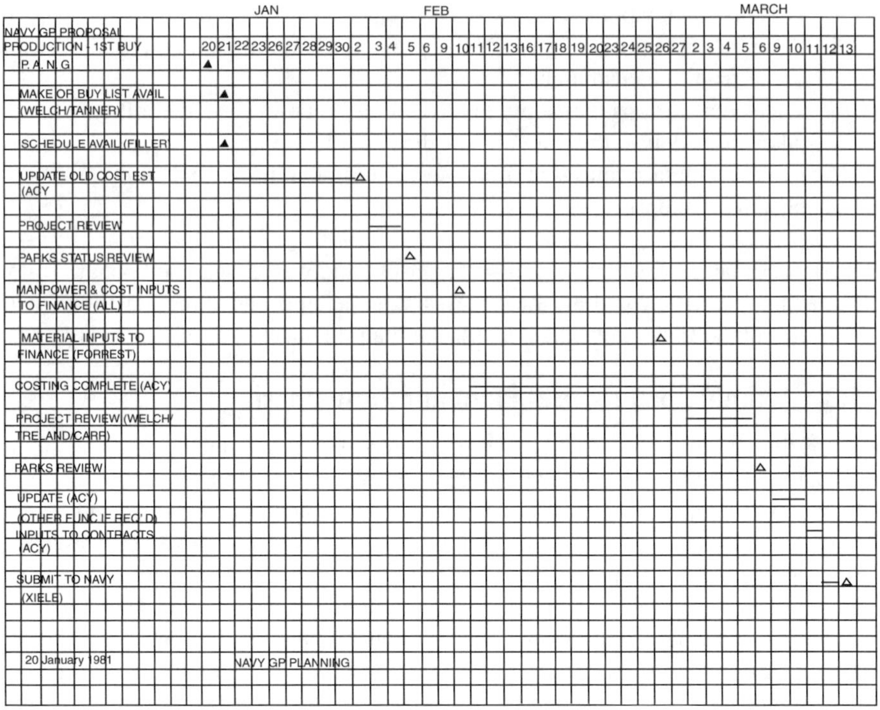

FIGURE 1.8 Cost estimate preparation plan.

to be expended in cost estimate preparation is the value of the new business acquisition. Management may review a preliminary estimate and decide that no further time or effort should be expended. The preliminary estimate is deemed sufficient to satisfy the requirements.

Proposal or Estimate Milestones and Schedule Plan

Key events in preparation of the estimate should be listed, and a time span assigned for accomplishment. Key milestones for most cost estimates of a manufactured product include:

Availability of engineering drawings and specifications
Completion of the engineering parts list and the manufacturing bill of material
Preparation of the make-or-buy plan
Completion of the listing of needed shop equipment and facilities
Completion of the preproduction planning
Preparation of the listing of needed shop equipment and facilities

Preparation of the tool list
Identification of long-lead procurement items
Completion of material and subcontract pricing
Completion of manufacturing labor estimate, including support labor
Completion of the total manufacturing cost estimate
Scheduled estimate checks, reviews, and approvals
Final pricing and costing
Submission to the prospective customer

Figure 1.1 shows other milestones, some of which may not be needed at all. In a small shop, the owner may have a relatively simple product line, allowing him or her to estimate the cost of new business with some degree of accuracy. The owner must make certain that he or she fully understands the bid requirements, and must cover the same proposal milestones as the larger organization, even though he or she does so in a less formal manner.

Management Control and Estimate Transmittal

As I. R. Vernon, in *Realistic Cost Estimating for Manufacturing* (1961), points out:

> Fast, economical and accurate estimates require proper management control of the estimating function. Management establishes the type of estimating department that will best serve company needs, and then formulates the procedures and administrative controls necessary for efficient departmental operation.

Such controls include the screening of incoming requests for bids or estimates, usually by a committee of top managers, to make the bid–no-bid decision. If the decision is to bid, they set up the proposal schedule plan, assign personnel and resources to prepare the cost estimate, and establish the administrative routings and controls for review and approval.

An essential management control of the cost estimating function is the identification and analysis of previous estimate cost deviations versus actual costs. Records of these deviations should be maintained and plotted to determine trends and to pinpoint areas of weakness in the cost estimating function. There may be many reasons for cost estimates that are too high, too low, or simply unrealistic. One of the first areas to examine is arbitrary cuts or reductions made by management, or unrealistic contingency factors applied to an otherwise sound cost estimate which results in others winning the award. Other reasons that estimates are too high include:

1. Being deliberately high to discourage business the company does not want
2. The tendency to be overly cautious with new products, processes, technologies, etc. that the estimator is not familiar with or does not understand
3. Preproduction planning that calls for more processing steps or more sophisticated tooling than is actually required
4. Poor material pricing practices, such as failing to consider price breaks on quantity purchases of material

5. Overestimating labor costs by assuming higher-than-normal start-up costs, a shallower learning curve slope than would actually be incurred, or by using loose labor standards in building the estimate

High estimates sometimes result in greater profits than anticipated, but may also lose business that could have been profitable. Other problems caused by high estimates are loss of customer goodwill, greater investment of resources in cost estimating because more estimates must be processed to book a smaller volume of work, and eventually being priced out of the market.

Some of the more significant reasons for low cost estimates include:

1. Incomplete or incorrect product design information, or a new design that "grows" after the estimate, in parts count, in complexity, and in design changes that occur after the production cycle has started
2. Higher labor costs than anticipated, possibly due to delays that resulted in production stoppages, higher-than-anticipated rework due to design or tooling problems, or poor planning in the preproduction phase
3. Higher material costs than anticipated, due to such factors as unplanned material price increases, design changes that cause material requirements to change, and higher-than-planned scrap and line losses

How the Estimated Should Be Transmitted

When the cost estimate for a manufactured product is transmitted to the potential customer, it should always be by written quote or other formal means of transmittal. The price and delivery may be given verbally, but should always be followed with the confirming written quote. The formal quotation should list and explain all assumptions and contingencies, and for how long a time the price given in the quotation is valid.

Qualifications and Caveats

The transmitted price and delivery quotation should always be given in a manner that is responsive to the customer's request for price inquiry. For example, if the price quotation is a bid on government work, specific cost breakdowns, usually by work breakdown structure, are requested in addition to end-item price and delivery. Failure to provide this breakdown results in a nonresponsive proposal.

Cost estimates that are developed by companies that do business with the government often require a government audit before a final contract award. Such audits can be very upsetting for the company and the estimating department that does not have good records, cannot show step by step how costs were developed in the estimate, or cannot show actual price quotations for major items to be purchased outside.

Final Price Negotiation

In most instances, the price and delivery quoted are either accepted by the customer, or the work is done by someone else. There is no final price negotiation. On the other hand, the larger the job in total dollar price, and the longer the production run, the more likely there will be a negotiation price. In such negotiations, profit margins and contingencies may be reduced in order to obtain the job. The estimated cost to actually do the work should never be part of the negotiating process.

The negotiating process is beyond the scope of this chapter, but it requires the best talent the company has in management and in cost estimating. They must have sufficiently detailed knowledge of the cost estimate and how it was prepared, be able to answer any questions, and face up to any challenge that is presented. A successful negotiation should result in a contract award very close to the price and delivery presented before the negotiations began. Additional negotiations may occur after contract award should there be a change in work scope, such as accelerated delivery, or an increase in number of units to be produced.

1.1.4 Bibliography

Vernon, I. R., *Realistic Cost Estimating for Manufacturing,* Society of Manufacturing Engineers, Dearborn, Michigan, 1961.

1.2 DEVELOPING THE MANUFACTURING PLAN

1.2.1 Review of Product Requirements

The manufacturing cost estimate must begin with a manufacturing plan. The thoroughness and accuracy of this plan determines to a large measure how good the estimate will be. The cost estimator must be able to concept a workable manufacturing plan for any product the firm manufactures. Such a plan must begin with an understanding of the product requirements.

As the cost estimator goes through the drawing package to do the preproduction planning, or what we know as the manufacturing plan, producibility problems may be apparent. In such cases, the problem should be noted, and if time permits, a cost trade study should be initiated. This is to determine if the desired change would generate sufficient savings to offset the cost of making the change. Producibility changes that take place after the drawings are released may not be as attractive as they appear initially, after the impact on schedule, tooling, retrofit, and the cost to change the engineering are all considered.

A checklist for reviewing producibility of an electronics product or assembly represents the kind of questions the cost estimator should be asking as he or she goes through the drawing package:

1. Does the dimensioning facilitate manufacturing, and are the tolerances realistic?
2. Is all marking and stenciling defined and visible?
3. Are assembly notes complete and definitive?
4. Is internal wiring critical? If so, is the location of the wiring specified?
5. Are test points and adjustments accessible?

6. Is harness development required? If so, can the harness be fabricated outside the unit, and installed later as a subassembly? Does the wire run list contain wire length information?
7. Does the design lend itself to mechanized or automated assembly techniques?
8. Does the design avoid the need for select-at-test component matching?
9. Are component parts accessible for assembly? Can all assembly and wiring be done without restriction?
10. Can required testing be performed without disassembling the unit?
11. If wire wrap is used, does the design facilitate automatic assembly?
12. Are standard connectors and assembly hardware used?
13. If circuit card assemblies are installed as part of the assembly, are they designed to plug in, or must they be wired in?
14. Are there mechanical loads such that printed circuit epoxy glass boards are in compression?
15. Has consideration been given to using printed circuit flex cable, or molded ribbon wire cabling instead of hard wiring of the assembly?

As a minimum, the cost estimator should review the drawings of a fabricated product to ensure that they:

1. Are dimensioned and have datum surfaces that are compatible with accepted machining and fabrication practices
2. Have sufficient stock allowances on castings, forgings, and stampings to provide for any mismatch or distortion that may result from heat treating
3. Have maximum allowable tolerances on nonfunctional features and characteristics
4. Have realistic tolerances on functional characteristics
5. Have adequate provision for clamping and locating
6. Provide sufficient clearance and access for the assembly of all component parts
7. Call out standard parts and materials, which can be processed and assembled using general-purpose machines, equipment, and tools

The estimator must next prepare a manufacturing bill of material from the parts list, and the make-or-buy plan that has been established. Next the estimator proceeds to concept the steps and sequences of manufacture for each part, subassembly, and assembly to be made in-house. Then, for each step in the process, he or she determines the machines, equipment, and tools that will be needed. Finally, he or she determines setup and run times for each step or sequence in the manufacturing process.

1.2.2 The Make-or-Buy Plan

Purchased parts and materials, as well as subcontracted items, can account for as much as 70% of manufactured product cost. It is therefore important that the initial make-or-buy analysis be made by a qualified cost estimating professional based on knowledge of the plant capability to manufacture or not to manufacture the items on the bill of material. For example, if the company is primarily an assembly house, then all fabricated and machined metal parts would be classified as "must-buy" items, in addition to hardware and bulk material items such as paints and solvents. Only those items that can be either made or bought will require a detailed analysis from cost and shop load standpoints. In larger companies, a formal make-or-buy committee with key people from all of the functions is chartered, chaired by the senior operations or manufacturing executive. This committee then makes final make-or-buy determinations based on recommendations from all disciplines concerned.

Must-Buy Items

Obvious "buy items" based on shop capability to perform the process of manufacture are the easiest decisions to make. However, management may decide to create the process or capability in-house to have a degree of control not possible when the work is placed outside. If the requirement exists for only a short time, this may unnecessarily commit company resources to a process capability that would stand idle much of the time.

Must-Make Items

There are usually a number of fabrication, assembly, and testing operations on any manufactured product that are critical to its manufacture and performance in the field. These are the operations and processes that must be done in-house to ensure product integrity and control. Such operations usually include product final assembly and test, and fabrication and assembly of close-tolerance or mating parts, among others.

Items That Can Be Either Make or Buy

These items are the ones requiring investigation and analysis prior to the decision to make or buy. Usually decisions on these parts and assemblies are based on cost, promised deliveries, and shop capacity or load. Cost trade studies are made, quotes are obtained from various potential outside sources, and the final make-or-buy decision is made either by senior management or the make-or-buy committee.

1.2.3 Outlining the Manufacturing Sequences and Processes

The operation process chart is the best way to clearly visualize all of the steps in the manufacturing process. This chart depicts graphically the points in the process where materials are brought into the process, and the sequences of all manufacturing, inspection, and test operations. This chart may also contain such detail and information as the standard time for each operation, production equipment and tooling required for each operation, and applicable process specifications.

With the completed operation process chart, the entire process of manufacture can be visualized. The process can then be reviewed and analyzed for optimum sequencing, alternative methods of fabrication, assembly inspection and test, and most efficient methods of production. Equipment and tooling requirements at each step in the flow can be readily envisioned and recorded.

The Operation Process Chart

The principles of operation process chart construction are shown in Figure 1.9. A preprinted chart format is not recommended, because of the wide range of use and application. Symbols used in constructing the operation process chart include:

F *Operation:* Occurs when an object is intentionally changed in any of its physical or chemical features, is assembled from another object, or is arranged for another operation, inspection, or test

G *Inspection/test:* Occurs when an object is examined or verified for quantity or quality in any of its characteristics

All steps should be listed in proper sequence for each part or component, working vertically from top to bottom. The major assembly or component is always shown at the far right, and all other components are allotted space to the left of this main component or assembly. The presentation is similar to that of a mechanized

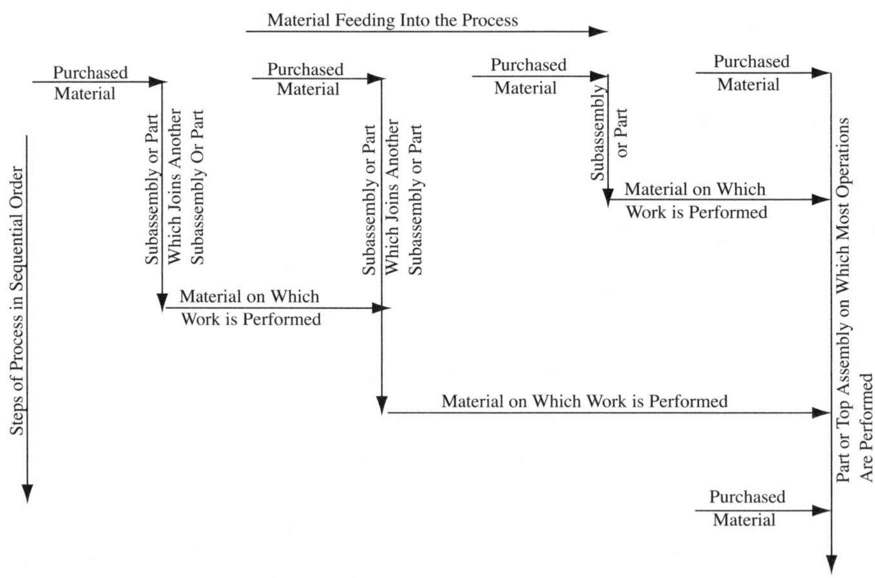

FIGURE 1.9 Principles of construction of an operation process chart.

assembly line, with parts and material, and subassemblies fed into the top assembly in proper sequence and at the correct point in the process. The operation and inspection descriptions should be brief, utilizing shop terminology such as drill, tap, ream, weld, assembly, solder, and test.

Figure 1.10 shows an operation process chart for a printed circuit board assembly. Time values, when assigned to each operation, should be broken down into setup and run times. Other useful or amplifying information, such as production machines and equipment used, tools required, and special process provisions for each operation, completes the operation process chart.

The completed operation process chart highlights material, operations, inspections, tests, and time. The visualization process needed to construct the operation process chart is fairly straightforward. The primary requirement, besides knowledge of the principles of chart construction, is a working proficiency or knowledge of the various processes involved and the ability to read and correctly interpret the requirements of the engineering drawings. The process of visualizing the steps of the product manufacture, and construction of the operation process chart, defines the manufacturing plan for cost estimating.

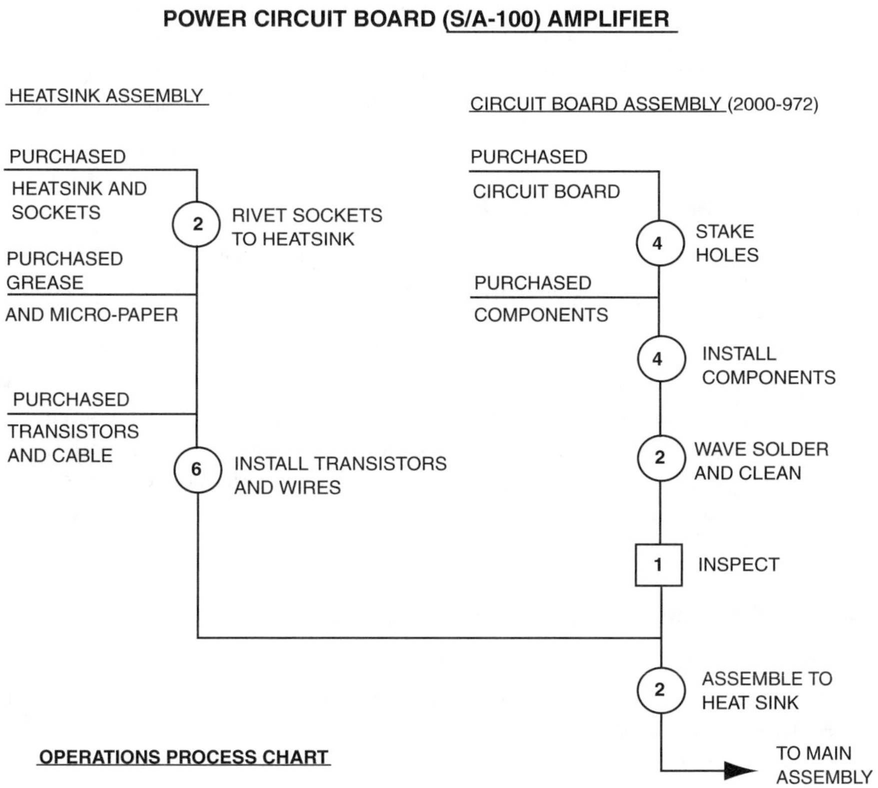

FIGURE 1.10 Operation process chart for a printed circuit board.

New Manufacturing Technologies

Advanced manufacturing technologies that are new to the company should be identified in the manufacturing plan for the cost estimate. These technologies can mean new opportunities for the company to reduce costs and gain a competitive edge. They also present the risk of cost overruns, and significant schedule delays, if they are not properly researched and understood by management at the time the cost estimate is prepared.

Some new technologies in today's world include laser welding and soldering, fiber optics, composite structures, ceramics, electron beam welding, and more. As a minimum, the cost estimate should identify the cost of new equipment and facilities for each new technology, the personnel skills required, any training that would be required for the existing workforce, the length of time that would be required to become qualified and proficient in the technology, any special safety or waste disposal requirements, and what long-term potential exists for work beyond the job that is being quoted. In addition, the possibility of subcontracting this work should be explored, especially if the need for the new process or technology is a short-term need. It makes little sense, for example, to set up a facility for coil winding—to invest in the winding machines, capacitance bridges, and Q-meters; hire engineers proficient in magnetics and qualified setup people; train operators in coil winding; and provide the facilities to varnish and encapsulate—when the coils can be purchased from a vendor who specializes in coil winding and has years of experience. This is true even if cost trade studies show that the coils would be produced cheaper in-house.

1.2.4 Equipment and Tooling Requirements

Capital and other production equipment costs to support the production of a new manufactured product often can be quite high. This is especially true when no comparable equipment exists within the company. A good example of this would be a metal finishing or plating line, which would also require waste treatment and disposal capability. Many times the cost of new equipment to manufacture a new product, or to continue manufacturing an existing product line and remain competitive, is such that it cannot be written off over the life of a single or even several product lines. In such cases, management may choose to invest in the new equipment with the knowledge that some risk may be involved, or it may decide to subcontract the work to outside sources.

Selection of Manufacturing Equipment and Machines

Once the process has been defined, the cost estimator must determine what equipment and machines will be required at each step of the manufacturing operations. In most cases this can be done quickly and easily, as the manufacturing process is centered around existing production machines and equipment, or standard equipment that can be readily obtained. Should it be decided to purchase new equipment or machines for the job being estimated, a decision must be made as to how to pay for this equipment. It may be chargeable to the product being estimated, and written off against products delivered. It may be charged to overall plant equipment depreciation, which would be a lower cost to the product being estimated but would increase the overhead charge against all other products in the plant. In many cases, a review of the make-or-buy decision is in order, and the new equipment cost and charges must be compared to buying some of the parts from an outside supplier. A complete understanding of the accounting practices in the manufacturing plant is an important requirement for good decision making.

Identifying Tooling Required

Tooling requirements are identified concurrently with the machine and other production equipment requirements. Working from the established tooling philosophy, the cost estimator can determine the tooling approach to take. He or she must know whether to plan the job around minimal shop-aid-type tooling, or a full hard-tooling-type approach, or even a no-tooling approach, by fabricating on setup. It is essential that this tooling philosophy or approach be fully defined at the outset. Once the tooling philosophy has been determined, tools required at each step in the process are defined and listed. A drilling operation, for example, may require a drill jig and a holding jig or fixture, whereas an assembly operation may require weld fixtures, holding fixtures, and/or brazing fixtures. A careful part print and assembly print analysis is required, with special attention given to the manufacturing step or operation, part or assembly shape and geometry, and peak planned production rate. In planning tooling requirements, the cost estimator should keep in mind that tools fall into one of several categories:

Special-purpose tooling
Tooling peculiar to the product it makes. This usually requires tool design, followed by special fabrication. It is good for the life of the life of the production run only. It can be considered a one-time, nonrecurring charge to the product, or can be prorated over the products delivered.

General-purpose tools
Tools that can be used in the production of many different products. No special design is required. This kind of tooling is commercially available and is much less expensive than special-purpose tooling. Some examples would include small hand tools, tool balancers, and standard assembly and motion economy devices. They may sometimes be charged to the product being estimated, but more often are considered part of the manufacturing overhead and need not be added to the product cost.

Expendable tools
Such items as cutters, end mills, drills, taps, broaches, and reamers. These tooling items are actually consumed over the life of the job or production run. They are usually considered part of manufacturing overhead, and are not charged to the particular job.

It should also be mentioned that the tapes, software, and programming required for numerically controlled (NC) equipment is also a tooling cost. Tooling costs and hence product costs can be kept as low as possible by following the rules outlined below.

1. Utilize standard tools and tooling components whenever possible.
2. When special tooling is required, utilize the design concepts of previously well-engineered tools.
3. When special tools and fixtures are required, use low-cost tooling materials in place of tool steel wherever possible.

1.2.5 Determination of Facility Requirements

The operation process chart, which describes the manufacturing process used to build the product, also defines the layout of the factory to support product manufacturing. The amount of floor space required is a function of the size of machines and of the equipment in the manufacturing process, the area required to work and service the machines and equipment, and the number of units of product in flow and in process at the planned peak production rate.

Once floor space requirements have been determined, the next step is to identify special facility requirements. These may include clean rooms; requirements for hoods, vents, and other similar openings in the roof or along outside walls; special power requirements; and requirements for sound and noise protection.

The operation process chart determines product flow and how dedicated areas would be laid out. If time permits, an overall block layout plan and most of the detailed area layouts should be developed. The primary purpose of this is to identify to management the impact on production facilities should the new job be won. No manufacturing plan should be considered complete until this has been done.

1.2.6 Putting the Total Plan Together

In summary, the manufacturing plan is the baseline for the manufacturing cost estimate. Identification of the process, the sequences of manufacture, the machines, the tools, and the facilities needed lead logically to the next steps in the cost estimate: the estimate of the setup and run times for each manufacturing operation, cost of material, etc.

1.2.7 Bibliography

Knox, C .S., *Cad/Cam Systems: Planning and Implementation,* Marcel Dekker, New York, 1983.

Maynard, H. B., *Industrial Engineering Handbook,* 3rd ed., McGraw-Hill, New York, 1971.

1984 N/C CAM Guidebook, Modern Machine Shop, 1984.

Practical Machining Principles for Shop Application, Machinability Data Center, Metcut Research Associates, Cincinnati, Ohio, 1981.

Tanner, J. P., *Manufacturing Engineering: An Introduction to the Basic Functions,* Marcel Dekker, New York, 1985.

1.3 THE CONCEPT-TYPE COST ESTIMATE

1.3.1 The New Product Concept

Perhaps the greatest challenge in manufacturing cost estimating is to determine the cost to produce a new product that is not fully designed. There may be little more to work with than hastily drawn sketches. Such a cost estimate may determine the advisability of continuing further design and development effort, or whether sweeping changes are needed to make the product producible at a low cost.

In the development of new products, the manufacturing cost must be determined from preliminary design concepts to evaluate whether one concept will be more economical to manufacture than another. This may be done for a commercial product, or in response to a government request for proposal early in the system evolution process. Figure 1.6 shows a design layout concept for an improved mortar round for the U.S. Army. There is sufficient information in this layout drawing to estimate the cost of this mortar round in production. This estimate must be as accurate as possible, since the decision to proceed with engineering development, and subsequent production of thousands of rounds, hinges on how well the manufacturing cost estimator does his or her job.

We next need to define the minimum product design information needed, and the form it should be in.

Product Definition

Before a significant investment is made in design and development of a new product, the product must show a real potential for sales, as evidenced by low estimated manufacturing costs, or because the new product wins by better performance in a competition with designs by competing companies. Companies that manufacture computer systems, or electronic devices or systems, or even commercial appliances and consumer-type products, may wish to select among several feasible new product design concepts to determine which would be the least expensive to manufacture. In such situations, sufficient information must be available to the manufacturing cost estimators to develop a valid cost estimate, yet the time and expense of developing and documenting a fully mature design is avoided.

In the high-technology companies, many new designs are concepted, yet few of these ever reach volume production. Anywhere from one to a few hundred units may be manufactured, and these units will change significantly in configuration between the first and the last unit built. In such companies, concept estimating is the norm for the manufacturing cost estimator. It is important that the estimator know and clearly understand what minimum engineering information is needed to prepare a valid concept estimate, and how to go about developing such an estimate. A concept estimate, more than any other kind, must of necessity draw heavily on the experience, ability to visualize, and ingenuity of the estimator.

Information Needed

Baseline design concept information can take many forms. It can range from reasonably good design layouts to freehand sketches on note paper. Often isometric or exploded view drawings are used in preference to design layouts. Figure 1.11 shows an exploded view of a fairly complex nutator assembly. This is an excellent way to show the parts and how they go together in an electromechanical assembly. A design layout would not provide sufficient information for the estimator to understand part configuration and how the assembly goes together. With this exploded view and a preliminary parts list, a reasonable manufacturing cost estimate can be developed.

Photographs of engineering models are also helpful, even though the final production design of the product may be somewhat different. A technique often used is to estimate the cost to manufacture the new product by similarity to existing products. A circuit card assembly, for example, may be 70% more complex and have greater parts density than a known circuit card assembly. A machined casting may be similar in configuration to one currently in production, but be only 10% as difficult to machine. In the manufacture of aircraft, the cost per pound of the airframe is one method used in determining the manufacturing cost of a new aircraft, long before the first drawing is released, yet the cost so estimated will come remarkably close to the actual cost of manufacture.

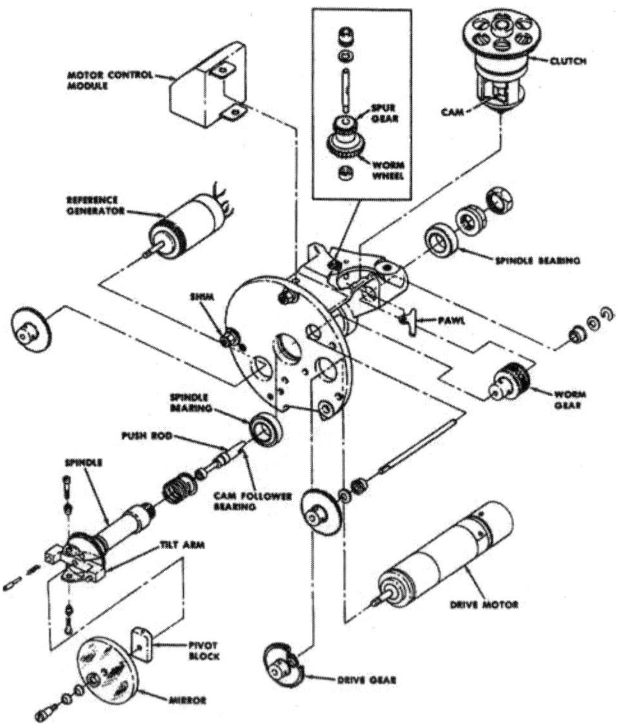

FIGURE 1.11 Exploded view of nutator assembly, showing sequence of assembly.

In addition to drawings, isometrics, sketches, and more, it is a mandatory requirement that there be a parts list or bill of material, reasonably close to the parts list that is released after final product design. This preliminary bill of material is used by the estimator to determine the make-or-buy plan, and as the basis for developing product material and subcontract cost. If major parts or assemblies are to be subcontracted, preliminary procurement specifications and subcontractor statements of work must be developed.

If preproduction units or models are to be built in a manufacturing shop rather than in an engineering model shop, the quantity to be built and the schedule dates when these units would be needed must also be included.

Product test requirements (if any), all specifications and other requirements the product must meet, and any new or unusual process requirements must be provided by the designer to the cost estimator. New manufacturing technologies must be clearly identified, with details of the process requirements as well as any new specifications and standards that would apply.

Finally, there must be a program schedule showing the major design and development milestones, including final release of manufacturing drawings. The schedule should also reflect the manufacturing start-up activities, including the preproduction and production planning milestones, tooling milestones, long-lead and other material procurement, the acquisition and bringing online of any facilities or production equipment, the low-rate initial production build, and the build-line up to rate (peak) production.

Producibility

The estimator's job is to learn as much as possible about the newly designed product, or product concept, and to assist the design team by providing cost estimates of the various design alternatives. The estimator must certainly point out design approaches that have caused manufacturing problems in the past. In short, the cost estimator, with his or her knowledge of manufacturing, should be an active and recognized participant in the design and development process. Cost trade studies should be made between design approaches such as castings versus forgings, sheet metal structure versus machined hog-outs, chemical process finish versus paint, or hardwired assembly versus flexible cable or ribbon cable.

1.3.2 Cost Estimate Development

Once the product has been defined sufficiently, and other necessary information such as parts lists, specifications, the plan for design and development, hardware requirements, and other vital information is available, the manufacturing cost estimator can begin to develop a preproduction plan, a make-or-buy plan, and a manufacturing bill of material. The purchasing department buyer or material estimator must immediately begin to price the buy and subcontract items on the bill of material, establish material procurement lead times, and identify any long-lead items. Any needed equipment and facilities over and above those available must be specified and priced.

The preproduction plan will identify equipment, facilities, and tooling required. Nonrecurring start-up costs are estimated first by the cost estimator, and include initial manufacturing engineering, process planning, and tooling costs. Fabrication, assembly, inspection, and test labor to actually build the product are estimated next. The cost of recurring support labor such as manufacturing engineering, liaison engineering, and tooling maintenance must then be added to the recurring touch-labor estimate.

Customer or internal delivery schedules will dictate peak rates for production, number of shifts, and multiples of equipment and tooling required. Time required for production start-up, and to progress from low-rate initial production to peak rate, will be determined by considering the material procurement, receiving, stocking, and kitting lead times; tool design and fabrication lead times; setup and shakedown of new equipment and facilities; and time required to hire, train, and build up the necessary workforce.

The manufacturing labor estimate should be completely documented, with all notes, backup data, and calculations carefully retained. All documents and data should be clearly labeled and dated and should include even very rough, handwritten notes and sketches. This file should enable the cost estimator and others to go to the file long after the estimate was prepared, with the backup data available to support it, and answer questions about the estimate that may arise. Such data are very useful in preparing estimates for similar products, or to update an earlier estimate.

In developing a concept estimate for a new manufactured product, it is absolutely essential that *all* of the ground rules and the assumptions used in preparing the estimate be fully documented. This information tells management and others the premises and constraints used in preparing the estimate, and goes far in explaining why things are as they are. Some examples might include:

1. The product baseline for the production estimate is the design concept as defined in the layout SK drawing dated 12/11/85.
2. Tooling and facilities are estimated to support a peak production rate of 500 units per month.
3. All work is to be performed on a single-shift, 10-hour workday, 5 days per week.
4. The estimate assumes the availability of the Harrison engine lathe and the DeVlieg mill.
5. All labor estimates were prepared using similarity to other products of similar size and configuration.

Similarity with Other Products

The similarity method is perhaps the most widely used way to estimate manufacturing labor costs when working with new product concepts. This method has the cost estimator select parts and assemblies that have been manufactured in the past that have features similar to the new product. Working from the actual or standard cost for the similar parts and assemblies, the estimator can develop the estimated cost for manufacturing the new product.

A circuit card assembly, for example, might be of the same overall size and shape as one currently in production, but the component density may be 50% higher. In this case, the assembly labor to load the new card may be 50% higher than the time it currently takes to load the existing production circuit card assembly. A machined casting may have some 25 drilled and tapped holes, while a new design casting may be similar in size, weight, and shape, but have only 5 drilled and tapped holes. The estimated reduction in time to machine the new casting might then be estimated at 20%.

The percentage that the new product differs from the similar known product is a judgment call on the part of the cost estimator. If there is any doubt as to the validity of the estimated percentage similarity, it should be verified with the shop supervisor who must actually do the work. Estimating by the similarity method is reasonably accurate, and enables the cost estimate to be completed rapidly. It is easy to support such estimates to management and in final pricing negotiations.

Weight, Volume, Parts Count

In the estimating of aircraft manufacturing costs, weight is a commonly used variable. Weight alone is seldom enough, and speed is almost always included as a second variable for aircraft airframes. According to the Rand Corporation report *An Introduction to Equipment Cost Estimating* (1969), one cost estimating procedure for aircraft uses all of the following in their parametric equations:

Maximum speed at optimal altitude
Maximum speed at sea level
Year of first delivery
Total airframe weight
Increase in airframe weight from unit 1 to unit *n*
Weight of installed equipment
Engine weight
Electronics complexity factor

In addition, the following characteristics were considered for inclusion as part of the estimating procedure, although they were not used:

Maximum rate of climb
Maximum wing loading
Empty weight
Maximum altitude
Design load factor
Maximum range

An airframe typically changes in weight during both development and production as a result of engineering changes. For example, the weight of one fighter-aircraft structure assembly varied as follows:

Cumulative plane number	Airframe unit weight (lbs.)
1–11	1456
12–116	1941

117–241	1541
242–419	9193

Since labor hours are commonly associated with weight to obtain hours-per-pound factors, it is important to obtain weights that apply to each production lot when airframe weights by unit are not available.

Depending upon the industry, similar factors such as weight, parts count, and other significant variables may be used to develop cost estimates.

Concept Estimate Preparation

The cost estimate for a new manufactured product may be developed in much the same manner as any other estimate, the difference being the degree of product definition and the lack of past actual costs to compare the current estimate with. Labor estimates for the fabrication, assembly, and test operations can be developed from standard times, from similarity to other products from weight or volume factors, or simply by professional judgment and experience on the part of the cost estimator. Estimates for support labor, tooling, facilities, etc. are developed as in any regular estimate. The primary difference in a concept estimate for a manufactured product and an estimate based on complete engineering documentation is the detailed listing of the ground rules, assumptions, and contingencies that were used in preparing the concept estimate.

1.3.3 Contingencies and Assumptions

In concept estimating, the contingencies and assumptions are the critical factors in preparing the cost estimate. This means that the estimate is only as sound and valid as the assumptions and premises used in its preparation. Examples of ground rules, assumptions, and contingencies are:

1. Assume the product baseline for estimating purposes is the current engineering concept.
2. Assume a peak production rate of 15 units per day, achieved 6 months after start-up.
3. All work is to be performed on a 5-day, 8-hr./day standard workweek.
4. No new machines or production equipment will be required.
5. Assume a minimal, shop-aid-type tooling approach.

In addition to the above, there should be a projected engineering change curve indicating number of engineering changes to expect on a month-by-month basis after the engineering drawings are released.

Parts and Complexity Growth

As a new design for a manufactured product evolves and develops, it tends to grow more complex, and the total number of parts that make up the unit or product tends to grow, sometimes exponentially. This is due to a number of things, including the transition from concept to fully designed product. This brings out details and features of the product not envisioned at the concept stage. Another may be a tendency to add functions not thought of at the concept stage, or it may be found that the original design approach simply will not work. Consideration of customer requests or preferences is often a factor.

Other factors that increase costs to manufacture are closer tolerances, a more expensive finishing process for the piece parts, or more elaborate testing of the finished product than was envisioned in the design concept stage. Rarely does the product get less expensive to manufacture as the design evolves. A concerted product producibility engineering effort during design and development will go far in eliminating much of this growth, but will never entirely eliminate it.

As a result of this tendency for parts count and complexity to increase, the manufacturing cost estimate must either contain a cost factor increase to compensate for this growth, or state clearly that the estimate reflects only the cost to manufacture the concepted product, and any growth in the design is a change that must be costed out later.

Design Oversights

In the design and development of any new manufactured product, there will be errors or oversights that do not become apparent until the product is in production, or in the initial stages of production start-up. It may be found that the product does not perform as intended, or that it fails to pass qualification testing. Perhaps the decision was made to proceed with production on management risk ahead of production unit qualification tests in order to meet tight production schedules. Such problems can have major cost impacts, and may even cause the new product to be taken out of production for an indefinite period. Other design deficiencies may be of a less serious nature, but nevertheless will impact production and production costs. The lesser design oversights can be predicted with some degree of accuracy and usually consist of the following:

1. Design changes that are requested by manufacturing to ease fabrication and assembly. These might include loosening of tight tolerances, relocating a tapped hole to make it more accessible, adding a missing dimension, adding notes to the drawings, etc.
2. Design changes required to make the product perform as intended, sometimes called "make play" changes. These might include modifying the configuration of a moving part, adding a ground wire, adding a fuse to a circuit, or adding a production test requirement where none existed previously.
3. Record changes, which incorporate many of the changes described above into the formal drawing package.

In some of the high-technology industries, and the aerospace industry in particular, design changes well into production are the norm, not the exception.

Samuel L. Young, in his paper "Misapplications of the Learning Curve Concept" (1982), states:

> Production is characterized by poor starting documentation. The first planes or missiles are modified many times before they are acceptable. Engineering specifications are modified after the fact to reflect changes made first on the hardware. Many programs are put into production concurrently with development effort on identical hardware. It is not unusual to find hardware being produced simultaneously on both development and production contracts with the former far more costly than the latter effort. The whole approach results in extensive rework of previous efforts to make the end-items perform to specification.

As indicated earlier, such design changes are predictable, and must be considered by the manufacturing cost estimator at the time of the original concept estimate. Such predictions should take the form of an engineering change projection curve.

Engineering Change Projections

Projections of engineering changes with time at the start of a newly designed product into production can be made with a high degree of accuracy based on experience with previous programs. These change projections are normally prepared by the product design staff, and take the form of a graphical curve, as shown in Figure 1.12. As would be expected, the number of changes would be high in the production start-up cycle and would gradually diminish with time until about the ninth month after initial drawing release, when engineering drawing changes would be minimal or nonexistent. The shape of this curve, and the time required to run its course, are functions of the

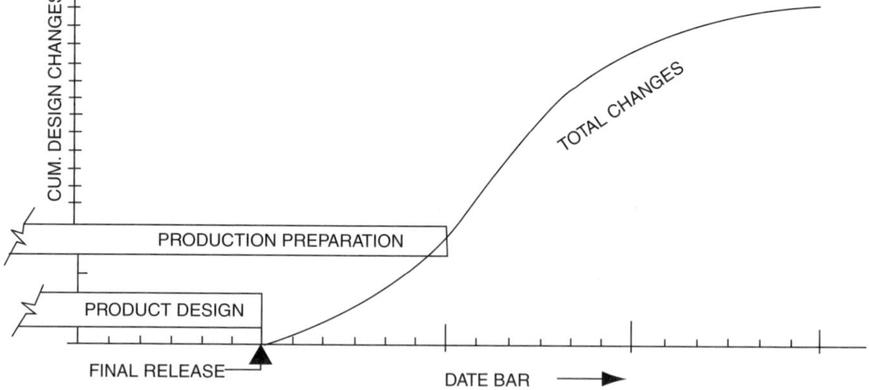

FIGURE 1.12 Design change curve.

product, the industry, and many other variables. That is why only the design department can draw this curve, based on knowledge of the firmness of the design at the time of drawing release, the condition of the drawing package, and the probability of major changes due to performance failures.

Engineering drawing-change-curve projections are a basic requirement if the manufacturing cost estimator is to develop a good estimate of manufacturing start-up costs.

They are especially needed if the estimate is based on concepts for a new manufactured product. If such curve projections have not been done in the past, there will almost certainly be resistance to providing them on the current estimate. In spite of this, both management and the manufacturing cost estimator should insist that these projections be provided from this point forward.

1.3.4 Examples of Concept Estimates

HAIL Mortar Round

The high-angle improved-lethality (HAIL) mortar round is being developed for use with the improved 11-mm mortar system for the U.S. Army. This estimate involves costing of the 1984–85 advanced development activities for the HAIL mortar round. Manufacture of rounds for the initial test phases, 70 rounds, will be done by a subcontractor. Then, during the advanced development phase, 765 rounds will be molded and assembled in-house for delivery to the army. A subcontractor will load the mortar bodies. Peak production rate will be 50 rounds per week.

The subcontractor who manufactured the initial 70 rounds will provide body molds for in-house manufacture of the 765 rounds. All other support equipment tooling will be provided in-house. The schedule allows 9 months to prepare in-house production facilities, and 6 months to manufacture the 765 rounds. Figure 1.6 is the design concept layout for the HAIL mortar round. The following ground rules and assumptions apply to this estimate:

1. Manufacturing labor standards were set at unit 1000 on a 15% learning curve, due to the developmental nature of the program.
2. Molding estimates are based on the subcontractor's current unit cost after approximately 50 units.
3. Supervision is estimated at 10% of assembly labor, and production control at 25% of assembly labor.
4. All estimated labor hours are based on an 8-hr. workday, 5 days per week.
5. All changes to the concepted product base line will be costed and funded separately.

Nonrecurring labor hours were estimated as follows:

Manufacturing engineer: 3295 hr.
Production control: 315 hr.
Tooling: 1587 hr.

Recurring labor hours were estimated to be:

Assembly: 1895 hr.
Manufacturing engineer: 2927 hr.
Production control: 433 hr.
Tooling: 919 hr.

Facilities requirements were determined to be as follows:

Inert manufacturing assembly: 5000 $ft.^2$
Pyrotechnic assembly: 900 $ft.^2$
Pyrotechnic bunker storage: 500 $ft.^2$

Capital equipment, including material handling and special high-cost tooling not covered by contract, is as follows:

Capital Equipment

	Order	Cost	Est. Weeks Lead
1.	Mold Press	41,200	12
2.	QC test, mold press	25,750	12
3.	Compounding equipment	20,600	21
4.	Electrostatic paint booth	9,000	32
5.	Lathe	1,000	32
6.	Cooling tower	15,500	21
7.	Curing oven	3,000	32
8.	Digital scale	5,000	24
9.	Air compressor	2,500	24
10.	Arbor press (avail.)	—	—
	Subtotal	132,550	—

The following high-cost tooling items are priced for subcontract build, but are covered by procurement.

1. Mold cavity inserts, 3 sets	$15,500	16
2. Mold	$40,000	16
3. Tooling for lathe (spin crimp oper.)	$8,000	16
Subtotal	$63,500	

Estimated material handling requirements are as follows:

1. Wire carts (6)	$3,000	16
2. ESD tote boxes (40)	$2,800	12

Subtotal	\$5,800
Total	\$201,850

All material requirements for manufacturing are to be provided by the engineering department, and are estimated separately.

Commercial Computers and Data Processing

The method of estimating a new product is shown in the outline below. The outline is a general guide and will vary depending on the complexity of the product or system, schedule, quantity to be produced, and technology.

Request design data
Prepare cost estimating outline
  Milestones
  Purchasing
  Original equipment manufacturing (OEM)
  Production
  Test
  Field engineering
Analyze material and labor inputs
Use independent judgment where voids exist
Prepare MLB (material, labor, and burden)
Determine learning curve base for each category:
Example:
  Material—cum. avg. or first-year block
  Labor—production cum. avg. or T_{100}
  Test—usually T_1

Extend MLBs, applying labor rate, published burden rates, and contingencies
Calculate cum. avg. and adjusted T_1 using published slopes for material and labor
Cum. avg. and adjusted T_1's are determined by applying a learning curve slope to each cost item for material and labor

1. Determine slope for each cost item from tables
2. Calculate cum. avg. and true T_1 for each cost item
3. Combine cost items by slope, separating material and labor
4. Total cum. avg. and true T1 for material and labor
5. Find adjusted T1 by dividing cum. avg. by true T1 and rounding off to the nearest percent slope, thereby eliminating fractions

The adjusted T_1 is the means of:

1. Calculating block averages
2. Calculating 75%, 100%, and 125% quantities
3. Simplifying MLB revisions

and results in the composite slope for material and labor used for monitoring costs after start of production

Calculate block averages using adjusted T_1's for labor and material

Document ground rules:

Preliminary review—for supervisor and contributors

First revision

Management review includes:

Pricing strategy

Maintenance

MLB review to the detail level

Example: How many PIBs?

AWW?

Unique costs

Note management recommendations and adjust final report accordingly

Submit MLB to office of pricing

1.3.5 Bibliography

Tanner, J. P., *Manufacturing Engineering: An Introduction to the Basic Functions*, Marcel Dekker, New York, 1985.

An Introduction to Equipment Cost Estimating, Rand Corporation, 1969.

Young, S. L., Misapplications of the learning curve concept, in *Learning Curves: Theory and Application,* Nanda, R. and Adler, G. L., Eds., Industrial Engineering & Management Press, Institute of Industrial Engineers, Atlanta, Georgia, 1982.

1.4 LEARNING CURVES

1.4.0 Introduction to Learning Curves

With the end of World War II in 1945, there was for the first time in history a mass of production cost data covering a wide variety of industries involved in quantity production. This included many different types of products, with some of the same products being produced simultaneously by multiple suppliers. There was an extreme focus for both government and private industry on building good products for a low cost at high production rates for an extended period of time. Stanford University was tasked to summarize this cost data, and to look for significant factors and trends. The university hired some of the better people from industry who had been actively involved in this production. One of the most significant things that came out of this study was the "learning curve" phenomenon. They discovered that, with the proper effort and motivation, the labor cost per unit continued to decrease as production quantities increased. Different products and process varied somewhat as to their slope, but they all seemed to match a standard logarithmic formula, of the form $Y = e^{-x}$.

For a manufactured product, the cost of value added (that is, labor hours per unit) declines predictably each time accumulated experience doubles. As an example, if the fifth unit built in a new process required 100 labor hours, and the tenth unit in the series required 75 hours, this 25% reduction represents what is commonly called a 75% learning curve. Similarly, if each doubling of experience brought about 20% reduction, the process would have a 10% learning curve. A learning curve is, in effect, a rate-of-improvement curve.

Labor reductions continue indefinitely as long as production continues. Such declines are not automatic; they require management and often capital investment. Learning curve effects can be observed and measured in any business, any industry, any cost element, almost anywhere. Key points about the learning curve concept are:

1. The learning curve itself is beyond question
2. The curve results mostly from management action and management learning
3. Costs will more surely decline if it is generally expected that they should and will
4. Full understanding of all the underlying causes is not yet available
5. The learning curve is so widespread that its absence is almost a warning of mismanagement or misunderstanding

Not all products have the same learning curve. Two main factors affect curve slope:

1. The inherent susceptibility of a process to improvement
2. The degree to which that susceptibility is exploited

The learning curve results from, and is a measure of, combined effects of both worker and management learning, specialization, investment, and scale. The effect of each is, at best, an approximation. The history of increased productivity and industrialization is based on specialization of effort and investment in plant, equipment, and tools, and so is the learning curve.

1.4.1 Types of Learning Curves

The Stanford and Crawford curves, as shown in Figure 1.13 and Figure 1.14, are two examples in general industrial use today. Though both curves embody the same basic principle, there are differences that can affect cost projections considerably. In addition, it is almost universal practice to assume that the learning curve is linear on log-log coordinates. A closer look, however, shows that long-cycle cost trends are not always straight. A typical S curve, as shown in Figure 1.15, is often found.

Stanford Curve

The Stanford curve shows that the cumulative average (the ratio of total accumulated time to total units) starting with the first unit is a straight line on log-log coordinates.

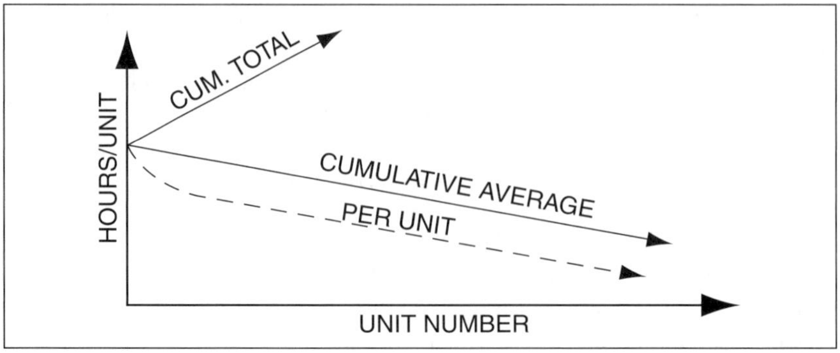

FIGURE 1.13 The Stanford learning curve. The cumulative average line is straight on log—log coordinates. The per-unit curve starts at a higher slope and soon parallels cumulative average as units are produced. This curve is often used in industrial applications.

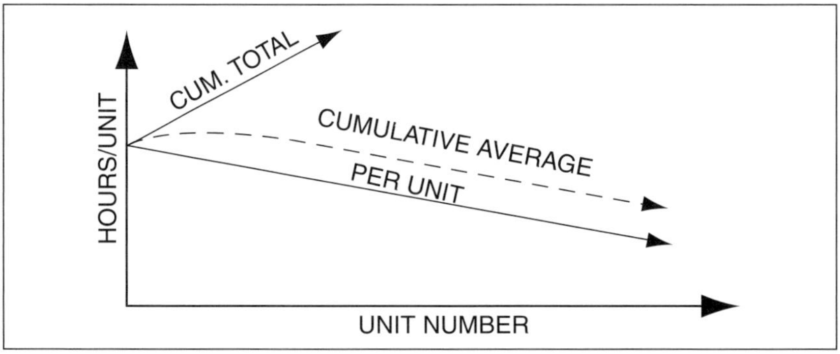

FIGURE 1.14 The Crawford learning curve. The per-unit line is straight on log—log coordinates. The cumulative average line starts at a lower slope and soon parallels the per-unit line. This curve gives higher projected values than the Standford concept.

The per-unit line is derived from this average and runs approximately parallel from about the tenth unit.

Crawford Curve

The Crawford curve assumes that the unit line is effectively a straight line on log-log coordinates. The cumulative average line is derived from the unit line and runs approximately parallel from about the tenth unit. The Crawford curve results in higher projected values than the Stanford curve. The Stanford cumulative average line gives a better averaging effect than the Crawford concept. The Stanford curve is consistent with historical data and minimizes inflated cost projections.

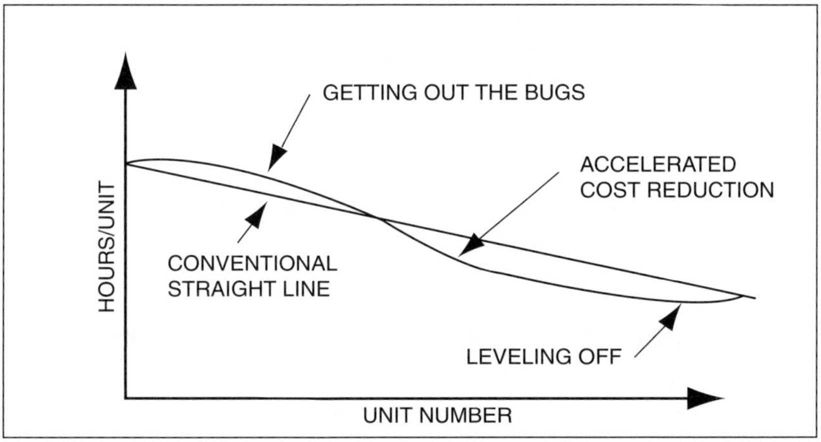

FIGURE 1.15 S-curve. The true learning curve shape may not be straight on log—log coordinates, though a straight line may approximate it well enough.

The S Curve

The S-curve upward bulge, as seen in Figure 1.15, signifies production cost penalties from time compression that occurs during new product introduction. The amount of penalty can vary among products and companies, because of differences in the degree of time compression of each. Such differences can be caused by any of several elements influencing new product introduction, such as product design, tools and facilities, worker capability, supervision, and support services.

1.4.2 Learning Curve Mathematics

The mathematical development of the learning curve is not complex. Correlation and other statistical methods show that a graph of real performance data (hours per unit versus cumulative unit number) can be described with relatively high significance by the following equation:

$$Y = AX^k$$

where
Y = labor hours to produce first unit
A = cumulative average time at any unit, X
X = unit number
k = learning curve slope

The equation describes the theoretical learning curve. It can be used to describe a learning curve unit time line, cumulative average time line, and total time line.

Though the equation will describe actual data, the mathematical exactness of the theoretical learning curve will not permit the unit time, cumulative average time, and total time lines all to be straight lines on log-log graph paper of the equation type described.

Learning curves, therefore, may be considered in these three classes:

1. The cumulative average time line and total line are of type $Y = AX^k$.
2. Only the unit time line is of that type.
3. All three lines are modifications of 1 and 2 above.

In practice, the learning curve is produced either graphically or in tabular form.

1.4.3 Practical Applications of the Learning Curve

The learning curve trend is a function of many variables. Factors that may affect curve slope include characteristics of the type of work to be done, program variations, and uncontrolled external factors. One of the more significant factors to consider in developing a cost estimate for manufacturing is the type of work. Different manufacturing processes can have different slopes. High-speed machining has a slope trend unlike that of conventional machining. Various assembly methods, wire harness fabrication, electronic assembly and wiring, final assembly, and testing all have different learning curve slopes. Consider the makeup of the end product with respect to different manufacturing processes in selecting learning curve slopes.

Also, to minimize distortion created by adding two or more learning curves with different slopes, segregate data as much as possible into appropriate subdivisions. If manufacturing processes such as machining versus sheet metal forming and fabrication cannot be separated, some other breakdown such as tail cone, center body, and/or forward body section might separate data of different process mixes well enough to minimize distortion.

Tooling and facility commitments affect both first unit cost and attainable learning curve slope. Influences on these commitments include the stage of product development, planned production quantities, lead-time constraints, limitations in available investment funds, and risk of not obtaining follow-on production.

For a minimum tooling approach, the expected first unit cost can be higher than if the tooling were designed for efficient rate production. Cost improvement for minimum tooling depends on improvement in shop personnel efficiency. If tooling is designed for efficient rate production, initial unit cost may still be high because of problems in shakedown. The learning curve should reflect rapid resolution of tooling problems. Speedy shakedown often results in a steep curve in early production, which is impossible with minimal tooling.

Automatic test equipment in many applications can bring large cost reductions in production. If the equipment has been fully debugged before coming online, its effect will show on the learning curve as a step reduction at the unit number where it starts to work.

Cause-and-Effect Relationships

Other factors affecting the learning curve include the following:

1. Changes from development to pilot production and to full-rate production.
2. Production rate constraints and dictated schedule changes. These are often caused by funding limitations, which bring program stretch-outs. Higher rates allow more units in work at a given time, allowing better labor utilization. Lower rates mean fewer units in work, and thus less efficient labor utilization. Lower rates also result in smaller lot sizes, increasing the number of setups required for the same number of units.
3. Number of setups. Detail parts fabrication incurs both setup and run costs. Setup costs occur each time a lot is fabricated, but are independent of the number of parts in the lot. Run costs are incurred each time a part is made. Average cost of parts in a lot (total cost divided by the number of parts made) depends directly on the number of parts made with each setup.
4. Production interruptions. Regardless of the cause, these result in a loss of learning.
5. Personnel changes and turnover.
6. Management learning and management action.
7. Timely resolution of manufacturing problems.

Pitfalls to Be Avoided

When applied correctly, the learning curve is one of many valuable tools available to the manufacturing cost estimator. However, some caution should be exercised in the selection and use of labor cost data in plotting the learning curve.

1. If a company receives a larger portion of its raw material in a finished state, its labor input per unit of product should decline. When this decline is merely the result of shifting labor input from the company to the plant of a supplier, it would be erroneous to consider the decline as a real reduction in labor hours. The same work is being performed in total, and no net saving has been realized.
2. It may be possible to generate a direct labor saving of 1000 hr. by spending 2000 hr. on additional tooling and manufacturing engineering. It is obvious that there would be no real savings as a result. It is clear from this that direct labor costs cannot be considered separate from the changes in the other elements of cost.
3. There is always the danger of false labor savings resulting from the reshuffling of accounting records. The learning curve plots direct labor only. What happens when greater use is made of supervisory labor? This labor is not the same; it is classified as indirect. If the emphasis is placed on direct labor savings alone, without considering indirect labor, a distorted picture of direct labor cost savings will result.

Finally, it should be stated that there is a practical limit to the amount by which costs can ultimately be reduced. A common practice is to assume that after unit 1000, the standard time for producing the product will be reached, and beyond that point the learning curve effect is negligible. While this is not entirely true, it is the point where learning and improvement is at a minimum. The simple selection of unit 1000 is recommended especially when estimating costs for job order manufacturing, or where there is a very high labor content per unit.

The selection of the point where the work will be performed to standard time may vary as a function of the number of days from start-up, rather than the number of units produced from startup. An example of this would occur in high-rate production shops, where as many as 1 million parts may be produced in the process of proving out special production tooling. This could be done at the rate of 100,000 to 200,000 parts per day—and the production parts could be produced at standard following the debugging process from day 1 (or, in this case, from production unit 1). Again, depending on the process, the tooling, and the factory, there may need to be a productivity factor applied to the standard. For example, routine operations of 90% productive to standard might be acceptable. In this case, the cost achieved would never drop below 111% of standard (standard ÷ 0.90 = 1.11). The reductions beyond this point would be due to changes in the process of doing the job, which would reduce the standard hours.

Graphical Techniques

Most learning curve applications in cost estimating involving the direct labor cost for a manufactured product are performed graphically using log-log graph paper. A learning curve slope protractor is shown in Figure 1.16, with the values of the slope coefficient for slopes from 60% to 97%. To use the protractor, the following procedure should be followed:

1. Plot the cumulative average time per unit, or time per unit on log-log graph paper
2. Fit a straight line through the data points
3. Overlay the slope protractor on the line, and find the learning slope in percent
4. Obtain values of the learning slope from the table
5. By extending the line, the time for unit 1 can be read directly from the graph

1.4.4 Application Examples

The learning curve is a primary tool of the Department of Defense and defense contractors in negotiating final prices for military hardware. In many applications, its usefulness in pricing decisions can extend even beyond direct labor costs. The curve is useful in make-or-buy analysis, especially when buying means ordering on a negotiated price basis.

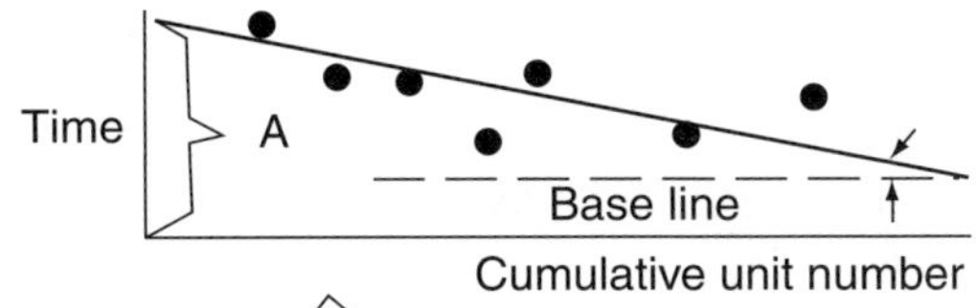

(1) Plot cumulative-average-time-per-unit or time-per-unit on log-log graph paper.

(2) Fit a straight line to data points

(3) Read time-for-one (A).

(4) Find learning percent from protractor.

(5) Obtain values of learning slope (B) from table.

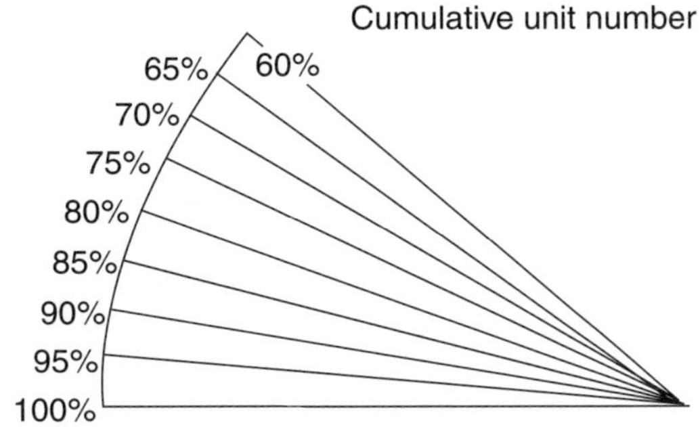

%	B	B+1	%	B	B+1	%	B	B+1	%	B	B+1
60	-0.737	0.263	70	-0.514	0.485	80	-0.322	0.678	90	-0.152	0.847
61	-0.713	0.287	71	-0.494	0.506	81	-0.304	0.696	91	-0.136	0.864
62	-0.690	0.310	72	-0.474	0.526	82	-0.286	0.714	92	-0.120	0.880
63	-0.667	0.333	73	-0.454	0.546	83	-0.269	0.731	93	-0.105	0.895
64	-0.644	0.356	74	-0.434	0.566	84	-0.255	0.748	94	-0.089	0.911
65	-0.621	0.379	75	-0.415	0.585	85	-0.234	0.764	95	-0.074	0.926
66	-0.599	0.401	76	-0.369	0.604	86	-0.218	0.782	96	-0.059	0.941
67	-0.578	0.422	77	-0.377	0.623	87	-0.200	0.800	97	-0.044	0.956
68	-0.556	0.444	78	-0.358	0.642	88	-0.184	0.816			
69	-0.535	0.465	79	-0.340	0.660	89	-0.168	0.832			

(6) Substitute values of A, B, and B + 1 to find learning curve.*

Learning curve	If unit times were plotted	If cumulative average times were plotted
(7) Time for Rth unit (R= 1,2,3...)	AR^B	$(B+1)AR^B$
(8) Cumulative average time for Rth unit	$\frac{AR^B}{B+1}$	AR^B
(9) Total time for Rth unit	$\frac{AR^{B+1}}{B+1}$	AR^{B+1}

* For best accuracy R should be greater than 10.

FIGURE 1.16 Learning curve slope protractor.

Problem

Given an estimated first-unit cost of 1200 hr., and an anticipated 11% improvement curve slope based on experience with similar products, calculate projected costs for the first 50 units and for 100 follow-on units.

Solution

This problem can be solved by using the cumulative average equation $Y = kx^{-n}$ after determining $-n$ by using the equation

$$-n = \log(p - 100) \div \log 2$$

Procedure

1. Determine $-n$, given that $p = 88\%$:

$$\begin{aligned} -n &= \log(0.88) \div \log 2 \\ &= -0.0555 \div 0.3010 \\ &= -0.1844 \end{aligned}$$

2. Substitute 1200 for k and -0.1844 for $-n$ in the cumulative average cost equation:

$$y = 1200 \times X^{-0.1844}$$

3. Solve for y at $x = 50$, and at $x = 150$:

$$\text{Cumulative average for 50} = 1200 \times 50^{-0.1844} = 513.2$$
$$\text{Cumulative average for 150} = 1200 \times 150^{-0.1844} = 496.3$$

4. Determine cost for 50 units:

$$\text{Total for 50} = 50 \times 583.2 = 29{,}160$$

5. Determine cost for first 150 units:

$$\text{Total for 150} = 150 \times 476.3 = 71{,}445$$

6. Determine cost of 100 units following 50 units by subtracting the equation in step 4 from the equation in step 5:

$$71{,}445 - 29{,}160 = 42{,}285$$

Another common application of the learning curve is in determining the slope of the curve, and the indicated first unit (T_1) value, when the hours for the first two blocks or lots of production units are known.

Problem

Given that the first 3 units cost a total of 2539 hr., and the next block of 4 units costs 2669 hr., determine the theoretical cost of the first unit and the improvement curve slope indicated. Assume that the difference in block size did not affect the trend.

Solution

The cumulative average costs through 3 and through 7 units can be readily calculated, and the cumulative average cost equation $Y = kX^{-n}$ can be solved to determine the first unit (k) cost, and the curve exponent ($-n$). The slope percentage equation $p \div 100 = 2^{-n}$ can then be used to determine the curve slope percentage.

Procedure

1. Determine cumulative average costs at 3 and at 7 units:

$$\text{At } X = 3, y = 2539 \div 3 = 846.3$$
$$\text{At } X = 7, y = (2539 + 2669) \div 7 = 5208 \div 7 = 744.0$$

2. Substitute these values into the equation $y = kX^{-n}$:

$$846.3 = k \times 3^{-n} \text{ (1)}$$

and

$$744.0 = k \times y^{-n} \text{ (2)}$$

3. To solve these two simultaneous equations, first divide Equation (1) by Equation (2) to eliminate:

$$(846.3 \div 744.0) = k \times 3^{-n} \div k \times 7^{-n}$$
or
$$1.1375 = (3 \div 7)^{-n} = 0.4286^{-n}$$

4. To solve for $6n$, take the logarithm of both sides of the equation:

$$\log(1.1375) = 6n \log(0.4286)$$

Using a scientific calculator, this becomes:

$$0.05595 = 6n(-0.36795)$$

5. Solve for $6n$:

$$6n = 0.05595 \div 60.36795 = 0.15206$$

6. Determine p:

$$p \div 100 = 2^{-n} = 2^{-0.15206} = 90\%$$

7. Substitute $6n$ into Equation (1):

$$846.3 = k \times 3^{-0.15206} = k \times 0.8462$$

8. Solve for T_1 cost (k):

$$k = 846.3 \div 0.8462 = 1000 \text{ hr.}$$

Problem

Assuming that the improvement trend in the previous problem will continue, calculate the cost of an additional 12 units.

Solution

The T_1 cost and n factor were determined in the preceding problem. The cumulative average method can be used to determine the cost of additional units.

Procedure

1. Given that $T_1 = 1000$ and $6n = -0.15206$, calculate the cumulative average through $3 + 4 + 12$ or 19 units:

$$y = 1000 \times 19^{-0.15206} = 639.1$$

2. Compute the total cost through 19 units:

$$\text{Total cost } yx = 639.1 \times 19 = 12{,}142.9$$

3. Calculate the cost of the last 12 of these units, given that the first 7 cost 5,208 hr., as determined in the preceding problem:

$$12{,}143 - 5{,}208 = 6{,}935 \text{ hr.}$$

Note that the block average cost for thee additional units is

$$6{,}935 \div 12 = 577.9$$

As in the previous problem, both of these problems can be worked without calculations by using the learning curve on log-log paper. It is often recommended that both techniques be used as a method of checking or verifying an answer.

Problem

A common problem often faced by the manufacturing cost estimator is that of constructing a learning curve from historical labor cost data, and then converting the series of learning curves for each cost center or department into a composite learning curve for the total manufactured product. Figure 1.17 shows how this can be done by using a weighted average to determine the composite curve.

This particular application involved taking historical labor cost data from development and initial low-rate production to determine composite learning curve slopes for projecting rate production labor costs to manufacture and test receivers and antennas. The weighted average is determined by calculating the percentage the individual department labor cost is of the total cost, then applying that cost percentage against the learning curve for the department, and then obtaining the composite curve slope by totaling all of the weighted slope percentages of the different departments. Figure 1.17 also shows this calculation.

1.4.5 Bibliography

Gibson, D.C., *Manufacturing Improvement Curves: Concept and Application,* McDonnell Douglas Corporation, St. Louis, Missouri, 1981.

Nanda, R. and Adler, G. L., Eds., *Learning Curves: Theory and Application,* Industrial Engineering & Management Press, Institute of Industrial Engineers, Atlanta, Georgia, 1982.

COMPOSITE LEARNING CURVE ANALYSIS

DEPARTMENT	LEARNING CURVE SLOPE (%)	LABOR HOURS ANT.	LABOR HOURS RCVR.	PERCENTAGE (%) ANT.	PERCENTAGE (%) RCVR.
Production	81	1,944	28,888	4.8	50.9
Quality Control	81	458	5,121	1.1	9.0
Magnetics	97	216	15,657		27.6
Mechanical Fabrication	81		1,683		2.9
Machine Shop	90	12,690	5,354	31.4	9.6
Antenna Lab	87	420		1.0	
Support Facilities	95	1,966		4.8	
Quality Control	95	592		1.5	
Product Test	95	1,486		3.7	
Integration Electronics	95	1,752		4.3	
SS Assembly	95	8,292		20.5	
Microelectronics Assembly	95	2,274		5.6	
Hybrid Assembly	97	4,261		10.7	
Top Assembly	87	4,088	12	10.6	
	TOTAL	40,439	56,715	----	----

$$\text{Receiver Composite Slope} = (.509)(.81) + (.09)(.81) + (.276)(.97) + (0.29)(.81) + (.096)(.90) = \boxed{87\%}$$

$$\text{Antenna Composite Slope} = (.048)(.81) + (.011)(.81) + (.314)(.90) + (.01)(.87) + (.048)(.95) + (.015)(.95) + (.037)(.95) + (.043)(.95) + (.205)(.95) + (.056)(.95) + (.107)(.97) + (.106)(.87) = \boxed{84\%}$$

FIGURE 1.17 Converting a series of element learning curve histories into one representative composite curve for the entire product.

Tanner, J. P., The learning curve, a line on labor cost, *Production Engineering,* May 1985.

Tanner, J. P., *Manufacturing Engineering: An Introduction to the Basic Functions,* Marcel Dekker, New York, 1985.

1.5 MATERIAL AND SUBCONTRACT ESTIMATING

1.5.1 Pricing the Bill of Material

In the majority of cost estimates for manufactured products, the cost of purchased parts, raw materials, commodity items, and items that are subcontracted constitute the biggest part of the cost estimate, often constituting 75% of the total cost of the product. The importance of accurately pricing the buy items on the bill of material cannot be emphasized strongly enough. The material estimator or buyer must thoroughly understand the potential pitfalls that commonly occur in estimating material costs:

1. Inadequate product specifications may result in prices for material at an incorrect quality level.

2. Incorrect delivery requirements may force the use of more expensive substitutes, or production changes that require more setups, or that cause production delays.
3. Incomplete product specifications may result in material estimates that do not cover the actual material costs involved in the final engineering design or product quality or reliability requirements.
4. Material price levels may change upward and exceed the estimated material costs. This is a particularly acute problem when estimating material costs for products with long manufacturing lead times, or for delivery at a future time.
5. Material price breaks may be anticipated and planned but not realized, because of delivery schedule changes or revisions to inventory policy.

The purchasing function is the primary support to the manufacturing cost estimator in determining material costs. For best results, direct quotes from vendors and subcontractors should be obtained. Where time to prepare the estimate is limited, purchasing must use cost history for commonly used purchased parts and material. If historical cost data is used, it must be factored for inflation and anticipated cost growth.

1.5.2 Estimating Standard Purchased Parts

Included in the category of standard purchased parts are items such as common hardware and fasteners, electronic components, certain types of bearings, gears, pulleys, belts, chain drives, electric motors, electrical connectors, wire, cable, clutches, batteries, power supplies, switches, relays, and similar items. These items can be purchased by ordering from catalogs or commercial specifications published by the manufacturer, and are usually available from stock inventory. They are sometimes priced from standard price lists provided by the manufacturer, with discounts or price breaks for quantity buys. In many instances such standard parts are available from local sources of supply such as distributors or manufacturers' agents or representatives. Figure 1.18 shows a typical line of standard fasteners available from one manufacturer.

The call-out on the bill of material for standard parts may be by the manufacturer's part number, by a military standard (MIL) number call-out, or by a number assigned by the company that is a drawing or specification of the item. MIL and company specification numbers allow the buyer to order the part from any manufacturer whose product meets the specification or MIL requirements.

Estimating the cost of standard purchased parts using cost history or manufacturers' price lists presents a fairly low risk in the pricing of these items. Only high-dollar standard purchased parts should be supported with a recent quote from the manufacturer or distributor. High-dollar items are those with a unit cost of $5 or more. Figure 1.19 shows a representative bill of material call-out for a latch on a computer-generated bill of material. This list summarizes the total requirement for a standard or component

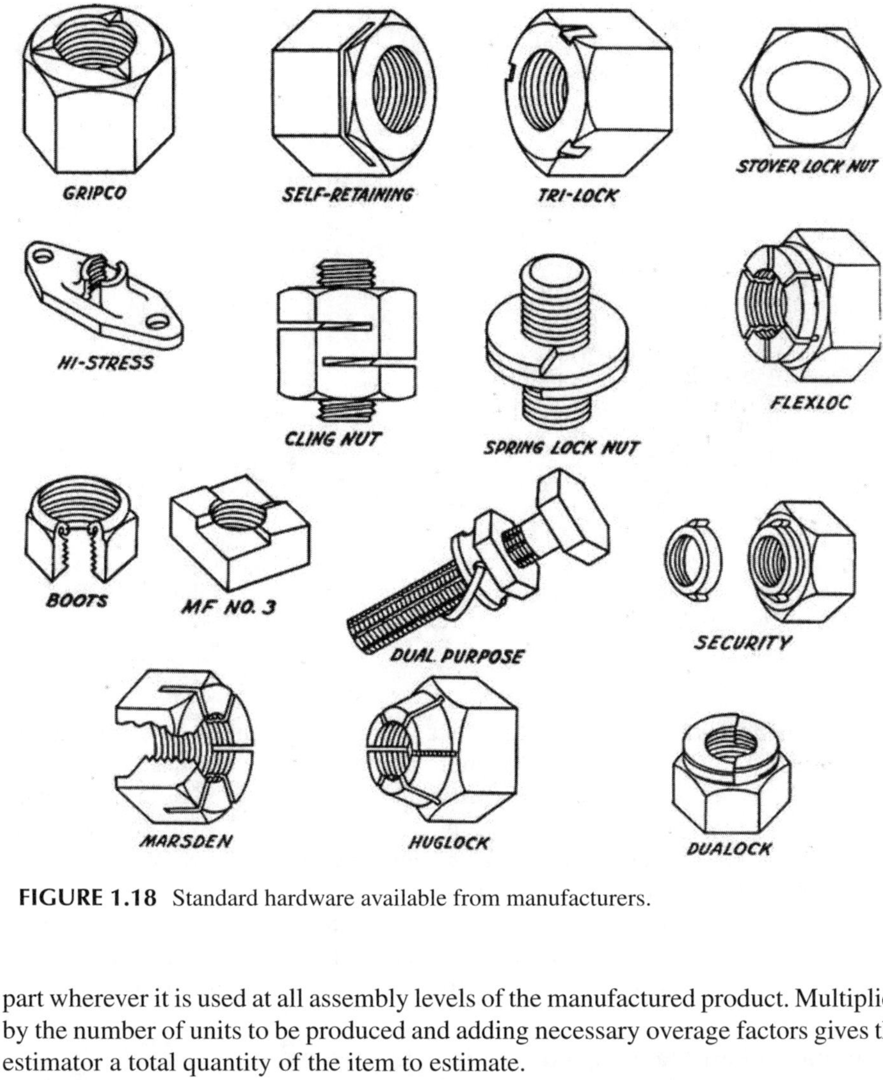

FIGURE 1.18 Standard hardware available from manufacturers.

part wherever it is used at all assembly levels of the manufactured product. Multiplied by the number of units to be produced and adding necessary overage factors gives the estimator a total quantity of the item to estimate.

There is much that the buyer can do to bring the cost of standard parts down below the list price level. He or she can, on his or her own initiative, improve costs by a number of tactics, including quantity buying, vendor price negotiation, material substitution, scrap reduction, etc. This should be taken into consideration in estimating the costs for standard purchased parts.

1.5.3 Estimating Raw Material Requirements

Much of the purchased material for any manufactured product must be ordered in the form of sheet, bar stock, or blocks of material; in spools of wire and solder; as

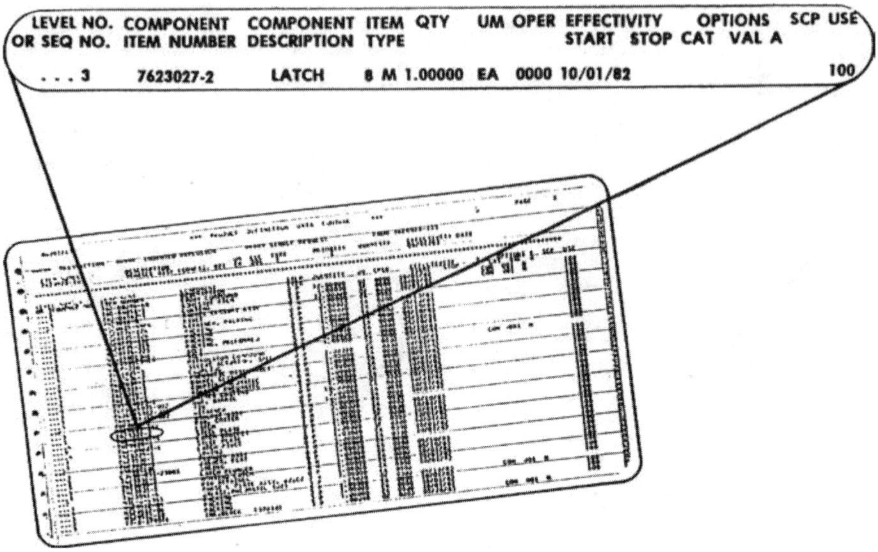

FIGURE 1.19 Representative computer-generated bill of material calling out a standard latch to be purchased.

drums of chemicals and solvents; in pint, quart, or 5-gal. containers; etc. The material estimator must calculate the correct quantities of this raw material and bulk supplies or commodities to buy in order to manufacture the required number of units with a minimum of scrap or waste. To do this, the material estimator must list every fabricated part to be manufactured in-house by quantity required, size, weight, overall configuration, and type of raw material needed. If the product has been manufactured previously, there is a good chance that this information is already available on the process sheet, methods sheet, or work instructions for the part in question. If not, the cost estimator or material estimator must determine the requirement.

Determining Raw Material and Bulk Item Quantities

The estimator analyzes each item to determine the amount of raw material that must be purchased to manufacture the required number of units. The amount of raw material for a given piece part usually comes from determining the weight of raw stock used per piece. A part that is machined would include the finished dimensions of the part plus the amount of stock removed by machining. The overall dimensions of such a piece of stock would be determined and the volume calculated. The volume is multiplied by the density of the material (weight per unit volume) to obtain the weight. A piece that is irregular in shape is divided into simple components, and the volumes of the components are calculated and added together to give total volume.

The total volume is multiplied by the density of the material to obtain the weight. The material cost is then obtained by multiplying the weight by the price per pound of the material. In the case of a sheet metal stamping fabricated from aluminum sheet stock, the procedure would be as follows:

1. Weight of sheet = gauge × width × length × density
2. Number of pieces per sheet = length of sheet ÷ length of multiple
3. Weight per piece = weight of sheet ÷ pieces per sheet
4. Piece part material cost = piece part weight × aluminum price/lb.

When estimating bar stock, the length of the piece, plus facing and cutoff stock, should be multiplied by the weight or price per inch of the stock diameter. Scrap, butt ends, chips, etc. that are lost in processing must be considered in the estimate. These losses vary from 3% to 10% depending on the job, current shop practices, and the material itself.

Bulk items such as spooled core solder can also be easily estimated by determining the weight of solder used for each solder joint, then multiplying by the number of solder joints to arrive at the total weight of solder used per assembly. When this number is multiplied by the total number of assemblies, the total weight of solder needed is determined. Using the table shown in Figure 1.20, the number of feet of the required type of solder needed can be readily determined. An additional 10% should be added to the calculated requirement to cover losses, solder required to tin the soldering iron, etc.

Diameter Inch	Area Sq. Inch	Lead Wire Ft. per Lb.	30/70 Solder Ft. per Lb.	40/60 Solder Ft. per Lb.	50/50 Solder Ft. per Lb.	60/40 Solder Ft. per Lb.	Tin Wire Ft. per Lb.
.032	.00080	254	297	310	325	340	396
.036	.00113	180	211	220	230	241	281
.040	.00126	161	188	196	206	216	251
.045	.00159	128	150	156	164	172	200
.050	.00196	104	122	127	133	129	162
.056	.00246	82.6	96.6	101	106	111	129
.063	.00312	65.1	76.2	79.4	83.3	87.2	102
.071	.00396	51.3	60.0	62.6	65.7	68.7	80.0
.080	.00503	40.4	47.3	49.3	51.7	54.1	63.0
.090	.00636	32.0	37.4	39.0	41.0	42.9	49.9
.100	.00785	25.9	30.3	31.6	33.2	34.7	40.4
.112	.00985	20.6	24.1	25.1	26.4	27.6	32.1
.125	.01227	16.6	19.4	20.2	21.2	22.2	25.9
.140	.01539	13.2	15.4	16.1	16.9	17.7	20.6
.160	.02011	10.1	11.8	12.3	12.9	13.5	15.8
.180	.02545	8.0	9.4	9.8	10.2	10.7	12.5
.200	.03142	6.5	7.6	7.9	8.3	8.7	10.1

Note: The number of linear feet per lb. of flux-core wire solder will be somewhat greater than the above figures for solid wire. Where the amount of flux is 1.1% by weight. linear footage is increased by 9% over solid wire, with 2.2% the increase is 15%, and with 3.3% the increase is 27%.

(Courtesy National Lead Co)

FIGURE 1.20 Number of feet of solder required.

When determining the amounts of raw material, bulk and commodity items, and the like required for the job, the material estimator must consider that material for stampings is bought by the sheet or reel, solder by the spool, paint by the gallon, and cleaning solvents by the 55-gal. drum. Certain adhesives, potting compounds, and bonding agents are bought in 5-gal. containers. In many instances, far more than the job requirement must be purchased because of this, often inflating the material cost for the job. This is especially so when there is no other use for the item except the current job. The offsetting consideration to this is that substantial price breaks are often available on these items when purchased in these amounts, and in these container sizes.

In foundry work, the manufacturing cost estimator calculates raw material costs as follows:

1. Determine the amount of metal required to charge the furnace by determining the ratio of finished casting weight to the weight of the metal charged into the furnace, based on previous experience, to determine the shop yield factor
2. Calculate the furnace charge per casting by dividing the casting weight by the yield factor

Shop yield will vary with different casting materials, and consideration must also be given to metal losses due to spills, oxidation, gate cutoffs, and overruns. A good rule of thumb is to add 10% of the casting weight for these losses. Any metal not consumed in the finished casting or lost is returned to the furnace for remelting. This remelt metal is determined by subtracting the sum of shop yield and metal lost from the amount of metal charged. By converting these values to percentages, with the amount of metal charged being 100%, we can express the amount of metal returned for remelting as a percentage.

In forgings and forged parts, the cost of material averages about 50% of the total cost. To determine material cost for a forged part, the estimator must first calculate shape weight using sketches or engineering drawings. The part can be divided into suitable geometric sections, the volume for each section obtained, and by adding the section volumes together, the total volume obtained. Multiplying this by the density of the forging material gives the shape weight of the forging, and adding 4% to the shape weight gives the net weight. The gross weight, which is the weight of forging stock required to actually make the forging, is determined next. This weight is found by adding material lost through flash, scale, sprue, tonghold, and cut waste to the net weight:

Flash is excess metal extruded as a thin section that surrounds the forging at the die parting line to ensure that all parts of the die are properly filled. Flash width and thickness varies with the weight of the forging. Flash also includes punch-out slugs from holes in the part. Punch-out slugs vary with the dimensions of the punched holes and the thickness of the section through which the holes were punched.

Scale is material lost due to surface oxidation in heating and forging. The amount of this loss varies with surface area, heating time, and material.
Sprue is the connection between the forging and the tonghold.
Tonghold is a projection used to hold the forging.
Cut waste is stock lost as sawdust when bar stock is cut to length by a saw, plus bar end loss from length variations and short ends from cutting the stock to exact length.

Gross weight is then calculated by totaling the percentages estimated for each of these factors and adding to the net weight. These percentages should be determined from historical data and experience in your shop, but should be approximately as follows:

Flash = 1.5%
Scale = 5%
Sprue = 7%
Tonghold (included as part of sprue)
Cut waste = 5%

The direct cost of forging material is then calculated by multiplying the gross weight by the cost per pound of the forging stock material.

Estimating Overages, Scrap, and Line Losses

Most firms have an established policy for estimating material scrap, overages, and line losses. Until such policies are established, or are determined from historical data, the following factors are given as good industry averages:

1. Purchased parts	
Less than $1.00 unit price	15%
$1.01 to $5.00 unit price	10%
$5.01 to $10.00 unit price	2%
Over $10.00 unit price	Determined by estimator
2. Forgings and castings	
Less than $100 unit price:	
Sand castings	10%
Pressure test castings	25%
Other forgings and castings	10%
Unit price $100 or over	5%
3. Raw material	
Titanium	10%
Aluminum, steel, magnesium, rubber (sheet, plate, bar, rod, or tubing)	25%

Extrusions	10%
Wire	35%
Miscellaneous (plastics, fabrics, tapes, etc.)	20%

It is important to point out that these factors may vary considerably among companies, even in the same industry. In the metal forming and stamping industry, sheet stock utilization can well determine whether the job is profitable, and what the percentages for sheet stock overages will be. These same considerations apply in the utilization of sheet material for printed circuit card fabrication. Careful planning of the position of the blanks on the sheet stock means maximum stock usage and minimum scrap.

Line losses of expensive parts and components can drive scrap factors very high if not controlled. A scrap tag system that requires the shop to complete a scrap tag for each part lost or damaged in assembly can do much to ensure that line losses remain low and under control. Such systems make the estimator's job easier and lead to lower material costs through lower scrap factors.

Estimating Material for Tooling

The cost of material for tooling is a significant part of the total nonrecurring cost for any new program. Tooling material requirements are taken from the bill of material or parts listing on the tool drawing, or are estimated from the tooling concepts envisioned by the manufacturing cost estimator. Included are such items as tool steel, drill bushings, quick-release clamps, hold-down buttons, and similar items for the fabrication of special-design tooling. Tooling material also includes standard perishable tooling, such as drill bits, cutters, end mills, punches, reamers, broaches, and other similar items that are consumed over the life of the program.

If possible, these materials should be calculated exactly from tool drawings and perishable-tool-usage experience. If such documentation or tool history is not available, then tooling material costs on previously built similar tools should be used, and perishable-tool usage estimated from experience on earlier jobs or programs that had similar requirements. It should be mentioned that the cost of special-design tooling can be held to a minimum by using inexpensive tooling materials such as wood, sheet metal, and aluminum instead of expensive tool steels to fabricate many tools such as holding fixtures, assembly jigs, and motion economy devices. Also, many times a standard, off-the-shelf tool or fixture with minor modification will do the same job as a special-design tool, at a fraction of the cost.

Tooling material may often be purchased directly by the tool design or manufacturing engineering organization without going through the purchasing, receiving, and receiving inspection groups, and as a result would not carry the normal material overhead or burden.

1.5.4 Long-Lead and High-Dollar Items

One of the most important tasks in the preparation of any cost estimate for a manufactured product is the identification of the long-lead and high-dollar material items.

This information is basic to the development of the production start-up schedule and will be a key factor in determining whether the requested delivery schedule can be met. The material items that are the high-dollar items in the cost of product material are the ones that need to be worked by the buyer and material estimator to obtain firm quotes, and if possible, competitive quotes. How well this is done will have a major bearing on whether a competitive and winning cost proposal is finally submitted.

If time permits, qualified backup sources of supply should be located for each and every long-lead item on the bill of material. This ensures that the failure of any long-lead supplier to meet delivery of an item of acceptable quality will not shut down the production line. There is a certain amount of risk in any cost proposal or estimate. Risk that the work will cost more than the estimated cost, risk that promised deliveries will not be met, and risk of not getting the work at all because it was awarded to a lower bidder, or one who promised a better delivery schedule, are the major risks in any cost estimate for a manufactured product. Anything that can be done to develop a winning, low-risk cost estimate should be a prime consideration.

Any item with a delivery promise of 6 months or longer can be considered long-lead. It is more difficult to define high-dollar items, except to say that any single item on the bill of material that costs significantly more than the others is high-dollar.

1.5.5 Material Learning Curves and Inflation

There are numerous ways to apply learning curve theory to the purchase of material. This normally means that the vendor's experience or learning should reflect in the cost of the material, much as the in-house factory labor declines in cost with experience in manufacturing the product.

When the NCR Corporation buys material from a vendor, the cumulative volume is measured in NCR's units of experience (measurement units of the cumulative volume), which differ from the vendor's units of experience. They claim that the following can be proven:

1. NCR's slope for a purchased material may be steeper or flatter than the vendor's slope. This is due to differing units of experience and other factors.
2. NCR's slope can be estimated from historical data, by a projection by comparison of experience with like products, or from the vendor's learning curve, when known.

Figures 1.21 and 1.22 show how a buyer's learning curve is related to a vendor's learning curve. Note that the buyer's slope changes from year to year. Given historical data for a year or more, the costs for the next 5 years can be estimated (see Figure 1.22). The amount of the year-to-year change decreases as the technology ages and depends on many factors: differing units of experience, the time point at which the buyer defined unit 1, the year-to-year vendor volume, and more.

Learning curves such as those shown in Figure 1.21 and Figure 1.22 reflect learning by the vendor only. The buyer can, on his own initiative, significantly improve

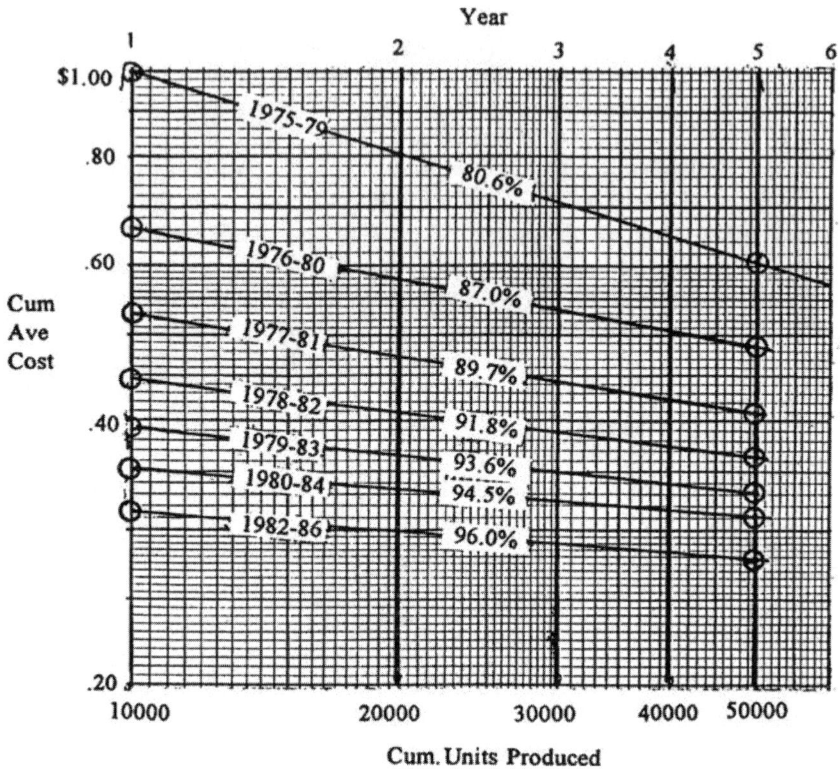

FIGURE 1.21 Buyers' cumulative average cost experience for a particular part number.

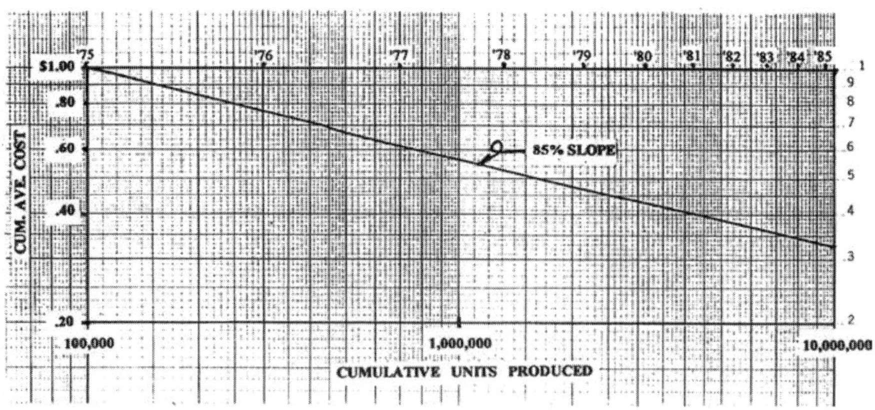

FIGURE 1.22 Suppliers' cumulative average selling price for a particular part number.

his material costs by a number of tactics, including quantity buying, vendor price negotiation, material substitution, and scrap reduction. Therefore, these curves represent a minimum slope achievable without any additional improvement by the material user. Further, even these minimum slopes are not automatically achieved unless the buyer goes after them and spurs the vendor to ride his learning curve.

While a firm's material slopes may be higher or lower than the vendor's slopes, in reality, a firm's slope depends mainly on its own actions, not on the vendor's actions alone. Material learning curve slopes are usually estimated by examining the history of past buys. Such costs include the average inflation experienced over the time period being analyzed as well as the improvement factors not attributable to vendor learning, such as quantity buys, better negotiation of prices, switches to lower cost vendors, etc. It is often claimed that material learning should be based on costs before inflation. This is undesirable and impractical because:

1. It is well known that there is a wide divergence of inflation rates among commodities. How does the material estimator determine the appropriate year-by-year rate of inflation for each commodity? Who can say what the real inflation rate has been for MOS or TTL devices, for example? A detailed analysis of vendor operations and input materials would be required for such a determination.
2. Even if inflation is removed, the learning rate is not the same as the vendor learning rate because of the differing units of experience, and because buyer as well as vendor learning is included in the applicable learning curve slope.
3. It has been shown that learning curves can be fitted to data that includes inflation, and fitted to the buyer's units of experience.

The factors that are important in determining material learning curve slopes include:

1. Validating the slope with historical data or by comparison with similar materials at the same technological age
2. Using known or forecast prices (reflecting current adjustments for inflationary surges, price wars, etc.) for the basic cost estimate that is used for entering the learning
3. Correcting the slope for likely enduring changes in the environment (a dying technology that will have little slope, a technology with surging usage that will have an increasing slope, a continuous relative increase or decrease in average inflation, etc.)
4. Correcting the historical slope for technology aging

Using the techniques described above can have a significant impact on how competitive the material estimate for a manufactured product really is. Material may be the largest part of the estimated cost of the product, and as such should be thoroughly analyzed.

1.5.6 Subcontract Estimating and Pricing

There can be many reasons for subcontracting part of the manufacture of a product, ranging from the use of the special expertise of the subcontractor to significant cost savings, usually in labor, by using a subcontractor. Whatever the reason, every consideration must be taken to ensure that the contractor will perform as promised at a reasonable cost.

With a subcontract, the outside source is doing more than providing material. He or she is providing material and a portion of the work to be performed on the manufactured product. He or she may provide a major subassembly, an operating part of the total system, or a specialized manufacturing process not available in-house. The subcontractor must provide a detailed cost estimate for his or her product and services, which become part of the total material estimate for the manufactured product. This cost estimate will detail the subcontractor's start-up costs, including such things as special tooling and test equipment, lead times required, manufacturing labor, material, burden, and profit. His or her delivery schedule should tie in directly with your production schedule, with deliveries feeding into your production at the point needed in the process.

Management and control of subcontractors is maintained by formal reporting requirements, visits to the subcontractor's facility, telephone reports, and status review meetings. Such techniques provide real-time communications and data flow to provide effective means of assessing subcontract status, early detection of problems, and initiation of any needed corrective action. The cost of providing this information should be included in the subcontractor's cost estimate.

1.5.7 Bibliography

Metals Handbook Desk Edition, American Society for Metals, Metals Park, Ohio, 1985.

Tanner, J. P., *Manufacturing Engineering: An Introduction to the Basic Functions*, Marcel Dekker, New York, 1985.

Vernon, I. R., *Realistic Cost Estimating for Manufacturing,* Society of Manufacturing Engineers, Dearborn, Michigan, 1961.

1.6 THE MANUFACTURING DIRECT LABOR ESTIMATE

1.6.1 Baseline Labor Estimating Techniques

Cost estimating for manufacturing can be done using any one or a combination of three general methods. The first method uses cost history and statistical methods. This technique, when properly applied, and where the required data exists, can be a reliable method of preparing the labor estimate. The problem with this method is that such data includes delays, shop inefficiencies, and time lost through a variety of reasons. Its use is recommended where standards do not exist, for budgetary quotes, and as a check against estimates developed using other methods.

The second method, using similarity with other products, assemblies, and parts, and/or using estimator experience, can be quite satisfactory when no previous production experience exists, when the product has never been in production before, or when the design is new and may still be in the concept stage. The accuracy of such estimates varies directly with the knowledge and experience of the estimator and the time allocated for preparation of the estimate. The completeness of the product definition also plays a large part in determining the accuracy of the estimate.

The third method is by using standard time data. In this method, all of the possible elements of work are measured, assigned a standard time, and classified in a catalog. When a specific operation is to be estimated, the necessary standard time values are added together to determine the total time. There can be little doubt that the use of standard time data is the most accurate and reliable method of estimating manufacturing labor. The use of standard time data promotes consistency among cost estimators, and requires little in the way of estimator experience with the work being estimated.

Historical and Statistical Data

Present and past costs may serve as starting points in preparing estimates for the future by recognizing the limitations outlined above and modifying them by forecasting conditions at the time the job will be in production. Cost estimates that are derived primarily by projecting past and current labor cost history are generally made only for guidance or planning purposes, and ideally should be followed by a detailed labor estimate based on standard time data. The manufacturing cost estimator must have a clear understanding of the principles of the cost accounting system providing him or her the data used to prepare the labor estimate.

Standard Data Use and Application

Standard time data is a compilation of all the elements that are used for performing a given class of work with normal elemental time values for each element. Without making actual time studies, the data are used as a basis for determining time standards on work similar to that from which the data were determined. Its use in manufacturing cost estimating offers certain advantages over other estimating methods. These advantages include:

1. It is far more accurate than any other estimating method. It is based on work content, rather than how much work is to be done and how long it will take to do.
2. It is easier to justify, because a series of individual elements and operations adding up to a given number of hours is easier to justify than one overall judgment of a given number of hours.
3. Standard data promotes consistency between estimates and cost estimators. Standard data, however, will show where there is a legitimate difference between similar equipment.
4. Estimator experience with the operation is not a requirement when using standard data. Such experience and knowledge, however, is extremely helpful.
5. Standard data coupled with learning curves can be used to estimate manufacturing labor for any production quantity. The cost estimator can use

experience to build an estimate in the 5 to 50 units range, but finds it difficult to estimate the same product at 1000 units. Standard data plus learning curves will cover the entire quantity spectrum.

The standards in this chapter are all based on standard data built up over the years by stopwatch time study, and using synthetic standards derived from MTM, MOST, Work Factor, and other predetermined time systems. Their application would be primarily in job order manufacturing quantities in the 25 to 2500 unit range. Figure 1.23 shows how the standard time data for the complete fabrication of any configuration of wire harness is organized and laid out for rapid determination of a standard time. It is simply a matter of determining the number of wires and wire ends, how each wire is terminated and routed as well as marked, entering the data in the appropriate space on the form, performing the extensions, and making the additions to arrive at the standard time for fabricating the wire harness.

Estimating by Similarity and Estimator Experience

The majority of cost estimates for manufactured products in the United States today are prepared by experienced cost estimators using their professional judgment. Their experience, based on detailed knowledge of their product, shop processes, and methods, is perhaps the most important single requirement for a correct estimate. New products are estimated much the same way, by a comparison of similar products with the new product being estimated. In many small shops the estimating may be done by the shop foreman or even the owner of the business. There is no attempt to use data from time standards, or any of the other, more sophisticated methods, to prepare estimates for new business.

Such techniques, if successful, may be all that are required. The shop may be small, the product may be produced in job order quantities, or the process may be highly specialized and very predictable. In such cases there is no need for a better estimating approach. As the firm grows larger and the cost estimating is done by more than one person, consistency among estimators may become a problem. As the product line becomes more diversified or the volume of production work grows, estimating judgment and experience exercised by one or two knowledgeable individuals may not do the job.

The ideal cost estimating situation is the experienced cost estimator using standard time data to develop and validate the cost estimate, and professional judgment and shop knowledge to determine whether the cost estimate is reasonable and attainable. The final questions are whether everything is included, and whether the estimate can be sold to management and supported in negotiations with the customer.

1.6.2 Preparing the Labor Estimate

This section provides examples and procedures of manufacturing labor estimating for a variety of different processes and basic operations covering the broad spectrum of manufacturing. Included are casting and foundry operations, metal forming

PART NAME ______ OPERATION ______

PROCESS PLAN DATE ______ MATERIAL ______

% ALLOWANCE ______ STD. S/U ______ RUN TIME ______

ESTIMATOR ______ DATE ______

HARNESS ASSEMBLY EH

DESCRIPTION	MINUTES	FR	TOTAL MINUTES
SET-UP			
DIT-MCO HARNESS TEST TE			
Prepare for Test/Lot	6.0		
RUN TIME			
PREPARE WIRE ENDS MANUALLY			
Clock in on DPI/Step	.60		
Fill out C.A.T. Card & Stamp/Job	1.50		
C.A.T. Cart, Stamp Only/Stamp	.04		
Find Wire Number on Plan/wire	.11		
Check off Wire Number on Plan/Wire	.10		
Cut with Diagonal Pliers/Cut	.07		
Cut with Nail Clippers/Cut	.05		
Prepare Pigtail Normal Wire/Pigtail	.85		
Prepare Pigtail "H" Film Wire/Pigtail	.63		
Comb out Shield (Incl. Cover Removal)/Shld.	1.27		
Untwist Conductors (2 & 3 Cond.)/Wire	.07		
Measure Conductor & Cut/Conductor	.21		
Measure Strip Lgth. & Mark/Conductor	.15		
Hand Strip-Auto Strippers/Conductor	.07		
Hand Strip-Knife/Conductor	.43		
Twist Strands/Conductor	.06		
Hand Tin with Soldering Iron/Cond.	.15		
*Hand Crimp Lug, Taper Pin, etc./Crimp	.20		
*Hand Crimp (2) Wires, 1 Lug/Crimp	.25		
*Hand Crimp (3) Wires, 1 Lug/Crimp	.29		
Install Ferrule/Job, Set-Up	.27		
Install Ferrule/Ferrule, Run Time	1.15		
Install Solder Sleeve/Sleeve	.60		
Install Vartex Over Pigtail/Pigtail	.59		
Install Vartex over Floating Shld./Shld.	.10		
Install Vartex over Ferrule/Ferrule	.61		

RUN TIME DESCRIPTION		MINUTES	FR	TOTAL MINUTES
Spot Tie/Each	Open	.26		
	Congested	.76		
Lacing, Starting Knot/Each		.26		
Lacing, Continuous Lock Stitch/Inch		.33		
Tywrap/Each	Open	.22		
	Congested	.31		
Install Tywrap thru Anchor/Each		.34		

SEARCH NUMBERED WIRES

NO. OF WIRES	MINUTES	NO. OF WIRES	MINUTES	NO. OF WIRES	MINUTES	NO. OF WIRES	MINUTES
1	.03	17	2.58	33	8.67	49	18.37
2	.08	18	2.85	34	9.17	50	19.09
3	.15	19	3.14	35	9.68	51	19.82
4	.24	20	3.45	36	10.21	52	20.57
5	.33	21	3.77	37	10.75	53	21.33
6	.44	22	4.10	38	11.30	54	22.10
7	.57	23	4.45	39	11.87	55	22.89
8	.71	24	4.81	40	12.45	56	23.70
9	.86	25	5.18	41	13.05	57	24.51
10	1.02	26	5.57	42	13.66	58	25.35
11	1.20	27	5.97	43	14.28	59	26.19
12	1.40	28	6.38	44	14.92	60	27.02
13	1.61	29	6.81	45	15.62	61	27.89
14	1.83	30	7.26	46	16.29	62	28.78
15	2.06	31	7.71	47	16.97	63	29.68
16	2.31	32	8.19	48	17.66	64	30.59

For a trunk containing more than 120 wires:

$$\text{TIME} = \left[\frac{(N+2)\quad(N-1)}{4}\right] + N \quad .0277\text{ Min.}$$

Where N = Number of wires in breakout

FIGURE 1.23 Wire-harness fabrication standard data.

Install **IT** Wire Marker (Non-Shrink.)/Marker	.20		
Mark Conductor, Cut, Strip & Twist	.38		
Wire Identification (Masking Tape)	.17		
*For Wires smaller than 10 gage For 10 Gage and Larger, Use Values Delow:			
Crimp Lug-Hydraulic Lugger/Lug	.69		
Crimp Lug-Bnch. Lugger 6-14 Gage/Lug	.77		
Crimp Lug-Bnch. Lugger 00-4 Gage/Lug	.80		
HANDLING			
Get Parts, Folder, etc./Job	1.8		
Get, Locate & Secure Plug or P/A to Harness Board/Plug or Plug Assembly	.70		
Get, Locate & Secure Sub-Assy. to Harness Board/Sub Assembly	1.02		
Get, Locate & Fit Item to Harness Assy./Item	.12		
Get, Locate & Secure Lacing Bar with 2 Nuts	.58		
Get & Locate Loose Wire to Harn. Bd./Wire	.14		
Remove Wired Plugs from Harn. Bd./Plug	.44		
Remove Wired Lug from Stud/Lug	.05		
Remove Wired Sub-Assy. from Harn. Bd./S.Assy	.74		
Remove Lacing Bar from Harness Board	.55		
Remove Harn.Assy. from Bd./Junct,or 10" Lgth	.09		
Place & Remove Complete Harness from Board fro Fit Check/Junct, or 10" Length.	.18		
Bag and Box Harness/Job	2.03		
Search Out Wire In Bag of - 30 Wires	9.91		
60 Wires	33.56		
90Wires	70.95		
120 wires	122.08		
ROUTING			
Route Wires Straight/Each 10" Length	.02		
Route Thru Bends or Breakout/Send or Brk.	.06		
Route Thru Vartex/Inch	.04		
Route Thru Transition/Wire	.18		
Route Thru Grommet or Hole, Open/Wire	.05		
Route Thru Grommet or Hole, Congested/Wire	.09		
Route to Clip Spring or Equiv./Wire	.03		
Dress Wires/Wire	.03		

WIRE PREP - ENDS 12 GA & UP

MEASURE, CUT & STRIP, PLUS SINGLE CONDUCTOR WIRE:	TIN	FREQUENCY	CRIMP (1) WIRE PER LUG OR PIN	FREQUENCY	CRIMP (2) WIRE PER LUG OR PIN	FREQUENCY	CRIMP (3) WIRE PER LUG OR PIN	FREQUENCY
Common Wire End - Open	.64		.70		.74		.79	
- Cong.	.82		.81					
Pigtall Shld. Norm.	2.08		2.13		2.16		2.22	
Float Shld. Norm.	1.58		1.64		1.68		1.73	
Pigtail Shld. "H" Film	1.86		1.92		1.97		2.01	
Float Shld. "H" Film	1.37		1.43		1.47		1.52	
TWO CONDUCTOR WIRE:								
Pigtail Shld. Norm.	2.78		2.90		2.99		3.08	
Float Shld. Norm.	2.29		2.41		2.50		2.58	
Pigtail Shld. "H" Film	2.57		2.69		2.78		2.87	
Float Shld. "H" Film	2.08		2.19		2.28		2.37	

MARK, CUT STRIP & TIN ONLY - SINGLE CONDUCTOR WIRE:	
Common Wire End - Open	.37
- Congested	.55
Pigtail Shield Normal	1.61

Values are per wire, vartex over shields included.

FIGURE 1.23 (Continued).

RUN TIME — DESCRIPTION — MINUTES — FR — TOTAL MINUTES

HOOK UP AND SOLDER WIRES

DESTINATION	SOLDER TIME ONLY MIN.	FR	1 WIRE/ TERMINAL MIN.	FR	2 WIRES/ TERMINAL MIN.	FR	3 WIRES/ TERMINAL MIN.	FR	TOTAL MINUTES
P.C. Pad	.05		.31						
Congested			.73						
Plug Pin 10—26G w/Weep Hole	.27		.64						
Plug Pin 10—26G	.27		.66		.85				
Congested			.92		1.54				
Plug Pin 4—8 G	.38		.84						
Congested			1.25						
Plug Pin 4/0—2 G	1.00		1.45						
Congested			2.74						
Turret, Island	.09		.48		.71		.94		
Congested			.79		1.33		1.89		
Turret, Buss	.89		1.28		1.41		1.54		
Congested									
Bifurcated, Isl.	.26		.64		1.08		1.49		
Congested			1.08		2.00		2.87		
Bifurcated, Bus	.89		1.28		1.41		1.54		
Congested									
Eyelet Terminal	.11		.50		.65		.81		
Congested			.76		1.08		1.41		
Hook & Other	.07		.45		.59		.73		
Congested			.64		.90		1.16		
Use Assist Tool			.06		.06		.06		
Tin Plug Pin	.18		.31						
Hand Solder Circuitry/Sq."	.09		.20						
Touch-up Circuitry/Sq."	.09		.19						

RUN TIME DESCRIPTION	MINUTES	FR	TOTAL MINUTES
COAXIAL CABLES			
Prepare One End	1.69		
Prepare One End & Install Connector	3.69		
Prepare Two Ends & Install Connectors/ Cable	7.37		
MISCELLANEOUS			
Insert Pinned Wire in Connector, Per Wire	.22		
Install Plug Seals, Per Seal	.09		
Install Prepared Lug Over Stud, Per Lug	.03		
Install Dust Cap, Each	.21		
Install Strain in Relief & Tighten, Each	1.56		
Remove Replace Potting Boot, Each	.36		
Install Plug in Vise & Remove, Per Lug	.16		
Install Common Points, Each	3.23		
Install Rubber Bands on Harness Bd., Each	.08		
Install Environmental Seal on Plug, Each	.54		
TORQUE MECHANICAL FASTENERS			
Change Socket on Torque Wrench/Socket	.16		
Set Required Torque/Gage	.05		
Torque Fastener/Fastener	.15		
Stripe Fastener/Stripe	.23		
POSITION HARNESS IN CHASSIS			
Harness Installation:			
Normal/Harness	.13		
Confined/Harness	.15		
Locate Straight Run:			
to 4" Per Run	.04		
Each additional 4" to 6" Per Run	.02		
Locate Bend of Breakout, Per Breakout	.06		

FIGURE 1.23 (Continued).

DESCRIPTION	OPEN MINUTES	OPEN FR	CONGESTED MINUTES	CONGESTED FR
ROUTE, PREP, TIN, HOOK-UP, SOLDER, CLEAN & DRESS WIRES				
Turret Terminal (Island) 1W	.90		1.39	
Turret Terminal (Island) 2W	1.56		2.53	
Turret Terminal (Island) 3W	2.21		3.68	
Bifurcated Term. (Island) 1W	1.07		1.68	
Bifurcated Term. (Island) 2W	1.93		3.20	
Bifurcated Term. (Island) 3W	2.77		4.67	
Eyelet Terminal (Island) 1W	.92		1.36	
Eyelet Terminal (Island) 2W	1.50		2.28	
Eyelet Terminal (Island) 3W	2.08		3.21	
Hook Term. or Other 1W	.88		1.24	
Hook Term. or Other 2W	1.57		2.10	
Hook Term. or Other 3W	2.00		2.96	
VARTEX & MARKER INSTALLATION				
Cut, Install & Tape Wire Marker (Shrinkable)			.23	
Cut, Install & Tape Wire Marker (Non-Shrinkable)			.20	
Install Bundle Marker (Shrikable)			.08	
Install Bundle Marker (Non-Shrikable)			.57	
Install Marker Plat/Marker			.62	
Slide Short Vartex Over 1 Wire/Piece			.04	
Slide Short Vartex Over Solder Connection			.05	
Install Short Vartex, Under 36", Over Wires (Non-Shrinkable)/Inch of Wire			.08	
Install Long Vartex, Non-Shrinkable/Piece			8.07	
Install Long Vartex, Non-Shrink./In. of Wire			.13	
Install Long Vartex, Shrinkable/Piece			.14	
Shrinkage Short Vartex, Per Piece			.15	
Shrinkage Long Vartex, Gun Handling/Piece			.14	
Shrinkage Long Vartex, Per Inch			.23	
Shrink Boot or Transition, Per Piece			11.36	
Wrap Vartex for Filter on Strain Relief, Per Piece			.17	
Cover Exposed Shield with Vartex & Tie/ Piece			1.67	

DESCRIPTION	MINUTES	
Reform & Locate Bend, Confined/Breakout	.12	
Dress Branched Conductors/Conductor	.06	
DIT-MCO HARNESS TEST TE		
Constant, Per Harness	3.51	
Connect and Disconnect, Per Connector	.51	
Test, Per Conductor	.10	
Self Check Equipment, Per Harness	.031 x Total R/T	

SUB TOTAL RUN TIME

PAGE #1 RUN TIME

ALLOWANCE

TOTAL STD. RUN TIME

FIGURE 1.23 (Continued).

and fabrication, machining operations, mechanical assembly, electrical/electronic assembly, painting, metal finishing, inspection, and testing.

Casting and Foundry Operations

The cost of manufacturing sand castings consists of material, foundry tooling, molding costs, core-making costs, grinding costs, cleaning and finish cleaning costs, heat treating or aging costs, inspection, and foundry overhead or burden costs. The most important factors in the cost of producing a sand casting include:

1. The cost of casting metal and weight of metal poured per mold
2. The melting cost
3. The method of molding (a function of the number of castings produced)
4. The type of pattern used (also dependent on number of castings produced)
5. The weight of the casting
6. The number of castings per mold
7. The number of cores per casting and per core box
8. The core material and weight
9. The core-making method
10. The number of risers used
11. The type and amount of finishing required

Estimating the cost of sand-casting material and how to estimate foundry tooling, including molds, patterns, mold boxes, cores, risers, etc. is covered in Subchapter 1.5. The cost of the foundry labor to pour the molds and to perform the grinding and the cleaning is estimated as follows:

If we assume that the automotive cylinder block in Figure 1.24 weighs 100 lb., and we realize a shop yield of 55%, a remelt factor of 40%, and a metal lossage factor of 10% of finished casting weight, then the following weights are calculated:

$$\text{Shop yield} = 100 \text{ lb.}$$

$$\text{Pouring weigh} = \text{finished casting weight} \div 0.55$$
$$= 100 \div 0.55 = 182 \text{ lb.}$$

$$\text{Remelted metal weight} = \text{pouring weight} \times \text{remelt factor}$$
$$= 182 \text{ lb.} \times 0.40 = 72.8 \text{ lb.}$$

$$\text{Lost metal} = \text{finished casting weight} \times \text{metal lossage factor}$$
$$= 100 \text{ lb.} \times 0.10 = 10 \text{ lb.}$$

The cost of the metal used in the finished casting is calculated from formulas as follows:

$$\text{Poured metal cost per casting} = \text{pouring weight per casting} \times (\text{labor and overhead} + \text{charged material cost})$$

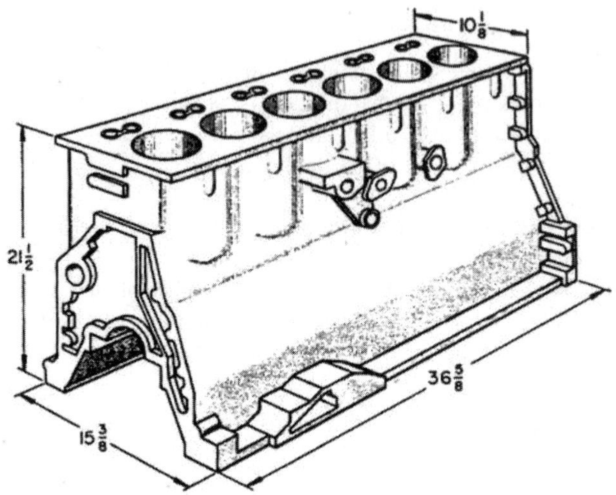

SAND CASTING OF AUTOMOTIVE CYLINDER BLOCK
Class 30, Gray Iron

COST ELEMENT	PERCENT OF TOTAL COST
Metal	13.6%
Core Material	6.8
Conversion	6.4
Molding	23.7
Coremaking	23.1
Grinding	10.5
Cleaning and Gaging	1.6
Finish Cleaning	5.1
Scrap Loss	9.2
Total	100.0%

FIGURE 1.24 Breakdown of sand-casting costs for automotive cylinder block.

Cost of metal in finished casting = poured metal cost per casting – (amount of melted metal × values of remelted metal)

If we assume that labor and overhead equal \$0.07/lb., charged material equals \$0.05/lb., and remelted metal equals \$0.03/lb.:

Poured metal cost per casting = 182 lb. × (\$0.07 + \$0.05) = \$21.84
Cost of metal in finished casting = \$21.84 – (72.8 lb. × \$0.03/lb.) = \$19.66

The estimated cost for cleaning and inspection of some 200 units of automotive cylinder blocks is \$530 for labor and \$25 for material. This means that the manufacturing cost for the 200 castings is \$4923. To this, of course, we would add foundry overhead or burden, general and administrative expense, and profit to arrive at the final selling price.

Metal Forming and Fabrication

Fabrication of sheet metal, whether steel or aluminum, accounts for most of the manufacturing today in the production of electronic chassis and cabinets. Many such structural units must withstand the unusual stresses of field military handling and shipboard shock from naval gunfire, so strength in design and construction is of paramount importance. In the sheet metal manufacturing shop today, there is usually a central principal machine, such as an N/C punch, which is programmed to punch out the various parts that make up the chassis or cabinet assembly in the flat, using sheared blanks of material. These punched-out part shapes are then separated from the blank by notching out the material still holding the part in the blank.

The flat piece parts are then formed to the required shape on a press brake and assembled by spot welding or riveting to complete the fully assembled chassis or cabinet. The first series of operations involves shearing the sheet stock of material into correct blank sizes for the N/C punch. The estimated standard times for power shearing of blanks are as follows.

Power-shear setup:

Set stops for front or rear gauging, test cut and measure, adjust hold-downs as needed
Standard time: 0.200 hr

Run-time analysis:

The maximum number of cuts required for a rectangular blank is four cuts for four sides. In actual practice the number of cuts is closer to one cut per blank, as each cut that is made frees at least one other blank. On the initial cuts, one cut is actually cutting the side of several blanks.
Standard time: 0.250 hr. average time/cut/piece

Unless otherwise noted, parts-handling time is included in the run time. Standard times indicated include either large, medium, or small part size.

The N/C punch takes the sheared blanks and performs the blanking, notching, nibbling, and hole punching necessary to shape the individual piece parts to their required flat shape. This is a machine-controlled series of operations. Standard time for machine, turret, and tape reader setup is the only operator-controlled time that is involved. Estimated standard times for N/C turret punch operations are as follows.

N/C turret punch setup:

Obtain punches, dies, and strippers called out on the tooling plan, and set up in turret of punch
Standard time: 0.400 hr. average

Average setup time (developed exactly by using basic minimum setup time):

Basic 0.0210 hours, plus installation and removal time per punch of 0.0175 hr., and time for punch orientation of 0.0185 hr. per punch

Run times for the N/C turret punch press:

Stock handling time of 0.0041 hr. per number of parts per blank, plus machine punch time of 0.00146 hr. per hit, multiplied by the number of hits per piece part

The cost estimator would have to estimate the number of hits by various punches as they progressively nibble out the part shape and punch any holes in the part.

Once the individual piece parts are fabricated in the flat on the N/C turret press, they must be formed to their correct shape. The press brake is one of the principal machines of the sheet metal shop. A wide variety of bending and forming can be done with standard dies. Most jobs are done on the press brake without changing dies. Positioning stops must be set, and space between the ram and bed must be adjusted for material thickness and die sizes.

Setup hours: 0.450 hr. plus an additional 0.200 hr. if dies must be changed
Handling and machine time for press brake bending and forming, including individual parts handling to and from the forming die:
- Small part: 0.200 hr./operation
- Medium part: 0.300 hr./operation
- Large part: 0.500 hr./operation

Machining Operations

Initial operations in the machine shop usually involve the sawing or cutting of bar stock, rod, or plate material into blanks for machining. The equipment is either a band saw or a power hacksaw. Very little setup time is required for changing jobs on either piece of equipment. Setup includes setting or resetting the stops, vise adjustments, and any blade changes. The time is:

Setup time: 0.150 hr./occurrence

Run time includes picking up the material from a tote or skid and positioning it to the saw, then removing the cut pieces and placing them on a cart or skid. The time values for this part handling time, covering part sizes from 1 in. to 3 ft., is:

Handle parts: 0.45 min./part

Machine times are given in minutes to cut 1 in. of metal of the indicated thickness:

1-in.-thick stock: 0.300 min.
3-in.-thick stock: 2.55 min.
6-in.-thick stock: 9.50 min.

The standard times given for lathe operations do not cover NC machine tools, only conventional machines. We will first consider turning operations on a Monarch 10-in. × 20-in. lathe or its equivalent.

Figure 1.25 gives standard machine shop formulas for the turning, milling, and drilling of machined parts. Parameters included are the cutting speed, spindle rpm, cutting time, rate of metal removal, horsepower required at the spindle and the motor, spindle torque, and milling feed per cutter tooth. Using these formulas, the cost estimator can determine machine run time and the size and power of required machine tool equipment.

Setup time for the Monarch 10-in. × 20-in. engine lathe includes obtaining necessary tools and gauges, installing the collet and/or chuck, setting in and squaring off the length of bar stock, and the time to check the first piece. Job teardown includes removing the collet or chuck and cleaning up the measuring equipment. Also included is the installation of the cutting tool and its removal:

Total setup time: 0.500 hr.

The handling time per part includes picking up a piece of cut-to-length tube or bar stock, installing and aligning it in the chuck, securing, checking concentricity, making the trial cut, and then checking the setup:

Standard time: 0.55 min./in.

Machine feeds for aluminum range from 0.002 in./min. to 0.030 in./min.

For milling machine setup time, assume a Milwaukee #2 milling machine or equivalent. Setup elements include obtaining necessary tools, cutters, gauges, etc. and returning those used on previous job; cleanup of the machine table, vise, or holding fixture; assembly and alignment; installation of the cutter to the collet; table adjustment for the initial cut; and first piece checkout:

Setup time: 0.600 hr./job

Part-handling time for milling machine operations includes the pickup and placement of the part in the vise or fixture, release of the part after work is performed, checking of the part, putting part aside to tote pan, and fixture cleanup for the next part:

Standard time: 0.400 min./part

Parameter	Turning	Milling	Drilling
1. Cutting speed (FPM)	$Sc = 0.262 \times D_t \times RPM$	$Sc = 0.262 \times D_m \times RPM$	$Sc = 0.262 \times D_d \times RPM$
2. Revolutions per minute (RPM)	$RPM\ 3.82 \times \frac{S_c}{D_t}$	$RPM = 3.82 \times \frac{S_c}{D_m}$	$RPM = 3.82 \times \frac{S_c}{D_d}$
3. Feed rate (in./min)	$F_m = F_r \times RPM$	$F_m = FtxN \times RPM$	$F_m = F_r \times RPM$
4. Feed per tooth	—	$F_t = \frac{F_m}{NxRPM}$	—
5. Cutting time (min)	$t = \frac{L}{F_m}$	$t = \frac{L}{F_m}$	$t = \frac{L}{F_m}$
6. Rate of metal removal (in.^{3}/min)	$R = 12 \times d \times F_r \times S_c$	$R = W x d x F_m$	$R = \frac{D^2 d}{4} x F_m$
7. Horsepower required at spindle	$HP_s = R \times P$	$HP_s = R \times P$	$HP_s = R \times P$
8. Horsepower required at motor	$HP_m = \frac{R \times P}{E}$	$HP_m = \frac{R \times P}{E}$	$HP_m = \frac{R \times P}{E}$
9. Torque at spindle (in.-lb)	$T_s = \frac{63{,}030\ HP_s}{RPM}$	$T_s = \frac{63{,}030\ HP_s}{RPM}$	$T_s = \frac{63{,}030\ HP_s}{RPM}$

Machine shop formula terms

S_c = cutting speed (ft/min)
D_t = diameter of workpiece in turning (in.)
D_m = diameter of milling cutter (in.)
D_d = diameter of drill (in.)
D = depth of cut (in.)
E = efficiency of spindle drive
F_m = feed rate (in./min)
F_r = feed (in./rev)
F_t = feed (in./tooth)
HP_m = horesepower at motor
HP_s = horsepower at spindle
L = lenght of cut (in.)
N = number of teeth in cutter
P = unit power (HP/in.^{3}/min)
R = rate of metal removal (in.^{3}/min)
RPM = revolutions per minute of work or cutter
T_s = torque at spindle (in./lb)
t = cutting time (min)
W = width of workpiece (in.)

FIGURE 1.25 Standard machine shop formulas.

Table advance, back-off, or adjustment time:

Standard time: 0.200 min./cut

Set table at proper position for work by moving up or down:

Standard time: 0.200 min./cut

Index dividing head:

Standard time: 0.150 min./cut

End milling beginning with rough profile (0.5-in. depth × 0.75-in. width cutter), followed by finish profile (same-size cutter):

Standard time: 0.120 min./in.

Surface or face mill using plain, helical, slab, shell, etc. end mills:

Standard time: 0.900 min./in.

Side milling, straddle milling, and slotting with side, half-side, and staggered tooth milling cutters:

Standard time: 0.070 min./in.

Corners, grooves, and slots:

Standard time: 0.060 min./in.

Drilling is one of the most common machine shop operations, and the time required for drill setup and run can vary significantly, depending on the type of drilling equipment, tooling and fixturing involved, and other significant variables. The following standard times for estimating are averages of many kinds of drilling operations in a variety of different industries. As with the time standards for engine lathe and milling operations, they should be used as a starting point until more accurate data is available for your shop conditions and your company.

Drilling setup includes obtaining necessary drills, taps, countersinks, gauges, etc., and time to return them when the job is complete. Also included is time to handle fixtures, jigs, and vises; make adjustments to the drill press, including changing feeds and speeds; make feed stop adjustments; install the drill in the spindle; and check the first piece:

Standard setup time: 0.300 hr./lot of parts

Part-handling time per hole includes time for moving the part from hole to hole and lowering the drill bit to the part surface. The cutting time per inch of depth should then be added to the handling time per hole:

Medium drill press, spindle rpm 500–2000:

Standard time: 0.300 min./hole

Heavy-duty drill press, spindle rpm 1–1000:

Standard time: 0.200 min./hole

Drill, tap, countersink time for 2-in. diameter hole:

Standard time: 0.300 min./hole (aluminum)
2.100 min./hole (steel)

Machine shop deburring is an operation that removes the thin ridges or roughness left on stock or machined parts after cutting, shaping, or drilling. It specifically removes forging flashings; mismatch material; scarfed or rough edges after welding, profiling, or sawing; and more. Burring is accomplished with a variety of serrated or gritted tools (files or router bits), which may be hand- or power-operated. The nature of the surface of the part, plus the degree of roughness, determines the proper equipment. Some machine shop burring tools include:

Rotary burr files or cutters
Drill motor split rod and emery
Hand files
Hand scrapers
Burring knives
Belt and disk grinders
Abrasive wheel or pedestal grinders
Sanding drums
Drill press burring tools

Table 1.1 gives standard times for general bench deburring operations found in most machine shops.

Mechanical Assembly

Modern welding processes provide a means of joining and assembling many different types of metal parts. Most metals can be welded if the right equipment and processes are selected. Available welding processes range from gas or electric arc fusion welding to spot and seam resistance welding to electron beam welding. Methods of applying the processes also vary widely. Workpiece thickness and composition usually determine the type of welding process that can be used. For example, the inert gas-tungsten arc (TIG) process is economical for welding light-gauge material, while the semiautomatic inert gas-metal arc (MIG) process is more economical for heavier-gauge materials and nonferrous materials such as aluminum.

Aluminum is not as readily welded as steel, but by using the proper alloy (1100, 5052, 6061) and cleaning prior to welding, the same results and efficiency can be obtained as with steel.

TABLE 1.1 Deburring Standard Times: Handling

GENERAL BENCH BURRING STANDARDS

Machine Parts - Forgings, castings, extrusions, shims, flates
Regular and irregular forms (aluminum mag. and steel)

Tools - Hand- and motor-operated burring devices (abrasive wheels, drums, belts, rotary and hand files)

(Types of burring methods and tools are listed below with their respective Time standards.)

I. Chart A -

Setup	.15

Sign in, study print, obtain tools, clean bench, and first piece inspection.

II. Handling time per part - P.U. and aside after each burr operation.
Chart B -

Complexity	S	A	C
Weight of Part Length of Part	½# 0 to 6"	½ to 3# 6 to 24"	3 to 10# 24" and over
Per Job:			
A. In–Hand			
Toss Aside	.0008	.0010	.0018
Stack	.0010	.0014	.0022
Pack in Box	.0019	.0024	.0030
B. In–Vise P.U. and Aside	.0014	.0022	.0024
C. Clamped to Table Remove and Aside	.0026	.0042	.0060

TABLE 1.1 Deburring Standard Times (*Cont.*): Types of Operations

Burring Time:

Includes burr edge and turn to next edge.

(Code: S=Simple, A=Average, C=Complex)

III.

Chart C -

	TIME FOR EACH 10 LINEAR INCHES (EDGE)					
Complexity Code	S	A	C	S	A	C
Weight of Part Length of Part	½# 0 to 6"	¼# to 3# 6 to 24"	3# to 10# OVER 24"	½# 0 to 6"	¼# to 3# 6 to 24"	3# to 10# OVER 24"
Material	Aluminum and Magnesium			Steel		
Burr Tool – Sanding Drum	.0009	.0012	.0016	.0012	.0015	.0021
Burr Knife	.0009	.0014	.0018	.0014	.0018	.0021
File-Hand	.0018	.0023	.0035	.0023	.0029	.0046
Drill Motor- Split Rod & Emery	.0012	.0014	.0019	.0014	.0017	.0023
Wire Brush Threads-10"	.0028	.0035	.0046	.0028	.0035	.0046
Lam. Cloth Burr	.0006	.0007	.0008	.0009	.0012	.0014
Bch. press – "No Burr" Tool – Pac Web	.0007	.0009	.0012	.0009	.0012	.0014
Web (each)	.0005	.0007	.0009	.0007	.0009	.0012
Form 1/32 Radius- Sand Drum	.0030	.0036	.0051	.0036	.0048	.0073
Form 1/32 Radius- Hand File	.0060	.0072	.0096	.0144	.0168	.0180

	TIME PER SQ. INCH			TIME PER SQ. INCH		
Length of Part	0 to 6"	6 to 24"	OVER 26"	0 to 6"	6 to 24"	OVER 24"
Material	Aluminum and Magnesium			Steel		
Surface Clean-up, Drill Motor & Split Rod- 250 / to 125 /	.0016	.0016	.0016	.0020	.0020	.0020
Sand Casting Surface- Sanding Drum	.0014	.0014	.0014	.0016	.0016	.0016
Hand Scrape Extrusion / Surface	.0012	.0012	.0012	.0016	.0016	.0016

TABLE 1.1 Deburring Standard Times (*Cont.*): Ferrous and Nonferrous Materials

	TIME PER INCH			TIME PER INCH		
Material	Aluminum and Magnesium			Steel		
Length of Part	0 to 6"	6 to 24"	Over 24"	0 to 6"	6 to 24"	Over 24"
Remove Forging Flash–Sand Drum 14 & 24ST.	.0072	.0072	.0096	.0080	.0100	.0120
Remove Forging Flash–Rotary & H-File 75ST.	.0140	.0140	.0180	.0150	.0160	.0200
Drill Press-Burr #40 Hole to 3" Diameter.	.0005	.0006	.0008	.0009	.0012	.0012
Machine Burr-Large Holes-per Circular Inch.	.0007	–	–	.0009	–	–
File (Rotary File or Wheel) to straight line: After Sawing	.0015	–	–	.0018	–	–
After Shearing.	.0006	–	–	.0009	–	–
Burr Ends (Rotary File or Wheel) after Sawing	.0005	–	–	.0009	–	–
Flange or Leg (Hand File or Wheel) after Milling or Profiling.	.0005	–	–	.0009	–	–

CALCULATION OF TIME PER PART

Burring:
1. Determine length of part and type of burring operation.
2. Multiply the burr operation from III (chart C) by standard per edge, inch, square inch or hole.

Handling:
1. Add handling time per length of part from II (Chart B) to burr time to obtain unit time.

Example: Sand Drum Burr, Part 8" long, one edge up to 10", ½ to 3lb.

Burr Time –	.0012
Handling Time per Job – Clamped to Table	.0042
Time per Part	.0054
Setup from Chart I –	.15

Setup time, including time for part handling, part positioning, and part alignment:

Standard time: 0.300 hr.

Run times, including normal preheating for aluminum, steel, or magnesium:

Weld 0.062 stock: 0.250 min./in.
Weld 0.125 stock: 0.400 min./in.
Weld 0.250 stock: 0.750 min./in.

Stress relieving using a heat-treat furnace is often required on some parts due to close tolerances or to highly stressed working parts:

Setup time: 0.100 hr./occurrence
Run time: 0.300 min./assembly plus oven or furnace time

Spot welding is defined as the welding of lapped parts in which fusion is confined to a relatively small circular area. It is generally resistance welding, but may also be gas-shielded tungsten-arc, gas-shielded metal-arc, or submerged arc welding. Spot welding is one of the most economical sheet metal assembly methods. It has the disadvantage of lack of structural strength, when compared to riveting, for example. A surge of electric current melts the two pieces of metal to fuse them together, and in so doing reduces the metal to the as-cast state, resulting in the weld having less strength than the surrounding heat-treated alloy. Setup of a spot welder requires installing and adjusting the contact points, current, and timing. Positioning and adjusting the welding fixture that holds the parts is also included:

Setup time: 0.400 hr.

Spot-weld run time includes time for parts handling and the moving of the assembly from spot to spot, and depressing the foot pedal to activate the machine weld cycle at each spot:

Parts-handling time: 0.200 min./operation
Welding time: 0.050 min./cycle

Standard times for the assembly of minor machine parts using bolts, screws, threaded inserts, and rivets, and also by adhesive bonding, eyelets, and nameplate assembly, are shown in Figure 1.26. Also shown are standard times for the positioning and alignment of parts for mechanical assembly. Setup time for these mechanical assembly operations is as follows:

Setup time: 0.400 hr.

Description	Per occurrence (hr)
Install spline nut	0.008
Install thread inserts (Helicoil-Keenserts etc.)	0.030
Install bolt/screw, washer, and nut	0.014
Install bolt/screw, washer, nut, cotter key	0.025
Install bolt/screw, washer, stop nut	0.025
Install bolt to nut plate	0.010
Install rivet (includes drilling)	0.008
Install eyelet (insert and swage)	0.017
Install name plate per screw (includes drilling)	0.015
Adhesive (application) per square inch	0.001
Hand stamp (per letter and/or number)	0.003
Rubber stamp (ink) per application	0.001

Position and align parts for assembly:

Number of parts per assembly (with fixture or handheld)	Hours
Two (2)	0.030
Three (3)	0.070
Four (4)	0.100
Each additional part over 4	0.030

FIGURE 1.26 Standard times for minor parts assembly.

Electrical/Electronics Assembly

In electrical/electronics assembly the work is very labor-intensive, and standard times for cost estimating and other uses should be based on sound engineering work measurement if at all possible. The standard times that follow meet this criterion, although, as before, standards should reflect operations in your plant or facility. Industry standard times such as these can often be tightened up through more efficient methods, tooling, and/or automation.

The standard data in Figure 1.23 provides standard times for most common hand-fabrication of wire harnesses and cables, including the hand-tinning and soldering of wire breakouts from a harness. The standard times that follow cover additional elements and tasks involved in the make-ready, work, and put-aside of job order electronic assembly operations.

Get parts and prepped wires from a workbench stack bin or wire rack:

Standard time: 0.025 min./occurrence

Get tools from workbench, such as pliers, cutters, wire strippers, soldering iron and solder, and aside tools to workbench:

Standard time: 0.050 min./occurrence

Finishing and Plating

There are numerous finishing operations, ranging from mechanical buffing and polishing to sophisticated metal chemical processes for passivation and surface protection. *Buffing* is a form of surface finishing in which very little material is removed. The sole purpose is to produce a surface of high luster and an attractive finish. Abrasives such as aluminum oxide, emery, ferrous oxide, rouge, pumice, and lime are applied to the rotating face of a buffing wheel. A composition containing the abrasive is pressed with the work against the face of the buffing wheel. The abrasive is replenished periodically. Table 1.2 provides standard data for calculating standard times for buffing operations.

Polishing is a term commonly used to designate that branch of grinding which employs various types of yielding wheels, cushion wheels, and flexible belts, the surfaces of which are covered or impregnated with some sort of abrasive. Polishing is sometimes referred to as flexible grinding. For finer finishing, emery cloth is employed in preference to an abrasive wheel in many instances. Table 1.3 gives standard data for polishing operations.

Spray painting, utilizing a waterfall paint booth as compared to a continuous conveyor that allows the operator to spray parts as they pass by, does require parts handling:

1. Handle parts to and from the turntable in the spray booth
2. Spray part or parts on the turntable
3. Handle parts to and from a drying rack, drying oven, or a heat bank conveyor

The major time variables include part size, part configuration, number of spray painting passes or coats, and viscosity of the paint or primer. Painting time values assume a part with at least four sides to be painted (no inside surfaces); part size is assumed to be a 30-in. cube.

Part masking and demasking covers applying masking tape by rough measure, tearing from a roll, and applying to the part. It also includes removal of the tape after the part is painted. All time values are averages for the purpose of quick, easy application to cost estimates. To set up for a new paint type or color, the following elements of work must be accomplished:

1. Prepare equipment, including paint booth, turntable, and spray gun
2. Secure paint and fluid tank
3. Thin paint if required, and transfer to tank
4. Secure air and paint lines, clean air lines, and attach spray gun

TABLE 1.2 Standard Times: Buffing Operations

Setup

Includes – Get tools, prepare work station, check first part, and record time.

Setup	.15 hr.

Run Time: (Handling plus buffing)

A – Handling Time Chart A – Includes pick up, positioning, & asiding of part.

Part Size in Cubic Inches	Handling Time – Hrs. per Part			
	Aluminum		Steel	
	Hand-Held	Holding Device	Hand-Held	Holding Device
To 12"	.002	.008	.003	.010
To 12" to 24"	.003	.008	.004	.010
Over 24" to 48"	.005	.012	.006	.015
Over 48"	.007	.012	.008	.020

B – Buffing Time Chart B – Includes periodic application of abrasive.

Type of Surface Finish Required	Hrs. Per Square Inch – Buffing	
	Aluminum	Steel
125	.0008	.0040
32 – 62	.0012	.0060

EXAMPLE:

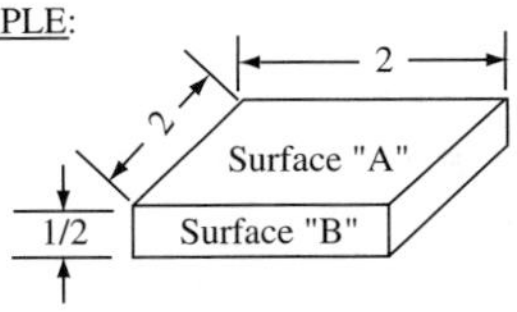

Material: Aluminum – 32 Finish
Surface "A" and "B" to be buffed-Hand Held

Size of Part – 2 Cubic Inches
Surface Area to be Buffed = 5 sq. inches

STEP I	– Setup			.15 Hrs.
STEP II	– Handling Time/Unit (from chart "A")	=	.002 Hrs.	
STEP III	– Buffing Time/Unit (from chart "B")			
	5 sq. inches × .0012	=	.006 Hrs.	
TOTAL RUN TIME (STEP II PLUS STEP III)		=		.008 Hrs.

TABLE 1.3 Standard Times: Polishing Operations

Set-up

Includes – Get tools, prepare work station, check first part, and record time.

Setup	.15 Hrs.

Run Time: (Handling plus polishing)

A – Handling Time Chart A – Includes pick up, position, & aside part.

Part Size in Cubic Inches	Handling Time – Hrs. per Part			
	Aluminum		Steel	
	Hand-Held	Holding Device	Hand-Held	Holding Device
To 12"	.002	.008	.003	.010
Over 12" to 24"	.003	.008	.004	.010
Over 24" to 48"	.005	.012	.006	.015
Over 48"	.007	.012	.008	.020

B – Polishing Time Chart B – Remove scratches and marks.

Type of Surface Finish Required	Flex Wheel and Hand Motor and Emery Cloth Unit Time per Sq. Inch – High Polish	
	Aluminum	Steel
125	.0010	.0017
62	.0014	.0025
32	.0025	.0037

EXAMPLE:

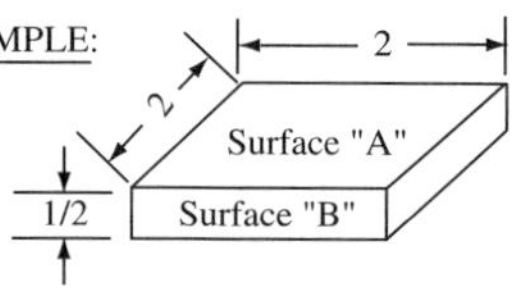

Material: Aluminum – 32 Finish
Surface "A" and "B" to be polished—hand-held

Size of part – 2 cubic inches
Surface area to be polished = 5 sq. inches

STEP I	– Setup			.15 Hrs.
STEP II	– Handling Time/Unit (from chart "A")	=	.002 Hrs.	
STEP III	– Polishing Time/Unit (from chart "B") 5 sq. inches × .0025	=	.0125 Hrs.	
TOTAL RUN TIME (STEP II PLUS STEP III)		=		.0145 Hrs.

5. Attach nozzle, adjust spray gun, and try out.
6. After job is complete, clean above items with solvent and put away.
   Standard setup time: 0.35 hr.

Detail spray painting standard time values for estimating include the following:

Parts-handling time: 0.35 min./part
Wash surface with solvent: 0.05 min./ft.^{2}
Spray paint (double this for 2 passes): 0.05 min./ft.^{2}
Brush paint areas not reached by spray painting:
  Standard time: 0.25 min./ft.^{2}
Apply masking tape: 0.12 min./10 in.
Remove masking tape: 0.26 min./10 in.
Assemble and remove masking plugs and stencils:
  Standard time: 0.10 min./plug

Inspection and Testing

Inspection labor is normally estimated as a percentage of total manufacturing labor. These percentages can range from 5% to a high of 14%, depending on the requirements; the criticality of the part, product, or assembly; and much more. Often inspection is performed by the production workforce, and time is allowed for this in the standard and the cost estimate. A safe rule of thumb, where no previous actual data may exist, is to use 7% of fabrication labor and 5% of assembly labor for inspection.

Test time is a function of the testing performed, the degree of test automation, whether troubleshooting is required, whether test data is required and must be recorded, whether the testing is destructive or nondestructive, whether it is mechanical or electrical, and a host of other considerations.

Test labor can be measured by standardizing the operation, applying elemental time standards, adding time study allowances, and adding a time factor for troubleshooting and subsequent retesting. Some of the electronics testing industry average time data is provided as follows for cost estimating:

Test setup and make-ready/teardown times:

Pick up and lay aside unit: 0.35 min.
Remove and reassemble covers: 0.95 min.
Hook up and unhook alligator clip: 0.20 min.
Hook up and disconnect plug or jack: 0.17 min.
Assemble test adapter/fixture: 0.30 min.

Test equipment adjustments:

Off and on toggle or rotary switch: 0.07 min.
Adjust Variac: 0.11 min.
Knob frequency adjustment: 0.13 min.

Scope adjustment for phasing, etc.: 0.35 min.
Observe scope and analyze pattern: 0.45 min.

Unit under test adjustments:

Circuit check with probe/voltmeter: 0.07 min./point
Adjust coil tuning slug: 0.30 min.
Adjust trimpots: 0.25 min.

Troubleshooting and retest will vary greatly depending on the maturity of the product design, the quality of the assembled unit and its component parts, and more. This can range from 15% to 50% of normal test time. The percentage used must be left to the professional judgment and knowledge of the cost estimator.

1.6.3 Bibliography

Hartmeyer, F. C., *Electronics Industry Cost Estimating Data,* Ronald Press, New York, 1964.

Matisoff, B. S., *Handbook of Electronics Manufacturing Engineering,* Van Nostrand Reinhold, New York, 1971.

Tanner, J. P., *Manufacturing Engineering: An Introduction to the Basic Functions*, Marcel Dekker, New York, 1985.

Vernon, I. R., *Realistic Cost Estimating for Manufacturing,* Society of Manufacturing Engineers, Dearborn, Michigan, 1961.

1.7 MANUFACTURING SUPPORT LABOR COSTS

1.7.1 Manufacturing Support Labor Requirements

Manufacturing support labor is a major cost driver in any cost estimate for a manufactured product. It includes the cost of the key functions needed to plan and tool for production of the item to be manufactured. Support labor includes manufacturing engineering, tool design and build, industrial engineering, supervision, and production planning and control. In some cases, the cost of support labor may be equal to or greater than the cost of the factory direct labor. In other cases, manufacturing support labor may be treated as an indirect cost and included in factory overhead. This is the case in many firms that do 100% of their business in the commercial or nondefense areas.

The amount of support labor required on any job is a function of the maturity of the product design, the technology or technologies involved, and the size of the company or program involved. In the aerospace and defense industries, support labor requirements are driven by government requirements for documentation, reports, and a host of other requirements that would not otherwise be needed. Manufacturing support labor may be estimated by first determining the tasks that must be accomplished, the time available to accomplish the tasks, and the level of support required.

Support labor costs are separated into recurring and nonrecurring, depending on whether the support provided is a one-time cost needed to plan and start up production or a continuing service to production over the life of the job. The design and fabrication of tools are nonrecurring support costs, but the maintenance and repair of the same tools are recurring costs.

1.7.2 Manufacturing Engineering

Manufacturing engineering includes the following activities and services:

1. Selection and design of manufacturing processes
2. Determination of sequences and methods for product fabrication, assembly, and testing
3. Selection and design of production equipment
4. Selection and design of tools and test equipment
5. Layout of factory buildings, machines, equipment, materials, and storage facilities
6. Determination of standard times for manufacturing operations
7. Selection and design of manufacturing systems and computer-aided manufacturing techniques
8. Manufacturing cost estimating, cost analysis, and cost trade studies
9. Manufacturing research and development
10. Review of product designs and specifications to ensure manufacturing producibility (or participation in concurrent engineering design team)
11. Management, coordination, and control of manufacturing operations

The primary recurring and nonrecurring tasks to be performed by manufacturing engineering must be identified, the staffing levels must be estimated for each task, and then these estimates must be man-loaded and time-phased to determine the estimated manufacturing engineering labor hours.

Development

Manufacturing engineering support during development is for the review of product design as it evolves through the various phases from concept to design layout, and to ensure producibility in its final form when released to manufacturing. Manufacturing engineers also plan the building of any engineering models or preproduction units that are to be fabricated by the manufacturing shops. At the same time, the manufacturing engineers work on and develop the preproduction planning that is implemented after the product is released to production.

This manufacturing engineering effort during the product design and development phase has been often overlooked in the past. It should be estimated and funded as part of the engineering development program for any new product. As indicated earlier, the best way to estimate this effort is to determine the task requirement and the staffing that is required to perform the task. The total manufacturing engineering hours are the staffing level times the number of weeks or months needed to complete the project.

Preproduction Planning

Preproduction planning starts with the production plan prepared as the basis for the estimate, and modifying it and adding details to update it as the actual requirements of the job begin to develop. This includes selecting and designing the process of manufacture; planning the sequences of fabrication, assembly, and test; selecting equipment and facilities needed; determining tooling requirements, methods, and factory layout; and anything else needed to decide how the product will be manufactured. The preproduction planning effort should begin during the product development phase, and the method of estimating this effort is the same as for development, by man-loading.

Start-Up and Production Shakedown

Start-up and shakedown involves the implementation of the preproduction plan and the initial low-rate production. This is followed by and concurrent with any modification or changes that might be required to debug the process and the tools. In this phase, tools and equipment are ordered, designed, built, and tried out. Work instructions, methods sheets, and operator visual aids are prepared. The factory layout and work flow are implemented, the operations are manned, and initial product manufacture takes place.

Figure 1.27 shows an example of detailed tooling planned for an orbital riveter to be used for a roll swaging operation on a circuit card. An example of detailed

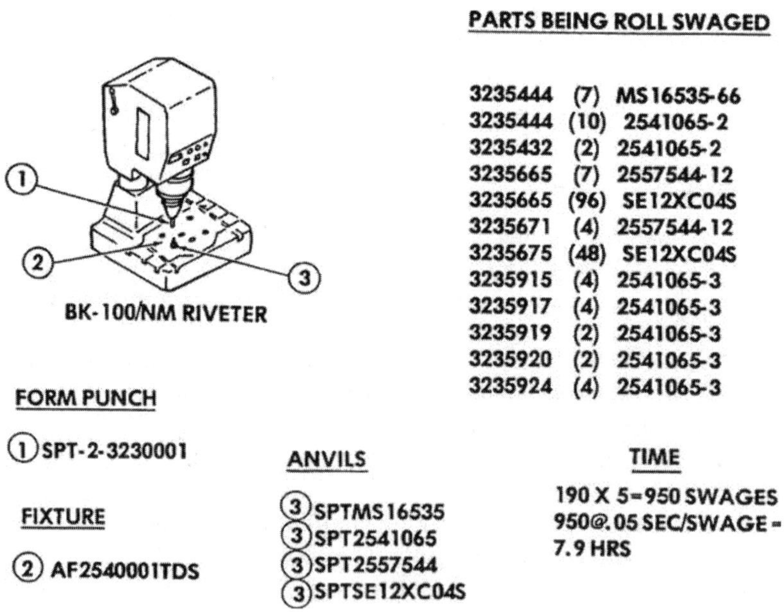

FIGURE 1.27 Orbital riveter tools, parts, and standard times.

operator work instructions is shown in Figure 1.28. This shows part of the instructions for assembling a gear train in an electromechanical subassembly. The man-loading technique for estimating manufacturing engineering man-hours described in the previous sections can also be used in estimating start-up and shakedown support in this phase. It is also possible to estimate the number of tools to be ordered, methods sheets to be written, visual aids to be prepared, etc., to prepare a more definitive budget. Manufacturing engineering hours for these tasks may be estimated by multiplying the estimated time per tool, methods sheet, or visual aid by the total number of each.

Manufacturing engineering support for start-up and production shakedown is part of nonrecurring manufacturing support cost. As such, it should be possible to define the tasks in sufficient detail to prepare an accurate cost estimate.

Sustaining Support

Sustaining support is the recurring manufacturing support effort provided by the manufacturing engineer after the job is in production. It includes resolution of the production problems that always occur from time to time, incorporation of any engineering design changes which may occur after production has started, and

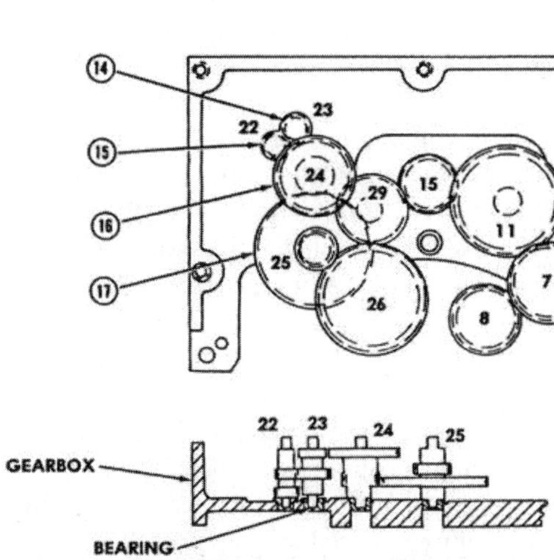

SEQ	PART NUMBER	DESCRIPTION	QTY	REQUIRED
14	39-00286-001	GEAR 23	1	POSITION TO HOUSING PER SKETCH
15	39-00287-001	GEAR 22	1	POSITION TO HOUSING PER SKETCH
16	39-00276-001	GEAR 24	1	POSITION TO HOUSING PER SKETCH
17	39-00275-001	GEAR 25	1	POSITION TO HOUSING PER SKETCH
18	—	—	—	SET HOUSING ASIDE AND POSITION COVER TO WORK AREA

FIGURE 1.28 Gear train assembly—operator work instructions.

improvements in the methods and tooling that may be realized after the job is on the floor and running. This sustaining manufacturing engineering effort is almost always estimated as a percentage of manufacturing labor. For a newly designed product that has never been in production, 10% of manufacturing labor is recommended. Again, this figure may be considerably higher or lower depending on the industry, the circumstances, and the degree of confidence in the product design.

For a product that has been in production previously and has a mature design, the sustaining manufacturing engineering support can be as low as 3% of manufacturing labor. The manufacturing cost estimator should review the history of similar jobs to determine the correct percentage for the job being estimated.

1.7.3 Tooling

Tooling can be a significant element of cost in the production of a manufactured product. Tooling is primarily a front-end, nonrecurring cost. The tooling philosophy established by management during the early phases of the proposal or cost estimate will largely determine what the tooling should cost. A heavy up-front tooling investment may ensure a low-cost product in production. A minimal tooling effort may still be the most cost-effective approach, depending on the circumstances and the length of the production run. Certain types of manufacturing, such as investment casting and ordinary sand casting, are by their very nature tooling-dependent, and as such will always request a large up-front investment in tooling. Other processes may not *require* a heavy tooling effort, but without an effective, up-front tooling investment, cannot be produced effectively in production. The length of the production run can be a big determinant in how much tooling is economically justified. A \$2000 assembly fixture amortized over a production run of 200,000 units adds only 1 cent to the cost of each unit of product, but a simple holding fixture, costing \$200 and needed for the assembly of a run of 200 units, adds \$1 to the cost of each and every unit.

Unfortunately, most companies spend far more on tooling than is really necessary. In the metal fabrication and forming industry, for example, it is possible to utilize standard shop tooling on many occasions to accomplish the same job that is often tooled with special-design tooling. It cannot be emphasized too strongly how important it is for the manufacturing cost estimator to review each drawing of a part or subassembly. Although the tooling required may appear obvious or apparent, another method may exist, or perhaps shop-aid tooling can be improvised to avoid this apparent expense. Tooling costs, and hence product costs, can be kept as low as possible by utilizing standard tools and tooling components whenever possible, and by using inexpensive tooling materials.

Tool Design

The design of special-purpose tools is estimated by determining how many tools are to be designed, then subdividing this number into tools that are highly complex, tools that are of intermediate level of complexity and detail, and those that are the most simple and straightforward. Average time values for each of the three categories are:

Highly complex: 24.0 to 40.0 hr./tool
Intermediate: 16.0 to 24.0 hr./tool
Simple design: 8.0 to 16.0 hr./tool

If a low-cost or minimal tooling approach is to be used, or one that utilizes little or no formal tool design documentation, then the cost estimator should figure 2.0 to 4.0 hr./tool to prepare a concept sketch or diagram to aid the toolmaker in building the tool. The time includes shop follow-up and liaison while the tool is under construction to clarify any details not clearly depicted on the tool drawing, and for the correction of any errors or design oversights.

Tool Fabrication

Tool fabrication includes the construction and tryout of the tool. The tool build hours almost always exceed design hours by an order of 3 to 1. In foundry tooling, for example, it includes building patterns, pattern plates, blow plates, and flasks, as well as various types of core-making tools.

Core-making tools make up most of the core estimate in foundry work. A pattern is set in the molding sand, and sand is packed around it to produce the impression into which the hot metal is poured to produce castings of a desired shape. Pattern plates separate the two halves of a pattern during molding, and flasks are the containers for the molding sand. Flask size, method of construction, and construction material are all key cost drivers in foundry tooling.

A core is a shaped projection of sand or other material inserted into the mold to create a cavity in the casting. Dry sand cores are formed separately and inserted after the pattern is removed, but before the mold is closed. A plan or layout of the cores in a casting is used to estimate the cost of core boxes, driers to support the cores during baking, blow plates, racks, special containers, fixturing including core-pasting fixtures, and ovens for core baking.

The method recommended for estimating core costs uses historical data. The number of acceptable cores made over a given period of time is divided by the cost of making the cores, which is direct labor cost only, thus giving the cost per core.

Tools to be fabricated for most manufacturing operations other than foundry, investment casting, die casting, etc. are estimated on an individual tool basis. That is, the cost of every tool is estimated individually. As indicated earlier, a good rule of thumb for the estimate of fabrication cost is to use three times the design estimate. Tool fabrication cost also includes tool tryout and any subsequent tool modifications that may be required as a result.

Tool design and tool fabrication costs are nonrecurring support costs in the total manufacturing cost estimate. They are concerned only with special-purpose tooling peculiar to a given product, but do not include standard tooling, or perishable tools such as drills.

Tool Maintenance

Another element of manufacturing cost that is often overlooked by the estimator is the cost of tooling maintenance. This is a recurring element of cost and includes everything from the initial tool setup and cleanup, if the job has been in production before and tooling exists, to repair, lubrication, painting, and other routine maintenance on operating tools used in production. This can be a major cost element if the production tooling includes automatic, special-design machines and their tools. It also includes the cost of tool preservation, and preparation for storage after the job is complete. Tool maintenance, if a direct charge, is estimated as a percentage of the total factory direct labor. It usually ranges from 3% to 5% of factory labor, depending on the degree of mechanization or automation of the process.

1.7.4 Industrial Engineering

Industrial engineering for manufacturing includes work measurement, standards setting, and maintenance, along with methods improvement, operations analysis, and cost analysis. In many companies this is an indirect function that is included in plant overhead. If this function is treated as a direct charge, the cost estimator should handle it by man-loading the industrial engineering personnel working on the job being estimated, and calculate man-hours by multiplying the staffing by the time period during which the services will be performed and converting this to hours. Industrial engineering support is a recurring cost. In many instances industrial engineering services are provided as part of the manufacturing engineering effort.

1.7.5 Supervision

First-line manufacturing supervision is often a direct charge in companies that produce manufactured products, especially in defense/aerospace industries. As such, it is a recurring cost to the job or program, and should be estimated as a percentage of the total factory labor hours on the job. Unless better information is available in your company, 10% of factory labor is the recommended norm for estimating supervision costs.

1.7.6 Production Planning and Control

Production planning and control includes manufacturing scheduling, material ordering, material control, material kitting, and issue to manufacturing, shop floor control, manufacturing status reporting and expediting, tool control, material handling and movement, work order release, packaging, and shipping. Even with today's sophisticated MRP software programs, the input and output is by people. The effectiveness of this function is critical to the success of manufacturing operations. Without an effective production planning and control system, manufacturing becomes a disjointed, uncoordinated group of activities. In a job order machine shop, for example, the scheduling and loading of the machines is critical to meeting promised delivery on all jobs.

Production planning and control labor is often estimated as a flat percentage of total factory labor. This percentage can range from 7% to as much as 15%, depending on the functions and degree of control exercised over manufacturing and production operations. It is also possible to estimate production planning and control labor directly by listing the functions, man-loading the staffing, and then time-phasing to arrive at total hours.

1.7.7 Bibliography

Malstrom, E. M., *Manufacturing Cost Engineering Handbook,* Marcel Dekker, New York, 1984.

Tanner, J. P., *Manufacturing Engineering: An Introduction to the Basic Functions*, Marcel Dekker, New York, 1985.

Vernon, I. R., *Realistic Cost Estimating for Manufacturing,* Society of Manufacturing Engineers, Dearborn, Michigan, 1961.

1.8 SUMMARIZING THE ELEMENTS OF COST

1.8.1 Cost Structure of the Final Manufacturing Estimate

In order to arrive at the final selling price of the manufactured product, costs beyond factory costs and profit margins must be determined. This cost and price structure for a manufactured product is shown in Figure 1.1. Prime cost is the sum of direct labor and direct material. Prime cost plus manufacturing overhead yields factory cost, or cost of goods manufactured. Total product cost is obtained by adding selling expense, general and administrative expense, engineering or development costs, and contingencies to the factory cost.

Several methods of pricing a product that is manufactured have been derived from the field of managerial economics. These pricing methods range from comparison of similar product prices to selling price based on total cost, with each product yielding the same percentage gross profit. Although many variations exist, there is usually some relationship between selling price and total cost. However, unit selling price is greater than total unit cost in all successful manufacturing firms. In cost estimates prepared for government audit after the contract has been awarded, the maximum profit is regulated by federal law.

In some companies the estimating department's responsibility for total manufacturing cost ends with establishing the direct labor, tooling, equipment, facilities, and material costs. The accounting group or marketing department adds wage and overhead rates from the various applicable cost centers, and performs the extension of the numbers to develop the final manufacturing cost estimate and price.

In structuring the final manufacturing cost estimate, care must be exercised in identifying those elements of cost that are recurring and those that are nonrecurring. In many instances the nonrecurring elements of cost, such as tooling, must be clearly identified to the prospective customer. Should a production contract result from the cost estimate, the tools so identified may become the property of the customer, to be delivered to the customer at contract completion. This mode of operation is fairly standard for contracts involving government work, and some of the major high-rate OEM companies.

If all of the nonrecurring costs are prorated into the final cost of the manufactured product, and no ownership of tools passes to the customer, it is still vitally important that management know and understand which costs are recurring and which are nonrecurring. Nonrecurring or one-time costs include the following:

Tooling that is specially designed and peculiar to the product but is not consumed in manufacturing the product, such as milling fixtures, drill jigs, and cable harness form boards

Manufacturing engineering services to do the initial and final production planning for manufacturing the product (this would include process plans, method sheets, factory layouts, workstation layouts, manufacturing bills of material, assembly parts lists, routings, and more)

Inspection plans and instructions, test procedures, test fixtures, and tooling peculiar to the product

Engineering and development costs incurred in the design of the product, including manufacturing drawings, product performance specifications, parts lists, and prototypes and models

Recurring costs include all costs that are contingent on the number of production units built, and include:

Labor and material to manufacture the product, including factory labor, inspection labor, and supervision and test labor, plus all materials and parts needed to build the product, even bulk and consumable items such as solvents, solder, paint, and plating chemicals

Perishable tools consumed in building the product, including cutting tools, drill bits, milling cutters, reamers, broaches, and similar items

Support labor, including manufacturing engineering, production control, engineering liaison, and other labor required to support the manufacture and assembly of the product on the production floor

Recurring and nonrecurring costs can be either direct or indirect. Production control labor, for example, would be both a direct charge and a recurring cost. Perishable tools are an indirect charge but also a recurring cost.

1.8.2 Computer-Assisted Cost Estimating

Without computer assistance, the manufacturing cost estimator must spend a great deal of time performing what are essentially clerical functions. These include researching data files, making many long and arduous arithmetical calculations and extensions, filling out spreadsheets and forms, running copies of originals, and much more. Not only is this very poor utilization of a skilled professional's time, it increases the possibility of error dramatically. With computer assistance, the cost estimating professional can spend the majority of time on basic estimating tasks that fully utilize experience and skill. These tasks include:

1. Obtaining a clear understanding of the requirements of the proposal or estimate
2. Concepting the plan and process of manufacture
3. Estimating the time required for each step or sequence of manufacture, including setups and run times
4. Determining equipment and tooling costs and alternatives
5. Calculating special costs, such as packaging and handling

Computer applications for manufacturing cost estimating also enable improved overall response time to customer and manufacturing needs, an overall improvement in cost estimate accuracy, and greater control and awareness on the part of management.

Computer systems are used in manufacturing cost estimating for data storage, to perform computations, and often to develop complete cost estimates. In data storage, actual cost data from previous jobs is much more accessible than when it is filed by conventional methods. The estimator can obtain process data, material costs, labor hours, and in some cases, burden rates to apply to the job being estimated. The computer can quickly and accurately perform any needed calculations, and prepare cost spreads and breakdowns needed to present the estimating data in the form and format required. Often, with proper programming and under the right conditions, the computer can do the entire estimate. Figure 1.29 shows an inexpensive computer cost estimating system for a small company built around a personal computer and a word processor.

1.8.3 Cost Estimating Examples

Many times the manufacturing cost estimating group is called upon to provide cost analyses, cost evaluations, and cost trade studies for the purpose of deciding which of several alternatives offers the greatest return on investment. The first example below determines the cost-effective breakeven point to change over from manual to automatic assembly of circuit cards used in the guidance system of a cruise missile. The study was critical in deciding whether the proposed investment in automatic assembly equipment could be justified, considering the number of circuit card assemblies that remained to be built on the contract. Such studies performed by the manufacturing cost estimator perform a vital function in the decision-making process.

The second example is a commercial amplifier assembly manufactured as part of a background music system by a small company. The amplifier assembly, consisting of two printed circuit card assemblies wired into a small metal chassis, is a standard product of this company. These units have been produced many times in the past, and all necessary processes, tooling, method sheets, and test equipment are in existence,

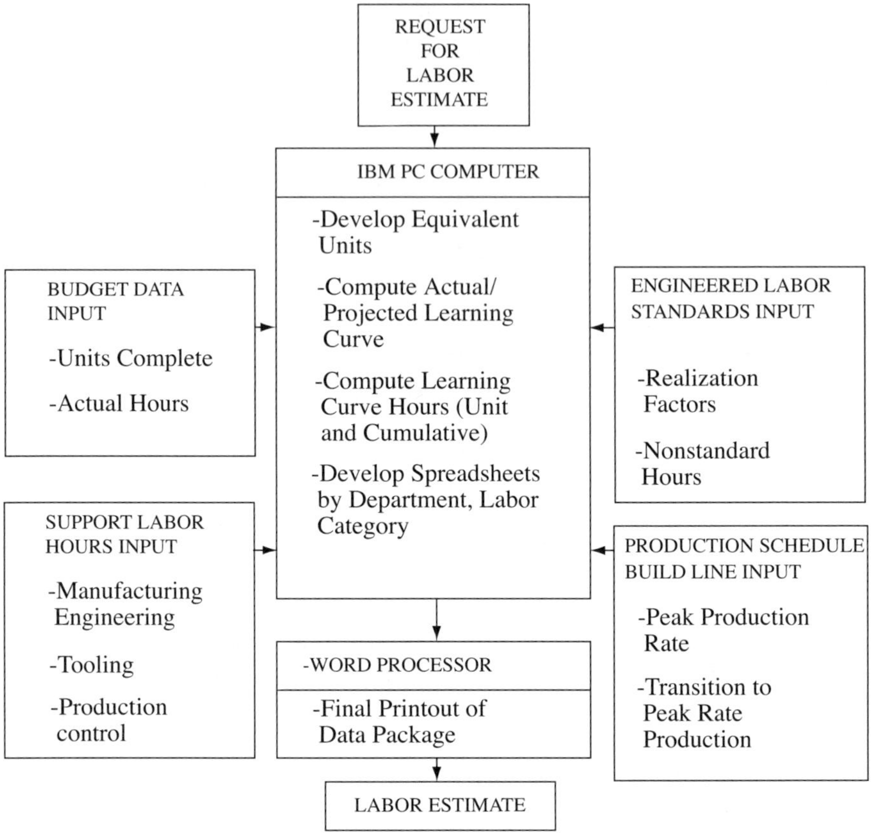

FIGURE 1.29 Computer-aided cost estimating system for small company utilizing a personal computer.

ready to be put to work. The company has been requested to quote on building some 200 additional amplifier assemblies. The last build of these units occurred 4 months ago and was for 150 units.

It should be noted that the examples given in this chapter tend to emphasize the development and analysis of labor costs. This is because the skills of the manufacturing cost estimator are built around estimating primarily factory labor cost. This should not make the estimating of material costs any less important. As indicated earlier, material can account for up to 70% of product cost.

Circuit Card Assembly Auto Insertion Cost Trade

Basic Assumptions

1. Eighteen boards per ship set, one hand-insert board not included
2. Fifty ship sets used as estimating quantity

3. Rework estimated at 3%
4. Dual-in-line package (DIP) inserter cost estimated at $79,000, with price increase by 5% in 6 months
5. Axial leaded components requiring strain relief to be formed on the automatic lead former, and inserted using Man-U-Sert. Some 34% of axial leaded components require strain relief in leads.
6. Loaded labor rates used (includes manufacturing overhead):
   Assembly @ $11.66/hr.
   Tooling @ $15.69/hr.

Nonrecurring Tooling and Programming Costs

1. Automatic sequencer: 18 programs × 7.0 hr./program × $15.69/hr. = $1,977.00
2. Variable center distance (VCD) auto inserter: 18 programs × 7.0 hr./program × $15.69/hr. = $1,977.00
3. Tooling: 4 tool orders × 1.5 hr./design × $15.69/hr. = $94.00
   175 hr. of tool fabrication × 4 tooling plates × $15.67/hr. = $8,630.00
4. Total tooling and programming = $12,678.00
5. Man-U-Sert inserter: 10 programs × 2.5 hr./program × $15.69/hr. = $392.00
6. Automatic DIP inserter: 17 programs × 7.0 hr./program × $15.69/hr. = $1867.00
7. Tooling: 4 tool orders × 1.5 hr./tool order × $15.69/hr. = $94.00
   50 hr. tool design + 125 hr. tool fabrication × 4 tooling plates × $15.69/hr. = $8,630.00
8. Total tooling and programming = $10,591.00

VCD Automatic Inserter Method Standard Times

1. Set up automatic sequencer: 0.050 hr.
2. Load automatic sequencer: 0.033 hr./reel
3. Automatic sequencer run time: 0.00065 hr./component
4. Set up VCD inserter: 0.083 hr.
5. Load and unload inserter: 0.0079 hr./board
6. VCD inserter run time: 0.00065 hr./component

Man-U-Sert Inserter Standard Method Times

1. Set up automatic lead former (including length and part number changes): 0.165 hr.
2. Automatic lead former run time: 0.00035 hr./component
3. Set up Man-U-Sert inserter: 0.067 hr.
4. Load parts trays, place on machine: 0.0065 hr./tray
5. Load and unload circuit board: 0.0079 hr./board
6. Run time to insert component, cut, and clinch leads:
   Axial leaded component: 0.0038 hr./component

DIP: 0.0085 hr./component
Radial leaded component (3 leads plus pad): 0.0106 hr./component

Automatic DIP Inserter

1. Set up automatic DIP inserter: 0.083 hr.
2. Load magazines: 0.0100 hr./tube/part number
3. Load/unload board assembly: 0.0079 hr./board assembly
4. Run time: 0.0006 hr./component

Equipment Setup Times

Sequencer: Load program from tape, enter count required into the computer, load reels of components into sequencer, set up the reels.

VCD automatic inserter: Load program from tape, load taped components onto reel, feed components into insertion head, verify sequence, assemble board holder to rotary table.

Man-U-Sert: Load program from disk, load trays into rotary bins, adjust board holder arms.

DIP inserter: Load program from tape, load magazines in proper sequence on machine, feed component into insertion head, verify insertion sequence, assemble tooling plate to rotary table.

Analysis

All components are currently programmed for the Man-U-Sert manual inserter. Some (592) components could be sequenced and run on the VCD automatic axial lead component inserter. This would leave 137 axial leaded components to be run on the manual inserter because of the requirement for strain relief when the leads are formed.

Some (299) components that are now inserted manually could be run on an automatic DIP inserter. Of the 19 different board configurations in the program, 11 are loaded on the Man-U-Sert, and 1 is totally hand-loaded.

Proposed Method

1. Automatic sequencer and VCD automatic inserter to be used:
   18 programs and 4 tooling plates would be required.
   All components (axial leaded) would have to be purchased on reels.
2. One Man-U-Sert program would be required.
3. Automatic DIP inserter to be used for integrated circuits:
   17 programs and 4 tooling plates would be required.

(The programming time could be reduced by half with the use of a program generator and host computer.) Seven different board configurations require work not covered in the analyses detailed above. These include adding jumper leads, piggyback components, removing circuit paths from the printed circuit board, and hand cleaning and soldering of the solder joints required to make these changes. The estimator's detailed analysis of this work is not shown, but a review of the bid-file actual working papers shows 6 additional pages of calculations for setup and run times for these three circuit board assembly methods. The costs in hours for each of the three assembly methods are:

Assembly Method	Setup time	Run time	Total time
Man-U-Sert	2.4870	294.68	297.167
Man-U-Sert + VCD	6.669	239.97	246.634
Man-U-Sert + VCD + DIP inserter	7.1150	134.67	141.78

These figures include lead strain relief, rework at 3%, and additional work described earlier.

The labor hours derived from standard times are then applied to a 15% learning curve to adjust for learning that would be realized for quantity increases. This curve, Figure 1.30, shows the actual learning curve currently being realized for the Man-U-Sert method of hand assembly.

Finally, the breakeven curves are drawn for both of the proposed new methods of assembly. These lines indicate the quantity of circuit card assemblies that would have to be produced using the proposed methods to pay back the investment required in machines, tooling, and software programs, plus methods and operations sheets required for new method implementation. These plots are shown in Figure 1.31, for the Man-U-Sert + VCD inserter versus Man-U-Sert only, and in Figure 1.32, which shows the DIP inserter + VCD inserter + Man-U-Sert versus Man-U-Sert only. The breakeven point for Figure 1.31 is 19,500 circuit board assemblies, and for Figure 1.32 is 95,000 circuit card assemblies. The second breakeven point in Figure 1.32 results if a DIP inserter is found to be surplus in another division of the company.

Job Order Assembly Cost Estimate

Basic Assumptions

1. Assume a 85% learning curve based on previous production experience
2. Standard time per unit attained at unit 1000
3. Some learning would transfer from previous production run, allowing point-of-entry on the learning curve at unit 12
4. All necessary processes, methods, tools, test equipment, and work instructions currently exist
5. Amplifier units will be manufactured in two lots of 100 units each
6. Material required is either already on hand or on order, thus allowing almost immediate start-up of production operations
7. Make-or-buy plan provides for buying all circuit cards, component parts, wire, connectors, metal chassis, cover, hardware, terminal boards, heat sinks, sockets, lugs, and consumables such as solder, grease, and paint
8. Off-line operations to be performed in-house include chassis paint, mark and drill, terminal lug installation on terminal boards, and all wire cutting, stripping, and marking operations

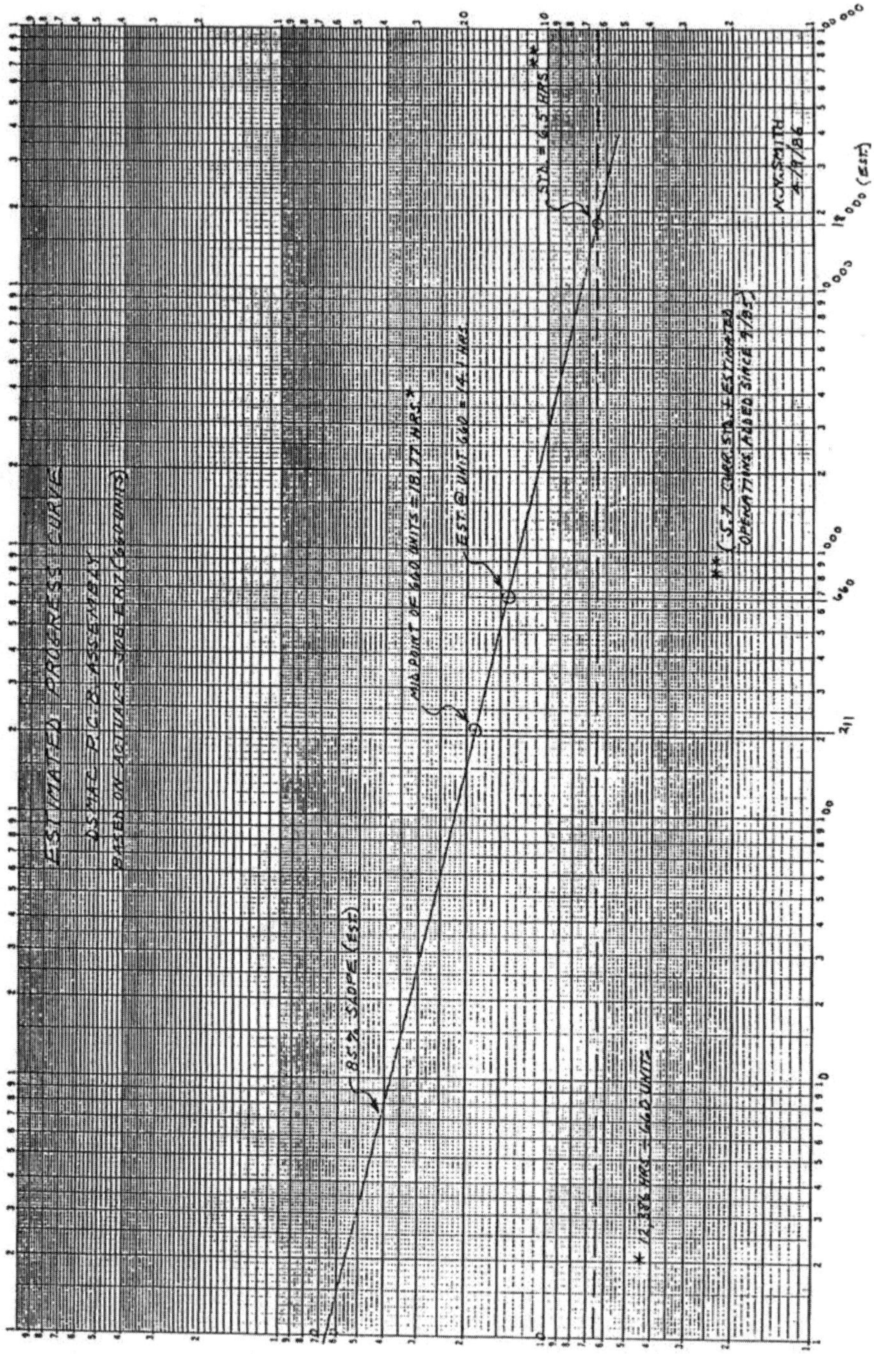

FIGURE 1.30

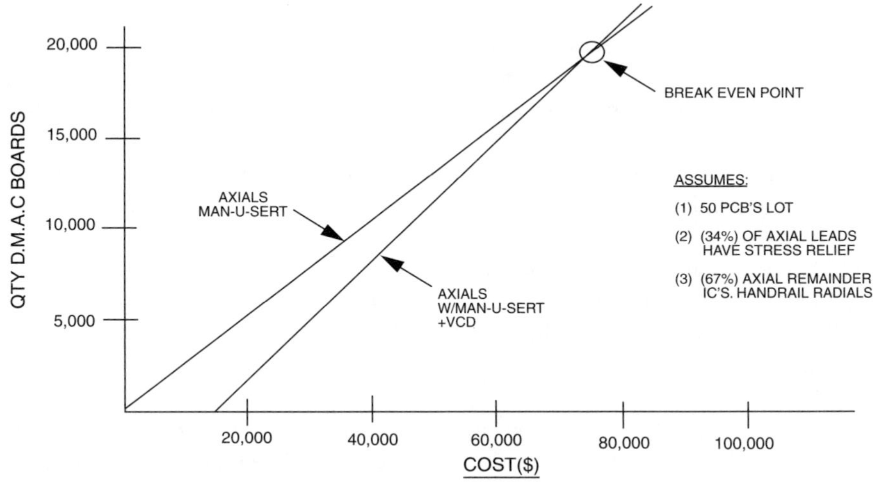

FIGURE 1.31 Breakeven cure when variable center distance (VCD) inserter is added to Man-U-Sert operation.

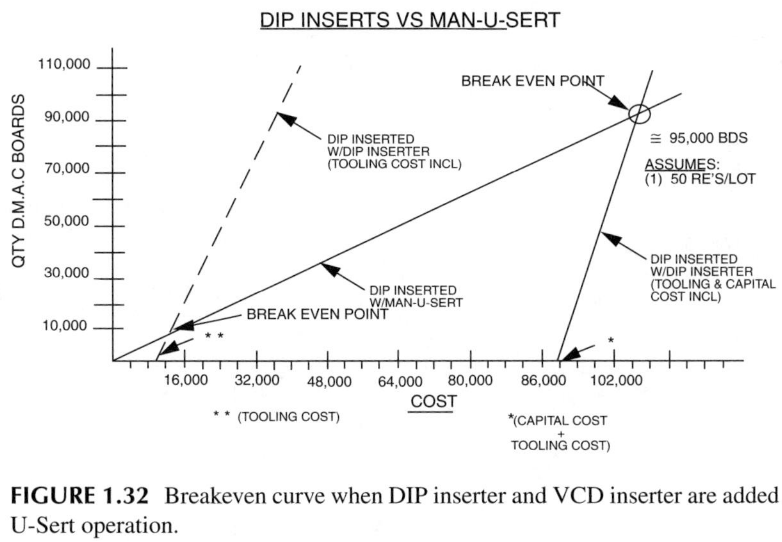

FIGURE 1.32 Breakeven curve when DIP inserter and VCD inserter are added to Man-U-Sert operation.

9. Standard times for all production operations currently exist, and are derived from original time studies of operations.

Preproduction Plan

The operation process charts shown in Figure 1.33 through Figure 1.35 fully describe the plan of manufacture for the amplifier units. These three operation process charts provide the information that is required to prepare the base labor estimate. By totaling the standard minutes shown on the charts for the various steps and then dividing by 60, we have the total standard time in decimal hours for the two subassemblies and the main amplifier assembly:

(2000–972) Circuit card assembly	0.65 hr.
(2000–973) Circuit card assembly	0.47 hr.
(S/A–100) Amplifier assembly	1.27 hr.
Total	2.39 hr.

If we assume that the units will be produced in two lots of 100 each, then there will be two setups in addition to the run time shown above. Setup time is estimated to be 4.50 hr. With two setups we would have a total of 9.00 hr. for setup and 478.00 hr. of run time for the amplifiers, giving a standard time of 2.44 hr. per unit for the 200 units.

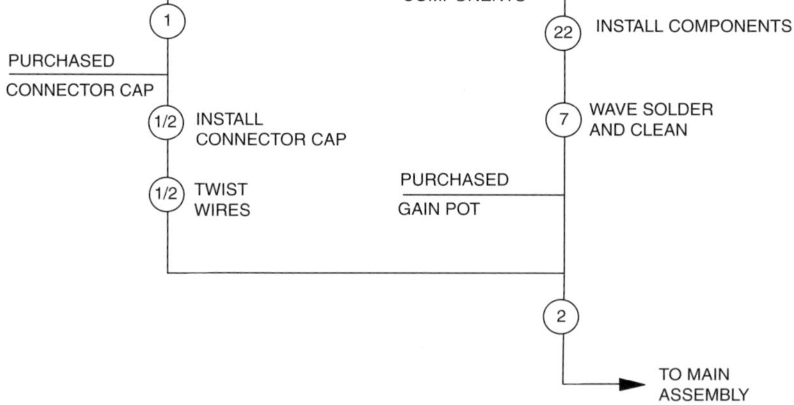

FIGURE 1.33 Operation process chart for main circuit board assembly.

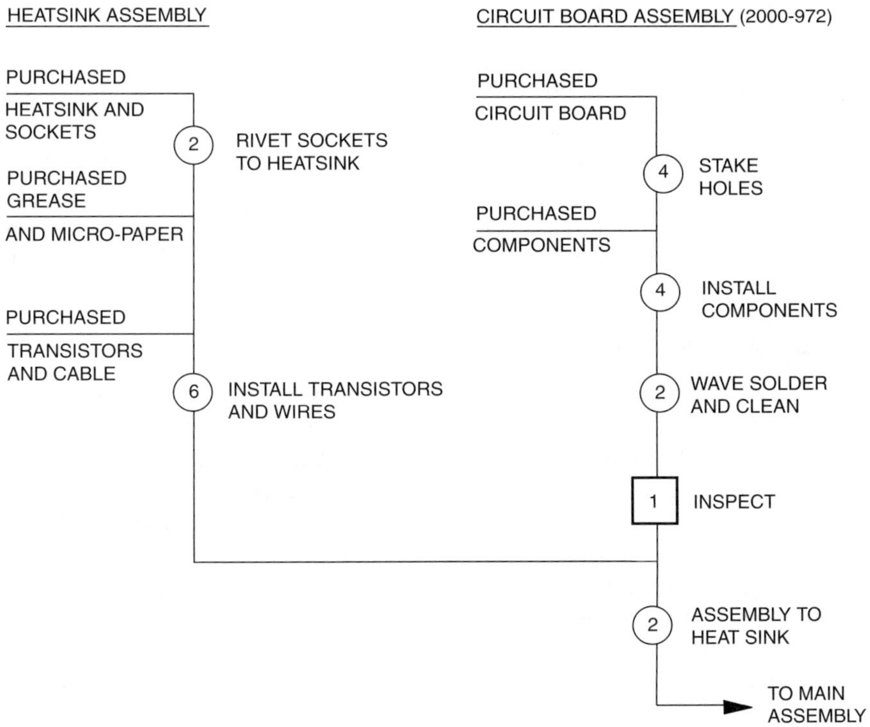

FIGURE 1.34 Operation process chart for power circuit board.

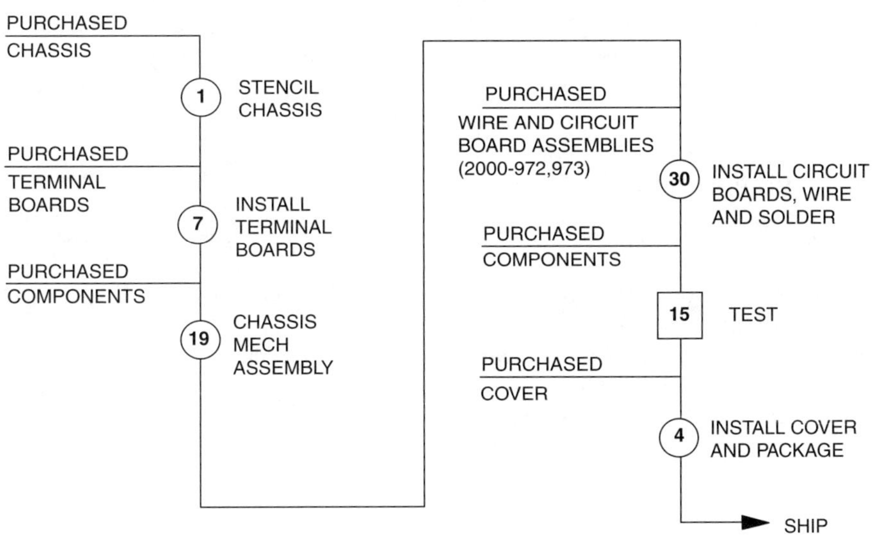

FIGURE 1.35 Operation process chart for main amplifier assembly.

Learning Curve Application

Previous production of the amplifier units followed a 85% learning curve, with standard being attained at unit 1000. Since the amplifiers have been manufactured previously, some learning would transfer to these units, allowing a point of entry on the learning curve at unit 12, giving an average time per unit of 5.15 hr. This represents what the run of 200 units will actually cost to produce. It should be kept in mind that the 85% learning curve slope is a composite slope which includes factory mechanical and electrical assembly labor, inspection labor, and test labor. If this customer has purchased units built earlier, he or she will expect a lower price this time.

Summary of Amplifier Costs

We next multiply the total labor hours by the hourly rate to obtain factory labor in dollars. The factory support labor is determined and multiplied by its hourly rate to determine support labor cost. Both of these labor dollar numbers are then multiplied by the factory burden or overhead rate to obtain indirect dollar costs, and are added to the factory and support labor dollars to obtain total dollar labor costs. Total material dollar costs are determined from the priced bill of material, and from the cost of available residual inventory already on hand. We then add material overhead or burden to give burdened material dollars. The sum total of burdened labor dollars and burdened material dollars gives the prime manufacturing cost for the units.

Selling price of the units to the customer is determined by adding the prime manufacturing cost to the general and administrative expense plus profit. This example is somewhat simplified in that no consideration is given to contingency factors, delivery schedule, or peak production rate requirements, all of which could add to the final selling price of the amplifier units.

After a careful check of all calculations, and review by management, the labor hours to manufacture the 200 units may be spread or time-phased over the planned build rate or schedule to determine monthly and even daily and weekly planned hourly expenditures. At this point, the selling price of the 200 amplifier units should be ready for transmission to the prospective customer.

1.8.4 Bibliography

Malstrom, E. M., *Manufacturing Cost Engineering Handbook,* Marcel Dekker, New York, 1984.

Matisoff, B. S., *Handbook of Electronics Manufacturing Engineering,* Van Nostrand Reinhold, New York, 1971.

Tanner, J. P., *Manufacturing Engineering: An Introduction to the Basic Functions*, Marcel Dekker, New York, 1985.

Vernon, I. R., *Realistic Cost Estimating for Manufacturing,* Society of Manufacturing Engineers, Dearborn, Michigan, 1961.

2 Product Tooling and Equipment

Richard D. Crowson

with

Clyde S. Mutter

2.0 INTRODUCTION TO TOOLING AND EQUIPMENT

Webster's dictionary defines the word *tool* as "any hand implement, instrument, etc., used for some work, or any similar instrument that is part of a machine." You could ask a hundred people what they think of when they hear the word "tool," and get answers anywhere from a screwdriver to a highly exotic instrument. The number and types of tools appear to be endless. The manufacturing engineering task in the tooling world of manufacturing is to understand all these tooling choices and select the right type of tool for each job. Think of the manufacturing engineer as the conductor of an orchestra, who must pick the correct instruments to produce the results that a critical audience is expecting

Planning and tooling are essential in bringing a product design into production. Tooling ensures proper fit and helps guarantee a repeatable process. Planning establishes the processes and establishes the timing at which the parts come together to make the finished product. Tooling and planning are two areas that can significantly affect the cost of a product.

Product tooling can range from inexpensive hand tools to highly sophisticated automatic assembly machines, from motion-economy devices for simple parts assembly to highly complex milling fixtures for precision machining. Product tooling can mean molds for die casting, or dies for extrusion. Tooling requirements for any manufacturing operation depend on the process requirements, the size and type of product, and the quantity of products to be manufactured. The ideas and principles discussed in this chapter apply to virtually any type of process or product.

When production levels are high over a period of several years, extensive and sophisticated tooling is easily justified. When production levels are low, or limited to a single run, the tooling must still perform its intended functions, but it is difficult to economically justify anything but the lowest costs. A $2000 assembly fixture amortized over a production run of 200,000 units adds only 1 cent to the cost of each unit of product, but a simple holding fixture, costing $200 and needed for the assembly of a production run of 200 units, adds $1 to the cost of each unit. The factory costs of low-volume or short-run products therefore may depend not only on the direct labor and material costs of production, but also to a very large degree on the costs of tooling. Tooling costs, and hence product costs, can be kept as low as possible by following the rules outlined below.

1. Utilize standard tools and tooling components whenever possible.
2. When special tooling is required, utilize the design concepts of previously well-engineered tools and fixtures.
3. When special tools and fixtures are required, use low-cost tooling materials in place of tool steel wherever possible.
4. Maintain close communication among the manufacturing engineer, the tool designer, and the toolmaker through all phases of the tooling program.

The planning and definition subchapter discusses the task of the manufacturing engineer, with the help of others, that decides the type of tooling or equipment to be used. This is a very important step and must be given much thought. If the product is new to the company, establishing a tooling philosophy is a good first step. The tooling philosophy is the guideline of the type of tooling to be used to build a product. Elements to be considered for setting the tooling philosophy include:

1. Quantity/rate product is to be built
2. Length of time product will be marketed
3. Estimated cost/value of product
4. Product design
5. Cost trade-off studies for tooling
6. Skill level of production workers
7. Profit margin required
8. Method of moving parts from station to station

The tooling philosophy and the production plan go hand in hand, with many people helping make the decisions.

Design, documentation, and control is an area that is often overlooked or minimized budget-wise by upper management. One of the reasons to invest in tooling is to have a repeatable process. Without control of the tool design and the physical tools, the process is in jeopardy. As the tools become more complex, the documentation becomes even more critical for building and maintaining them.

Tool construction and maintenance is as important as the design. There is nothing worse than having an expensive tool in the toolroom for repair. It should be on the production line making money. Building the tool with proper materials and workmanship, plus a good preventive maintenance program, saves both time and money.

Selecting the proper hand tool or other standard tool to do a job is a process all manufacturing engineers need be aware of. The cost of automating some operations may be very expensive, whereas the hand-eye coordination of a skilled worker, with the proper hand-tool selection, can perform the same function at a much lower cost, with better reliability and the added advantage of flexibility. A tool should never be designed if it can be purchased off the shelf.

Proper coordination during the planning phase should define the type of assembly tooling required. This is followed by refining the many small details that must be considered.

The assembly task may require only a minor holding device, or a large, complex piece of equipment to maintain alignment of key features of the product. Consideration must be given to the load/unload portion of the assembly job, as well as the movement of parts to the workstation, and the subsequent transport of the assembled product to the next position. Special machines and high cost go hand in hand. Special machines should never be ordered or designed without a great deal of thought and deliberation. However, a good special machine can make a very significant contribution to the success of a company. The ability to envision the machine function and process, being aware of the latest in off-the-shelf machine components such as part positioners, sensors, computer controls, etc. will be a great help in knowing what can be easily achieved. Computerized simulations of the workplace, and the overall factory flow of the product, may be valuable assets in this field.

Inspection and test tooling is another area where off-the-shelf equipment has made great strides. In the last few years, the number of instrumented equipment modules that are easily incorporated into a tool has greatly increased. This reduces the cost to make tooling for inspection or testing requirements into state-of-the-art, easy-to-maintain, and quick-to-build equipment.

2.1 PLANNING AND DEFINITION

2.1.1 Introduction to Planning and Definition

During the production planning phase for the manufacture of any new product, or the continued production of existing lines of products, it must be decided what approach will be used for tooling. In many instances, the process itself dictates the tooling requirements, leaving little room for deviation. This is the case for foundry and investment casting operations, as well as die casting, injection molding, and similar processes. In the high-rate production of automobiles and appliances, tooling is the key to successful production of a quality product, and little is spared to ensure that the tooling is appropriate and will do the job.

However, many companies spend far more on tooling than is necessary. In the metal fabrication and forming industry, for example, an experienced manufacturing engineer can specify standard shop equipment on many occasions to accomplish the same job that is tooled with special-purpose tooling.

This is also true in metal-removal operations. In the past, precision machining often required elaborate fixtures to ensure accuracy and productivity. Modern numerically controlled (NC) equipment may require only a holding fixture in addition to any special cutters, drills, or form tools. It is important to review each product drawing and ask yourself, even though the tooling approach *seems* apparent, what other methods exist, and whether standard tooling can be modified or improvised to avoid this expense.

2.1.2 Tooling Preplanning

The tooling philosophy should reflect a low-cost yet fully adequate tooling approach. In an increasingly competitive world, the company that is successful is the one that goes a step beyond the standard tooling approach, and finds ways to do things better and cheaper. This requires the knowledge, experience, and resourcefulness of the manufacturing engineer. The tooling preplanning should determine how the low-cost tooling philosophy will be carried out.

Reviewing the Product

The first and most important phase of any new tool or fixture design is the preplanning accomplishment by the manufacturing engineer. The manufacturing engineer must review the entire fabrication or assembly operation, define the tool or fixture requirements, and consider such factors as:

How the part or assembly will be held or clamped in the tool or fixture

What orientation and accessibility the tool must provide the machine or operator so that the work may be performed most efficiently

What relationship the tool will have to the workplace (if an assembly tool), parts bins, hand and power tools, and other equipment required for the operation

The spatial relationship between the product, part, or assembly and the operator to provide an efficient and comfortable working position for that operator (assembly tool)

To understand the importance of tooling preplanning, it is necessary to explore in more detail the tooling requirements, how the tool is conceptualized, critical tool features, tool design standards, and the importance of maintaining a standard tool inventory.

Defining Tooling Requirements

The requirements for the tool must be thoroughly understood by the manufacturing engineer, since this is basic to all else that now follows. The manufacturing engineer must thoroughly understand *why* a tool is needed, what *function* it will perform, and what *process* parameters it must support. The first step is to prepare some type of process flow chart on the assembly steps, followed by preparation of the parts-fabrication process sheets.

In the injection molding field, as an example, an understanding of all the conditions that prevail during the molding operation should ensure that necessary precautionary measures are incorporated into the mold design to safeguard the expected features in the product. The function of an injection-mold type of tool is to receive molten plastic material ranging in temperature from 350 to 750°F at pressures between 5,000 and 20,000 psi. In the injection process, the plastic comes from a heated nozzle and passes through a sprue bushing into feed lines, or runners, and then through a gate into a cavity. The cavities are maintained at lower temperatures, at which solidification takes place. The range of temperatures and pressures depends on the type of plastic material. The plastic is held in the cavity of the mold for a prescribed time, until full solidification takes place. At this point, the mold opens, exposing the part to the ejection or removal operation. From the general concept of the molding operation, it is recognized that it is important to design a mold that will safely absorb the forces of clamping, injection, and ejection. Furthermore, the flow conditions of the plastic path must be adequately proportioned in order to obtain, in cycle after cycle, uniformity of product quality. Finally, effective heat absorption from the plastic by the mold has to be incorporated for a controlled rate of solidification prior to removal from the mold. The tool designer must incorporate all details that are conducive to good molded parts.

Investment castings are made in both small lots and very large lots, and range from the simplest design to extremely intricate configurations. Many different methods of tooling are possible. If the requirements are small or moderate, a single-cavity tool may be sufficient to make the patterns (Figure 2.1). The pattern rate from a single-cavity, simple die can vary from 20 to 50 parts per hour. A highly automated single-cavity die with hydraulic openers and compressed-air pattern removal can produce as many as 250 patterns per hour.

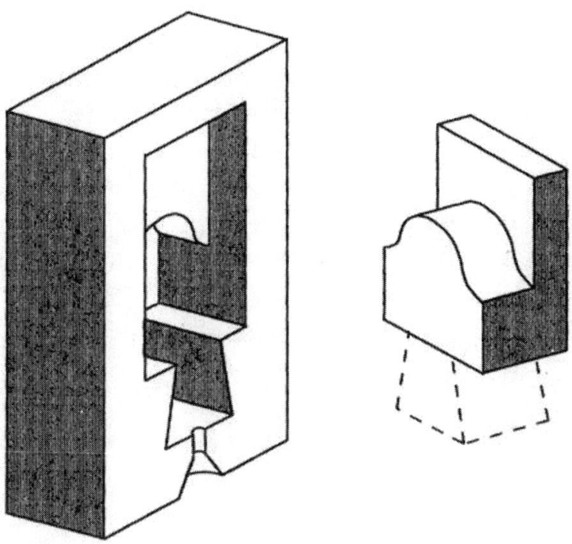

FIGURE 2.1 Single-cavity mold for investment casting patterns.

Automated dies, machined from steel or aluminum, are used when thousands of patterns a month are necessary. Once this tooling is ready for production, it is expensive to make changes. For this reason the product designer must be sure of the product design before the manufacturing engineer proceeds with development of high-rate production tooling. If an area of possible design change is known, the manufacturing engineer should be involved so that provisions for such changes can be incorporated in design of the die at the original concept stage. An alternative to this traditional approach is to include rapid prototyping in the process.

A well-thought-out tooling concept for a special tool to open the insulation track in aluminum window sash bars is shown in Figure 2.2. This was an especially troublesome problem in the manufacture of aluminum window frames, as the extruded shape of the sash bar was such that the insulation track was necked down at the sash bar cutoff operation, requiring an extra manual operation with a screwdriver to open the track. The new tool cut the time for this operation in half. The concept sketch tells the tool designer exactly what kind of tool to design and what the tool is expected to do.

Figure 2.3 shows a concept for a lazy Susan fixture to hold and turn large window frames while caulking and insulation strips are applied to hold the glass panels in place. Previously this required two operators to physically pick up the frame and turn it 180° on the workbench. Now one operator can turn the window frame by depressing a foot pedal, without leaving the workplace.

An excellent way to assist the manufacturing engineer in developing a tooling concept during the preplanning phase is to provide a manual that catalogs, illustrates, and describes tooling that was successful in the past. Such a manual also shows applications

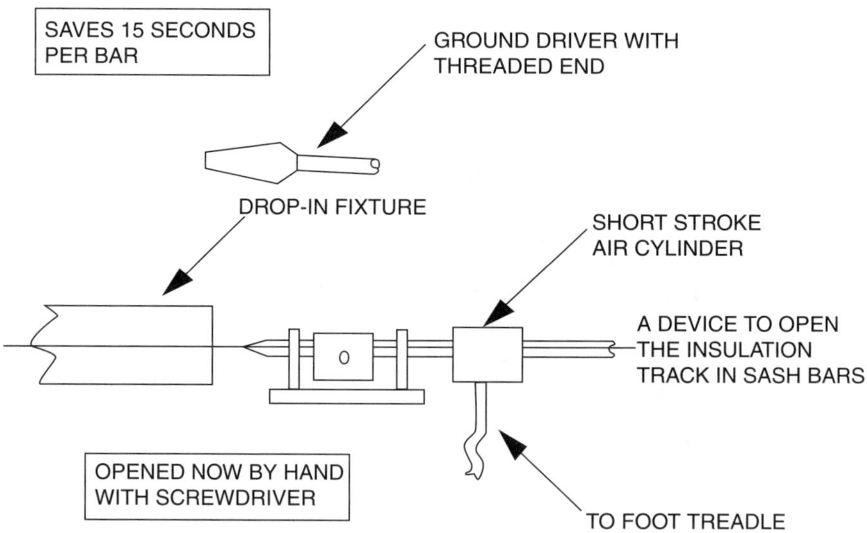

FIGURE 2.2 Tool for spreading windows sash bars.

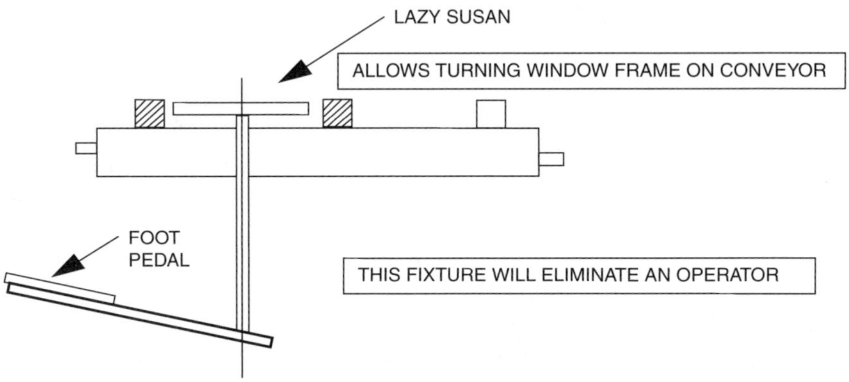

FIGURE 2.3 "Lazy Susan" tool for turning large windows with a foot pedal.

of standard tools and tooling components that the company has in its tooling inventory. These would include such items as small vises, stops, quick-acting clamps, ball joints or swivel fixtures, rotary fixtures, air cylinders, small electric motors, punch-and-die sets, and so on. Reference to such a manual provides the manufacturing engineer with ideas that might help with current tooling requirements, and could uncover a design or application of a previously used tool that might be used with minor modification for the new tooling application.

Communication

It is one thing for the manufacturing engineer to know the tool required for the fabrication or assembly operation, but often another thing to adequately communicate these concepts to the tool designer. Concept sketches such as those shown in Figure 2.2 and Figure 2.3 go a long way toward conveying this information, as does the written tool order used in many companies. Nothing takes the place of direct verbal communication between the manufacturing engineer and the tool designer. The tool designer must understand how the operation is to be performed and must discuss with the manufacturing engineer precisely what is required of the tool. The tool designer must learn critical dimensions and features, as well as desired tool locating points and surfaces, and stops and clamps required. The expected life of the tool in terms of anticipated total production of the product should be discussed, as this is the basis for the degree of durability required of the tool. For example, tooling made of wood or plastic components in place of tool steel can, in some cases, provide the required durability for low-volume, short-run production and reduce the total cost of tooling by 40% or more. In many shops, the tool designer prepares the concept sketches as part of the tool definition process. This aids in the mutual understanding of the tooling requirements, the advantages and disadvantages, a preliminary cost estimate, and schedule impact of the various choices available for an operation.

2.1.3 Rapid Prototyping

A few years ago, product design for plastic parts, and for cast or forged metal parts, was a long, drawn-out exercise. The design engineer would have a master machinist hog-out a part from a drawing or sketch. The part would be evaluated and changed, a new part made and evaluated, etc. In the meantime, the manufacturing engineer and the rest of manufacturing waited for the product design. Today's rapid prototyping techniques have changed this process. Now, using computer-aided design/computer-aided manufacturing (CAD/CAM), models can be made quickly. The flexibility is available to adjust the design and see the results in record time without affecting schedules or cost. Manufacturing can now proceed earlier with expensive hard tooling, without concern about last-minute engineering changes. As an added bonus, the models can be used to prove-out other tools.

Armstrong Mold Corporation (East Syracuse, New York) is one firm that specializes in rapid prototyping. Armstrong has the ability to quickly produce prototype plaster mold metal castings and cast thermoset plastic parts—often within 1 to 2 weeks for simple parts and 1 to 2 months for complex parts.

The latest rapid prototyping techniques include stereolithography, laminated object manufacturing, and CAM hog-outs from solid stock, as well as traditional pattern making and craft skills. With all of these skills and methods available, premium-quality parts can be produced in a short time at relatively low cost. Some of the advantages of rapid prototyping are:

- Paperless manufacturing (working directly from CAD files)
- Functional parts in as little as 1 week
- Produces models that can be used as masters for metal castings and plastic parts in small or large quantities
- Fit and function models available to detect design flows early in development
- CAD file formats available include:
  - IGES, DFX, CADL, STL
  - Floppy diskettes: 3.5 or 5.25 in.
  - Tape: mini data cartridge, DC 2000 style

Stereolithography

Stereolithography can be used to create three-dimensional objects of any complexity from CAD data. Solid or surfaced CAD data is sliced into cross sections. A laser that generates an ultraviolet beam is moved across the top of a vat of photosensitive liquid polymer by a computer-controlled scanning system. The laser draws each cross section, changing the liquid polymer to a solid. An elevator lowers the newly formed layer to recoat it and establish the next layer's thickness. Successive cross sections are built layer by layer, one on top of another, to form the three-dimensional plastic model.

Laminated Object Manufacturing

Laminated object manufacturing (LOM) is another method of creating three-dimensional objects of any complexity from CAD data. Solid or surfaced CAD data is sliced into cross sections. Thin plastic sheet materials are laminated one on top of another. A single laser beam cuts the outline of each specific layer. The process continues until all layers are cut and laminated, creating a three-dimensional, woodlike model.

CAM Hog-Outs

Three-dimensional objects can be cut from a variety of solid stocks (metal, resin, plastic, wood) directly from two- or three-dimensional CAD data using multiaxis CNC machining centers.

2.1.4 Tooling for Special Machines or Robots

In recent years, in addition to special machines, the robot has become a very useful tool for manufacturing. It has found its place in material handling, spot welding, seam welding, spray painting, machine tool tending, and numerous other applications. In all these cases, sensors are employed to exclude uncertainties—that is, to detect them and react appropriately. In the context of assembly, these uncertainties can be traced to two factors: uncertainties from tools (feeders, fixtures, and machines) and uncertainties in the parts themselves (manufacturing tolerances). There are two basic ways to handle these uncertainties in a manufacturing environment:

1. Avoid all uncertainties in the assembly planning phase
2. Detect uncertainties with the aid of sensors

The first possibility is the traditional approach, which gives rise to inflexible, expensive, and time-consuming systems, especially in the construction of specialized feeding devices. Such a solution makes quick product changes uneconomical because of long changeover times. A compromise between these highly sophisticated and therefore expensive systems and conventional grippers can sometimes be achieved by mounting some sort of compliant device between a conventional gripper and the robot. These compliant devices consist of two metal plates connected by elastomeric shear pads, and are known as remote center compliance (RCC) devices. RCC devices can compensate positioning faults resulting from reaction forces in the assembly phase.

2.2 DESIGN, DOCUMENTATION, AND CONTROL

2.2.1 Introduction to Design, Documentation, and Control

Many factors enter into the proper design of jigs, fixtures, and other tools that are not apparent to the casual observer. A finished tool may seem to be nothing more than rough castings or built-up pieces of steel with a few seats and straps to hold a part in place. The study of the manufacturing engineer to provide a proper sequence for fabrication or assembly, the care with which the tool designer lays out those straps to hold the work without springing it, and the accuracy with which the toolmaker sets the seats and bores the holes do not show on the surface. The real value can be easily determined, however, when the tool is put into use.

It will be found that tool design is not a haphazard, hit-or-miss proposition. Each manufacturing concern must decide upon the quality of work that it wishes to produce, and it must adapt certain standards of design and tolerance that will produce that quality. This subchapter will explore the duties and constraints of the tooling functions that are responsible for designing, building, and controlling the tooling in modern industry.

2.2.2 Control of Tooling

After the decision has been made to design and build a tool, the next step is to actually perform the work. Some type of system is needed that controls the tool through every aspect of the tool's life. A tool order is the normal way that companies authorize a tool to be built. A common system or procedure for processing the tool order is shown in Figure 2.4. In larger companies, much of this may be done by computer, but the steps are essentially the same. The tool-order form will vary from company

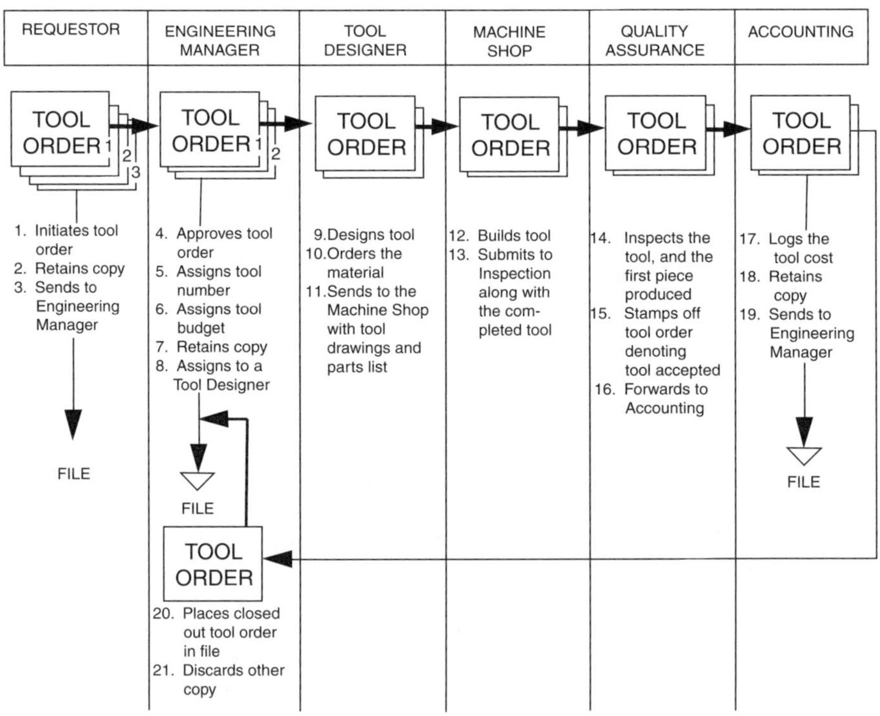

FIGURE 2.4 Typical tool-order control system.

to company, but the data required is quite consistent. The tool order will contain the following information:

1. Description of the tool function
2. Product part number to be used in the tool
3. Production rate the tool will support
4. Quantity of tools required
5. Life expectancy of tool
6. Special conditions the tool will see (hot, cold, wet, etc.)
7. Safety considerations
8. Expected skill level of user

Tool Orders

The tool order is the primary document used by the manufacturing engineer in most companies to formally authorize the design and fabrication of tools. Figure 2.5 shows a tool-order format for a small to medium-sized company that contains all of the necessary information blocks to convey information to the tool designer for the design of the tool. In addition to conveying information, it is a formal work-authorizing document for design and toolroom work. It is used by accounting to

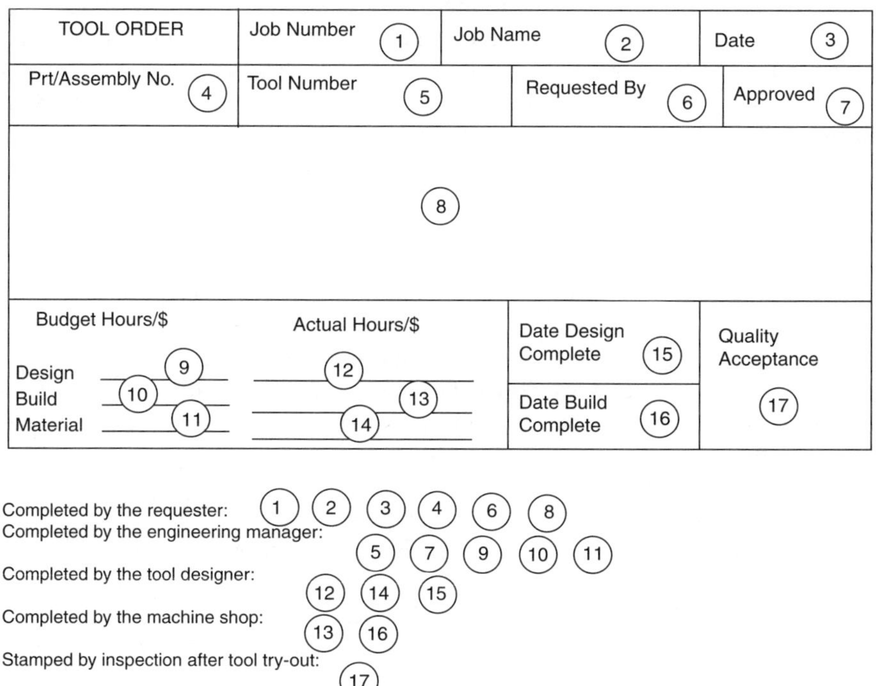

FIGURE 2.5 Tool-order format.

identify and collect costs by individual tool for a given job. All tool-order forms should be similar to this form.

When the tool drawing is completed and checked, the manufacturing engineer should review the design and accept the design concept. At that time, he or she should estimate the cost of building the tool and review the economics of the tool and the manufacturing operation to determine whether a more or less elaborate tool would be more economically feasible. On certain projects, the budget for tooling may have been established as the driving factor. In other cases, the product fabrication or assembly cost may permit spending more money on tooling.

Although the tool drawing should be so complete that no other communication with the toolmaker is needed, a few minutes of explaining the function of the tool is usually time well spent. If the toolmaker knows what is expected of the tool, he or she can normally do a better job of building that tool, and can often offer suggestions for improvements or cost reductions during the design phase.

Upon completion of the tool, both the tool designer and the manufacturing engineer should observe the tool tryout or, if shop conditions permit, try out the tool themselves by performing one or more complete operations with production parts. Any required debugging or modification can be accomplished before starting manufacturing operations. All changes and modifications made during tool build and tool tryout should be recorded so that the tool master drawing can be updated to agree with the final configuration of the tool.

Once a tool is complete through tool fabrication, tool tryout, and final acceptance, control of the tool is still required. All tools require some form of maintenance and may have to be recalibrated periodically. Do not expect the production floor to remember to do these tasks. When the tool control system works properly, the *system,* with recall tickets or orders, will do the job.

2.2.3 Design Considerations

Once the necessary concept and descriptive information have been transmitted to the tool designer via the tool order, concept sketch, and verbal communication from the manufacturing engineer, it becomes the tool designer's responsibility to communicate adequately with the toolmaker. The primary medium of that communication is the formal tool drawing. Beyond following standard drafting practices, tool designers should consider the following to assist in communications and reduce tooling costs:

1. Tool drawings should describe the tool completely and also specify optional construction in noncritical areas wherever possible
2. Tool tolerances should be specified as loosely as possible, with every attempt made to avoid extremely tight tolerances
3. Tool drawings should be dimensioned completely, so that scaling is not required by toolmakers
4. Noncritical screw and dowel locations should not be dimensioned; instead, a note such as "locate approximately as shown" should be indicated

5. A complete material list should be included on the first sheet of the tool drawing
6. A list of standard sizes and shapes of materials stocked in the toolroom should be available to each tool designer, and materials should be specified from this list whenever possible
7. Clamps, stops, locating pins, rest buttons, and hold-down screws should be located so that the part or assembly will not be distorted or have its surface damaged
8. The tool drawings should specify marking the tool number, part number, and any other numbers to allow positive identification of that tool in the future
9. The current revision letter or number of the part or assembly should be noted on the tool drawing and also on the tool

The details of tooling design depend on several factors, such as cost, quantity of work, and the ingenuity of the designer. There are some suggestions to bear in mind, with differences of opinion as to whether they are listed in order of importance:

Simplicity
Rigidity of clamping devices
Sequence of operations
Interferences in the tool itself
Interferences with the machine on which used
Clearances for work and hands
Avoidance of chip pockets
Locating points corresponding with dimensions on part drawing and with locating points on other tools for the same part
Convenience and speed in operation
Accuracy of work produced
Durability
Economy of construction
Stock sizes of material used

In making the drawing, it is necessary to consider the accuracy as to scale, the correctness of the projections, proper representation of the fixture and the work, and its reproduction qualities. The dimensions on the drawings must be accurate and sufficient without unnecessary repetition, legible, and contain the tolerances in understandable form. Dimensions that are out of scale should be underlined to call attention to that fact. Drawings must also contain information or specifications as to materials, finish, the kind of heat treatment (if any), finish of surfaces by grinding or otherwise, necessary notes, instructions for marking part numbers, and the title of the part to be used in the tool.

After all the general rules of design are covered, more decisions are to be made. Some companies require every detail of a tool be drawn, which adds to the design time, but reduces the fabrication time by allowing many people, including a machinist, to work on the tool at the same time. When duplicate tools or replacement details are to be made, we can be assured they are all alike. Other companies prefer the design to be as basic as possible, with only a few key dimensions provided. This saves design time, but requires the tool to be built completely by one toolmaker.

The most important consideration in design of tooling for production is the product. The dimensions, finish, quantity, and ultimate use of the part in the completed product will usually dictate the processes available to achieve the desired results This process definition will then aid in determining the machine or other equipment to be used with the tool. The following are some examples of the choices and the relatively complex factors involved in defining the tooling requirements—and therefore the details of the tool design.

2.2.4 Points to Be Considered in Design

Patterns

Where patterns are to be made there are four major considerations:

Economy of construction
Ease of molding
Equality of sections to avoid unequal shrinkage
General appearance

Wherever possible, stock patterns should be used.

Drilling

A part requiring 14 operations of drilling, reaming, countersinking, and counterboring to be done in one jig can be handled in several ways. The operations can be performed on two eight-spindle gang drills, on three six-spindle machines, or even on four four-spindle machines by passing the job from machine to machine and returning it to the original station on a conveyor. This, of course, requires several duplicate jigs and is one method for large production. The same jig can be used on any one of the machines by carrying out eight, six, or four operations, according to the number of spindles, and by finishing up with a change of tools in quick-acting chucks for the remaining operations. Any of these methods may be said to have good points in that all operations are performed in a jig at one setting—all holes being necessarily in proper relation with each other. One method ties up machine tools, and the others expend labor hours. If, for the same part, two or three jigs are made for the eight- or six-spindle machine, a more flexible arrangement results. It is possible to group the larger holes in one jig and the smaller in another, gaining time by using a faster speed for the small holes. With modern NC machines the drill jig may sometimes be replaced with a simple holding fixture, where the accuracy and repeatability is supplied by the machine.

Press Blanking

In the metal fabrication industry, there are many choices available for stamping out the flat pattern of a sheet metal part. As shown in Figure 2.6, a simple rectangular blanking tool can be made from four pieces of low-cost hardened steel measuring 6 in. by 6 in. Two of the four pieces will act as a base for spot welding the punch and die portions, and for mounting the two guide pins. One other, which will become the punch portion, is cut to the exact size of the hole it will punch. It is then spot welded to the 6-in. by6-in. base. The remaining piece is now sawed or punched, leaving a cutout in the piece of the desired size plus 0.007 in. This becomes the die, and the 0.007 in. is for die clearance (clearance will vary according to the material to be blanked). This piece is now mounted to the other base, again by spot welding. Two guide pins are inserted in opposite corners, or in the center of the cutout in the female portion of the base. Two holes of the same diameter as the pins, plus 0.003 in. for clearance, are added opposite the pins in the other base, which is the punch. Slide one onto the other, using the pins for alignment, and the blanking tool is complete. It can then be mounted in a standard die set, which maintains alignment and provides for attachment to the punch press. The guide pins can now be removed.

This tool is capable of piercing 800 large holes per sharpening. It can be sharpened twice before being discarded. It would take 6 hr. to make this economy tool. For comparison, a tool with the same function made in a tool shop by a more standard method using more durable materials and a dedicated die set would cost many times more, plus a number of weeks for delivery. This tool would have great value, in that similar tools could be made in any good sheet metal shop by a journeyman mechanic. The part tolerances would be about the same as for a permanent tool. Under normal conditions, this would be plus or minus 0.005 to 0.010 in. The real differences of tool life, automatic feeding, and the like may not be as important for the part as lead time or initial cost.

Brake Forming

On many press-brake bending jobs, solid blocks of urethane and other elastomers can be used to replace conventional machined dies. The savings, of course, are great. The

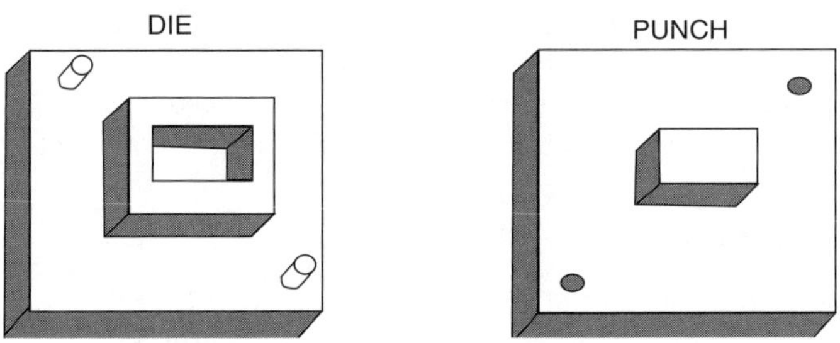

FIGURE 2.6 Simple blanking tool.

value of the urethanes lies in their ability to transmit kinetic energy. Urethane, unlike rubber, is virtually incompressible. If a blank is forced into a block of the material by a punch, the energy of the punch is transmitted almost evenly in all directions. If the material is confined by a retainer, the energy is reflected back by the retainer walls, since it has no place else to go. It will force the blank against the punch with uniform pressure. The advantages of urethane tooling can be summarized as follows.

1. Low tool cost (urethane is normally used in block form with minor relief to direct metal flow; it can be machined with ordinary shop equipment)
2. Minimum setup time
3. Extreme flexibility
4. Nonmarring of surface finish
5. Excellent part definition when properly used
6. Production of sharper bends or smaller radii than with metal dies
7. Sharp reduction of springback and wrinkling
8. No compensation needed for variation in stock thickness

Urethane is now accepted as a standard die material, and its use is standard practice in most progressive sheet metal shops. Techniques of this type may eliminate designing and building dedicated punch-and-die sets to do the same job.

2.2.5 Documentation

It was only a few years ago that tool documentation meant a decision between a drawing on Mylar or vellum, and using pencil or ink. While there are still occasions for manual drafting, most tool design is accomplished using Unigraphics, CADD, AutoCADD, or one of several other well-known CAD programs. Drawings are now received electronically, and tool design data can be sent directly to N/C machines to make the tool. Newer and better programs are available every day and most are well within the price range of even the smallest company. It is important to remember that while the computer is an aid to design preparation, the same human thought processes are required for good design. Computers can be very good tools when used correctly, or expensive toys when not. A computer cannot improve the design of a tool; it just makes it easier for the designer to work. One important consideration for using CAD in the design of tooling is the ability to make changes or modifications. The basics of a good design are still in the hands of the designer. Standards have been established on how the basic designs are arranged, to allow the toolmaker to be able to read a drawing without misinterpreting the design intent.

Arrangement of Views

For drawings in orthographic projection, the third-angle system, known in Europe as American projection, has been in practically universal use in the United States for many years and is continued as the American standard. A brief discussion of this practice will be based on sketches of the object shown as Figure 2.7. In third-angle

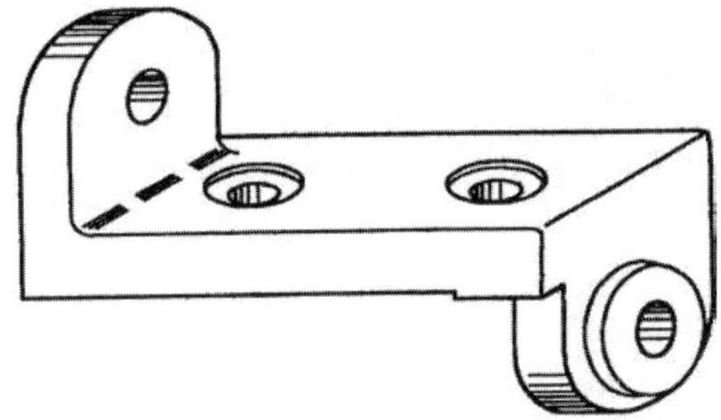

FIGURE 2.7 Basic isometric sketch of a part.

projection the top view is placed directly above the front view, and the right-side view to the right of and facing the front view (Figure 2.8).

Figure 2.9 shows the relative positions of the six possible principal views of an object sometimes needed to describe a part or tool: front, top, right side, left side, and bottom. A bottom view, or "view looking up," can be used to advantage instead of a top view when the shapes or operations to be shown are on the underside of the part. For example, in a normal punch-and-die drawing, the arrangement of views would be as in Figure 2.10, with the view of the bottom of the punch placed in the position of the bottom view and the top of the die in the position of the top view, each facing the front view. In case of lack of space, this arrangement can be modified by placing the drawing of the bottom of the punch to the right of and in line with the top view of the die, as if it were turned over from the top view (Figure 2.11). In drawings where any such arrangements of views are employed, the views should always be carefully titled to aid in reading.

For objects for which two side views can be used to better advantage than one, these need not be complete views of the entire object, if together they describe the shape of the object (Figure 2.12). Only those views should be drawn that are absolutely necessary to portray the shape of the part clearly. Often two views will suffice, and many cylindrical parts may be portrayed adequately by one view if the necessary dimensions are indicated as diameters.

After the design is documented, a decision is made on whether tool usage and maintenance instructions are required. Complicated tools and machines always require these two items as part of the documentation package.

2.2.6 Tool Construction

Tooling is controlled and tracked per schedule by the tool-order system. It is important to have this system to ensure that the tooling arrives on the production line at the correct time. Another item needed is the actual cost of the tool. If the tool is fabricated by a vendor, this information is available from the purchase order. When the tool is fabricated in-house, it is much more difficult without a good system and

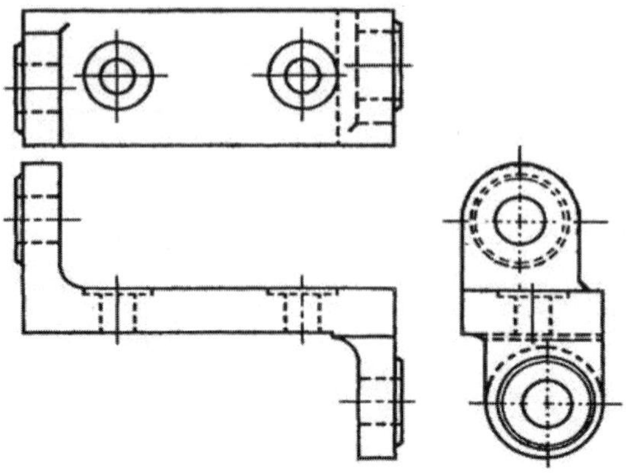

FIGURE 2.8 Normal third-angle projection.

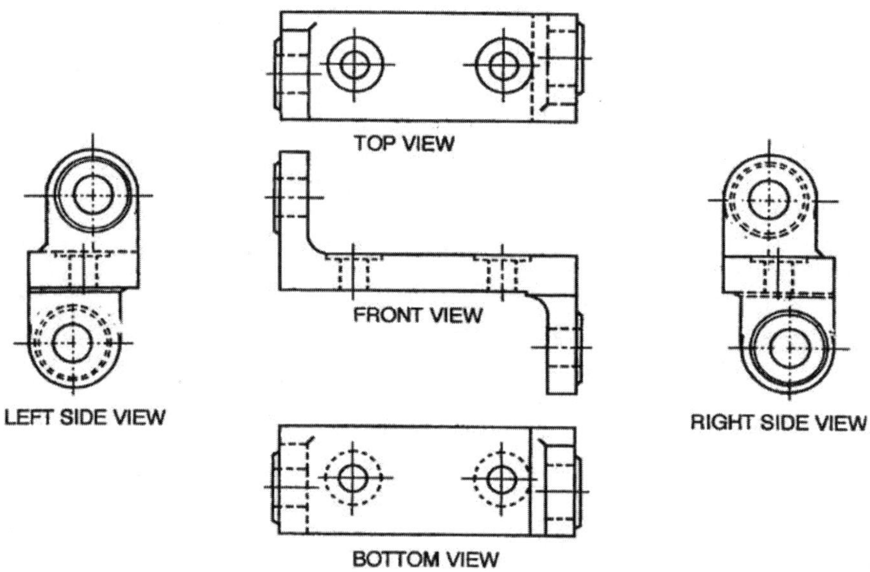

FIGURE 2.9 Additional views of a part are sometimes needed.

people willing to use it. Listed below are a few reasons that the cost information is valuable.

1. As costs are collected against each tool, they should be compared to the estimates. We can see if the estimates were correct, or if the estimating technique needs to be corrected up or down. This exercise should be never-ending, and over the years the estimates will become quite good.

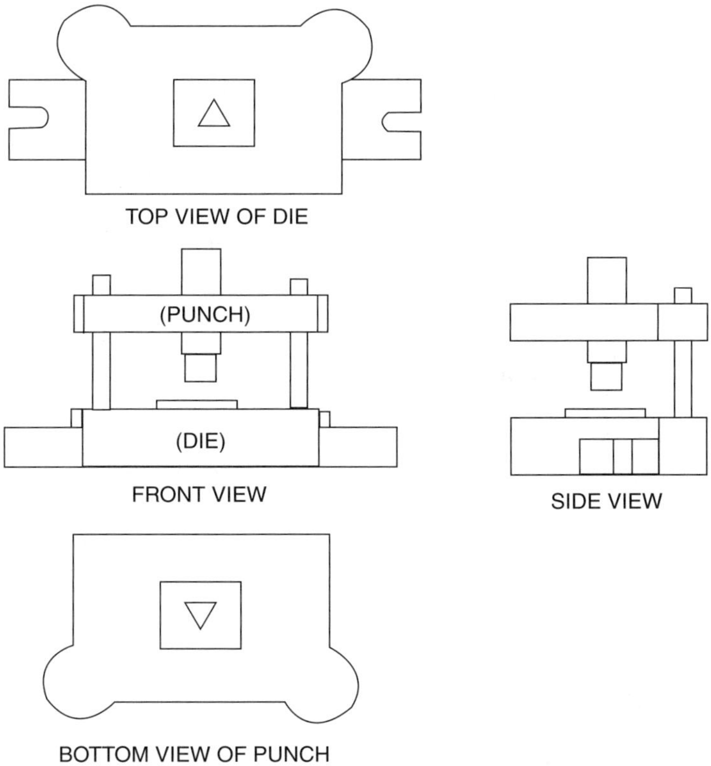

FIGURE 2.10 Normal arrangement for punch-and-die drawings.

2. Knowing the real cost of tooling is important when duplicates of the same tool are required. Estimates of the cost can then be made with a great deal of confidence.
3. Having a large file of actual cost data to support your estimates will prove invaluable, since future cost estimates can be made with a great deal of confidence.
4. The data may also be used to help determine how competitive your in-house tool fabrication shop is versus having the work done by a vendor. This can be done by comparing costs on similar type tools, and the results are often valuable.
5. Knowing the true cost of the tools on a production line is necessary to determine product cost.
6. Tools are tangible company assets and may play a role in the tax structure of a company. The actual dollar value of company tooling can be depreciated each year, or sometimes expensed in the year the tool cost is incurred. This difference may be important in future bidding for the same product.

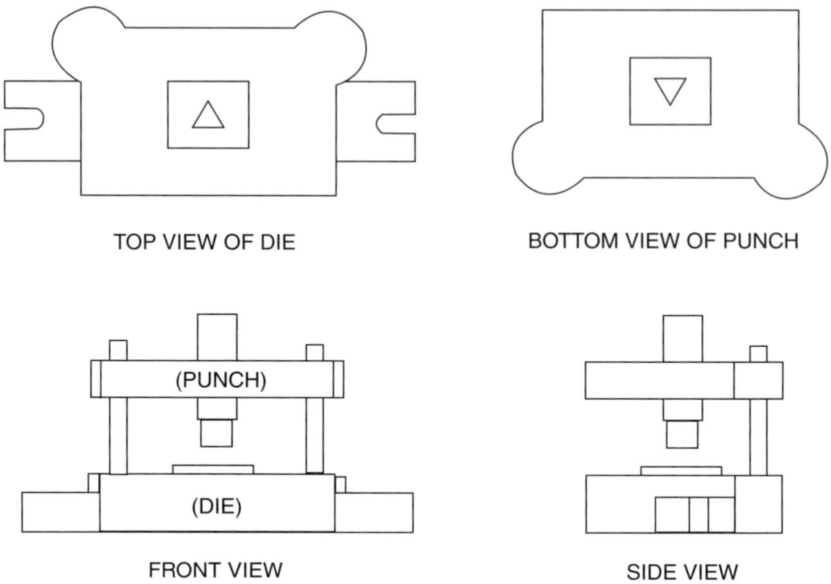

FIGURE 2.11 Alternative punch-and-die drawing arrangement.

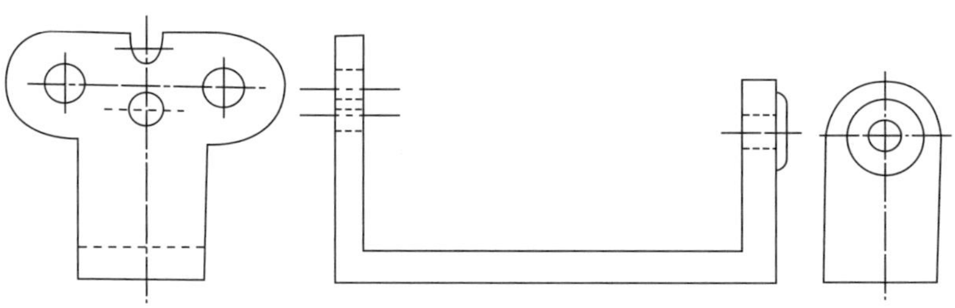

FIGURE 2.12 One or two views may be sufficient to describe an object.

2.2.7 Acquisition of Special-Purpose Machines

Acquisition often becomes a major task, whether the machine (or robot) is to be used in a fabrication shop or in assembly of the product. Most special machines are designed and built by companies that specialize in this type of work. In the author's experience, the simpler, more straightforward machines may be built in-house, but usually not the larger, more complex ones.

The task of bringing a new machine online may logically be divided into two phases. The first phase is prior to placing a contract with the outside firm, and the second phase is after placement of the purchase order.

Most of these machines are quite costly. Their timely installation and subsequent performance is very important to the success of the product cost, delivery schedule—and perhaps the success of the company. Therefore, all levels of company management will require accurate and timely reports on the progress of the project.

Activities Prior to Purchase Order Placement

Activities prior to order placement are probably the most important activities, requiring the most analysis and in-house user coordination. Some things to consider are:

- Assign a project manager or task leader
- Make a detailed schedule of activities
- Prepare good flow charts of the operations
- Coordinate with product design, tool design, quality, safety, maintenance, and training experts
- Involve the production shop and the purchasing agent
- Prepare a detailed machine specification
- Send the specification to several suppliers
- Review the supplier proposals for technical approach, price, and delivery time
- Select a supplier
- Coordinate with the supplier to agree on final changes to the technical statement of work, price, delivery, etc.
- Place purchase order

Note that most of these same concerns would apply if the user plans to design and build the machine in-house.

Activities after Order Placement

Close coordination with the machine builder is usually valuable during final design, construction, debugging, and prove-out. Travel cost should be added to the price of the machine during evaluation. Items to consider during this period include:

- Update master schedule
- During design, consider the maintenance problem
- Watch for safety issues
- Make sure that any computer software is documented
- Check final drawings for "as built" configuration
- Review the training manuals and maintenance manuals
- Insist on accurate status reports during build
- Plan on debugging run at the supplier
- Provide plenty of prove-out parts. These must be *exactly* representative of the production parts
- Maintenance personnel should observe the final operation and prove-out at the supplier
- Final acceptance should be in your plant after final prove-out runs

Never pay out more money in progress payments than justified by the machine progress

2.3 CONSTRUCTION AND MAINTENANCE

2.3.1 Tool Construction

Proper construction materials and tool fabrication are established by the tool design, but it always helps to perform a liaison with the toolroom. Listed below are a few topics that should be reviewed as tool construction progresses:

Is the tool too heavy for its use? (Too light?)
Has the environment the tool will see been addressed (expansion/contraction, rust/corrosion)?
Have all the wear points been addressed?
Will the tool hold up on the production floor?
Will there be any high-maintenance details?
Should the tool be painted?

After fabrication of the tool, the following items should be checked:

Have all sharp edges and corners that could injure the operator been removed?
Do all the moving parts function as designed?
Are there any high-maintenance details for which spare parts should be ordered?

2.3.2 Tool Maintenance

After spending large amounts of money and time on the design and construction of a tool, it may be ignored until it breaks. Most companies have systems, procedures, or policies on tool maintenance, but often the system fails. The tool is turned over to production, where the driving goal is to produce an item, and there seems to be little time for anything else. The missing principle all must believe is that keeping a well-functioning tool online is more important than continuing to use a tool until it stops, and then wait for repairs.

Before a tool is released to the production floor, the manufacturing engineer, tool designer, and quality engineer should decide the type of maintenance and frequency required. Many tools will require very little maintenance—perhaps a quick trip through the toolroom for a cleanup. Tools that require some sort of regular maintenance should be the focal point. A computer can be one of the best tools when setting up a maintenance system. Establish the maintenance that needs to be done and the time frame (weekly, monthly, etc.), and program the computer to extract work tickets at the proper time for maintenance. Tools requiring lubrication are the ones most neglected. A lubrication chart, which shows the correct type of lubrication needed, how often, and when the task was performed, should be attached to each tool. Any number of systems can be established, but without discipline and management's complete backing, none will give the desired results.

2.4 HAND TOOLS

2.4.1 Introduction to Hand Tools

Hand tools are tools that most of us are very familiar with. We all have used a screwdriver, wrench, hammer, etc. Visiting a hardware store or thumbing through a hand-tool catalog will remind you that there are hundred or different types of tools. A well-stocked tool crib or use of a catalog library will save time and money. Use all the resources at your disposal to find the right tool for the job before having a special one designed and built. Hand-tool salespeople in your area can be helpful if given the chance. The wise engineer uses this cheap resource to full advantage.

2.4.2 Hand Tools

Many hand tools are now air-, electric-, or pneumatic-driven. Proper setup on the production floor is essential for proper function and tool life. For air tools, you must ensure that the tool obtains the correct air pressure and oil mixture; electrical tools need proper grounding and correct voltage; and pneumatic types need proper line pressure. Working with experts from the fields of industrial hygiene and industrial safety is recommended to assist in the selection of new hand tools. Ergonomics should be included in this selection. Plant layout and facilities engineering are also involved in providing the correct infrastructure to support the use of various hand tools. A maintenance schedule with a recall system should be set up to clean, functionally check, calibrate, and replace worn parts.

A computer is the perfect device to use when setting up a maintenance system, and in scheduling maintenance requirements. At the time of purchase, most power hand tools come with an instruction booklet that provides care and maintenance information. This information may be entered into the computer along with a date-recall program. When the maintenance schedule time is reached, the computer will print out a recall ticket with all the data required. This system will help plan the spare parts required and prevent any unnecessary duplication. It can also help in the planning for future overhead cost requirements by allowing you to see the condition and age of the equipment presently available.

2.4.3 Tool Cribs

An inventory of commonly used standard tools and tooling components should be built up specifically to meet the requirements peculiar to your shop and your plant operations. This inventory would include standard cutters, end mills, broaches, reamers, drills, taps, and similar metal-cutting tools if your manufacturing operation is machining. Also included in an inventory for a machine shop would be a variety of different chucks, collets, and hand tools such as files, scribes, deburring tools, vernier calipers, and micrometers.

Tooling components such as drill bushings, quick-release clamps, tool balancers, and small vises would also be included. The manufacturing engineer should determine what the standard tooling inventory should contain, and should specify the items to be stocked, from whom they should be ordered, and how many of each to order. Expendable tools and cutters should be placed on "max-min," which means that when the actual inventory of an item reaches a certain minimum quantity, an order is immediately placed for a predetermined number of the items to bring the inventory back up to the desired level.

Inventory

The standard tool inventory is usually stocked and issued from a central tool crib. This tool crib can be a part of the manufacturing engineering department, but more than likely is under the production control department's jurisdiction. The tool crib will also stock items such as shop rags, work gloves, lubricating and cutting oils, C-clamps, machinist's scales, safety glasses, and shop aprons.

If the plant operations include structural and mechanical assembly, the standard tool inventory will include welding rods, standard clamps and holding fixtures, ball-peen hammers, power nut runners, combination wrench sets, socket wrench sets, screwdrivers, adjustable wrenches, pliers, clamping pliers, rubber mallets, power screwdrivers, rivet squeezers, and welding and brazing equipment. If operations include electronic assembly and wiring, the standard tooling inventory will include soldering irons and soldering iron tips, solder pots, temperature-controlled soldering irons, soldering aids, heat sinks, spools of cored colder, side-cutting pliers, needle-nose pliers, long-nose pliers, chain-nose pliers, assorted small brushes, and plastic squeeze bottles. Also included would be such items as holding fixtures and power arms.

2.5 TOOLING FOR PARTS FABRICATION

2.5.1 Introduction to Tooling for Parts Fabrication

Producing parts for the assembly line may seem rather straightforward. Looking a little deeper, you will find it very complex. A production part can have as many different and varied process steps as a complicated assembly operation. The cost of producing these parts will also have a major impact on the final cost of the assembly being produced. The manufacturing engineer must always be on the watch to reduce or eliminate any machining or process steps. A good understanding of all machine shop equipment, methods, and practices is a must. With this knowledge as a base, sound decisions can be made. One of the worst problems on the assembly line is late or poor-quality production parts.

2.5.2 Study the Part to Be Made

When you are reviewing a part for fabrication, look at it as the most important part in the assembly. Learn everything there is to know about the part. Listed below are some good questions that should be answered before deciding on tooling:

1. Why was the part designed the way it was?

2. What does it interface with?
3. If there are attach holes or mating fits, has a tolerance analysis been performed?
4. What are the total number of parts that will ever be needed?
5. Is the part slated for redesign in the future? If so, when?
6. Will the part be sold as a spare or replacement part?

2.5.3 Tool-Planning Process

With an understanding of the history and future of the part, you can plan the fabrication process as economically as possible. The plan should start when the raw stock is picked and continue to the delivery of the part to the production line. Break down each process and list all the equipment available to you to perform the process. Cost trade-off studies may be performed to decide whether a machined part requires a conventional or an N/C machine. Each piece of equipment requires different types of tooling, setup, and machine time. For a sheet metal part, the choice may be between a die or forming on a press brake and other assorted hand operations. Once the cost trade-off studies are completed and the equipment decision is made, the fabrication plan for the part can be finalized. Using the plan, discuss with the quality engineer where the best inspection points will be for the part. Critical holes or surfaces may need to be inspected long before the part is completed, or some type of in-process quality control plan established.

2.5.4 Ordering the Tool

The tooling required for the part can now be ordered. Each tool order will describe in detail which operations have already been performed, what operation is to be done, and which machine the tool is to be used on. Most tooling for parts fabrication is straightforward when the tool designer uses good design practices. Examples are backup of the part to eliminate chatter during milling and locating the part in the tool from targets or key starting positions the way the design engineer dimensioned it on the product drawing.

Punch Press Dies

Dies are another story, since most dies are sent out to be built by a die shop. Tool drawings are usually not provided. Extra care must be given to this type of tooling. Even after finding a die shop with a good reputation, you will still need to ensure that all necessary data is given to them to get your desired results. Writing a total "scope of work" is a good way of doing this. This will not only help the vendor understand the task but will also ensure that the purchasing department knows what your wishes are. The following is an example of a scope of work for a combination die (courtesy of McDonnell Douglas Florida Missile Plant).

Scope of Work for Combination Die

1.0 Scope

This specification covers the requirements to design, build, ship, install, and debug, a combination die to fabricate Flat Pattern Blank of AWM8020 (Det. 3) Rev. "F".

2.0 Design Requirements

2.1 Format drawings are not required. Vendor to furnish prints of all shop drawings with strip layout, die stations and regrind instructions.

3.0 Approvals

3.1 MDMSC-FMP required approval points are:

3.1.1 Vendor to supply (25) sample parts for MDMSC-FMP and approval prior to die shipment.

3.2 Final approval will occur only after an in house tryout.

3.2.1 Vendor to assist in initial setup of die at MDMSC-FMP with in house tryout.

3.2.2 Final approval will be given from 3 parts, which meet flat pattern specifications, that are taken from a random sampling of 10 parts.

4.0 Tool Requirements—General

4.1 Die must be constructed to provide maximum safety for the operator and meet or exceed:

4.1.1 MDC TFIM 70.101.

4.2 All components are to be of first line quality and conservatively rated for the given application.

4.3 Die to be identified as "PROPERTY OF MDMSC-FMP" and "CDAWM8020–3TDS".

4.4 Die to be capable of producing a minimum of 78,000 parts.

4.5 Die to have quick change punches with supplier part number.

4.6 Die set to be 4 post ball bearing precision.

4.7 Die to have miss-feed detector.

4.8 Die to have stock pusher.

4.9 Die to have scrap chopper and chute.

4.10 Die to have spring loaded strippers.

4.11 Feed arm 6" high to bottom of material.

4.12 Strip/coil stock alum. alloy 6061T-6, 0.032 $\times$ 7.5 wide.

4.13 Top of die shoe to conform to sketch on Page 3.

5.0 Tool Description

5.1 The tool is to be a progressive die to produce flat pattern to print revision as stated above.

5.2 The part is to be run length wise thru the die.

5.3 Production rate 300 min. per week.

6.0 Press Description

6.1 Type: Komatsu 110 ton press.

- 6.2 Model: OBS 110–3
- 6.3 Serial No.: 11813
- 6.4 Stroke: 4.0
- 6.5 Shut Height: 14.96
- 6.6 Slide Adjustment: 3.94
- 6.7 Slide Stroke: 5.97
- 6.8 No. of Slide Stroke: 32–65 SPM

7.0 System Specifications

- 7.1 MDMSC-FMP will supply part material required for tool tryout. The Vendor should notify MDMSC-FMP when material will be required.

2.6 TOOLING FOR ASSEMBLY

2.6.1 Introduction to Tooling for Assembly

The techniques described in this subchapter are concerned with minimizing the cost of assembly within the constraints imposed by the design features of the product. The specific focus of this section is to discuss the assembly tooling required to support the assembly process selected. During the normal course of developing a new product design, the manufacturing process must be developed progressively and concurrently. Chapter 1 describes a preliminary manufacturing plan that is required in order to estimate manufacturing cost and pricing. This plan should be finalized as the ultimate product design is in work. Information that was used to develop the manufacturing costs, including quantity, production rate, delivery schedule, and the like, should be utilized in establishing the tooling philosophy. The factory layout, as described in Chapter 4 and Chapter 5 of Volume 1 also influences tooling requirements.

Assembly cost is determined at the product design stage. The product designer should be aware of the nature of assembly processes and should always have sound reasons for requiring separate parts, and hence longer assembly time, rather than combining several parts into one manufactured item. Each combination of two parts into one will eliminate an operation in a manual assembly workstation, and usually an entire workstation on an automatic assembly machine.

The assembly process should be developed early in a program, since the lead time for assembly tooling may be greater than the time needed to make the detail part fabrication tools. Also, in a normal program, the final piece-part assembled configuration may be the last design drawing completed. The assembly process will often dictate the critical parameters needed of some of the key parts, and will therefore influence the tooling required for parts fabrication.

2.6.2 Assembly Tooling Systems

Assembly tooling can be divided into automated or manual systems. Production process flow charts should help define the system. Assembly tooling can also be grouped by requirements for performing work operations, making checks or tests, or transportation and movement. Special requirements may consider painting or other finish processes. As the assembly process matures, the need for duplicates of some of the assembly tooling will become apparent. This may be the result of "choke points" in the assembly line. Subchapter 5.1 discusses the technique of computer simulation of production assembly flow. Manual assembly differs widely from automatic assembly due to the differences in ability between human operators and any mechanical method of assembly. An operation that is easy for an operator to perform might be impossible for a special-purpose workhead or robot.

2.6.3 Manual Assembly

In manual assembly, the tools required are generally simpler and less expensive than those employed on automatic assembly machines. The downtime due to defective parts is usually negligible. The direct operator labor portion of manual assembly costs remains relatively constant and independent of the production volume. Manual assembly systems also have considerable flexibility and adaptability. Sometimes it will be economical to provide the assembly operator with mechanical assistance in order to reduce the assembly time. Assembly tooling for manual operations will require careful consideration of ergonomics. The problem of repetitive operations causing poor worker performance and injuries is well understood today, and is becoming an increasingly important factor in both the processes and the tooling provided. The selection of pneumatic squeezes and other power tools should be considered as the assembly tool develops, since the solution may lie partly in the assembly tool design and partly in the ancillary equipment selected.

Concept sketches of the planned assembly process and assembly tooling are an important part of understanding the details involved in setting up the production line. Computer simulation, as described in Subchapter 5.1, is a valuable technique for analyzing the assembly line. Choke points, requirement for duplicate stations, and the like will become apparent. Options can be simulated, and final decisions made as to the final production process. Manual assembly can be performed by an operator at a fixed workbench, or by multiple operators utilizing some sort of transfer device to pass work progressively down a line. Parts may be provided in tote pans or supplied by single-purpose parts feeders (Figure 2.13).

All resident production experience should be involved in the assembly process. The product design, quality, shop supervision, safety, industrial hygiene, and other elements should provide suggestions and recommendations from their point of view. The manufacturing engineer often acts as a coordinator, or team leader, in developing the optimum process required for a world–class manufacturer. With this information and process definition, the assembly tool designer now has a good basis for actual design of the assembly tooling.

2.6.4 Automated Assembly

Automated assembly can be broadly broken down into special-purpose machines, which have been built to assemble a specific product, and programmable assembly

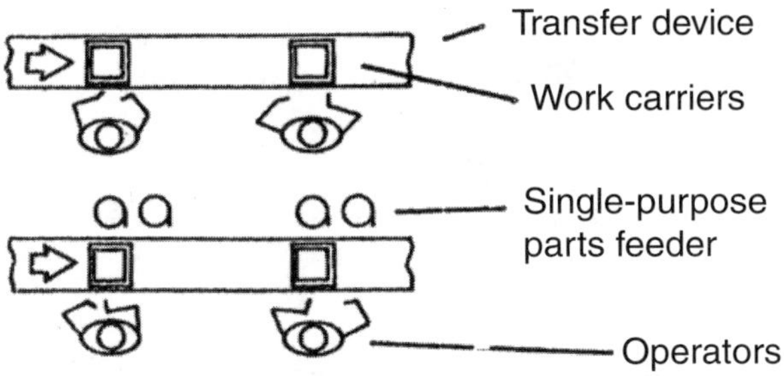

FIGURE 2.13 Manual assembly systems.

machines, where the workheads are more general-purpose and programmable (see Figure 2.14).

Special-Purpose Machines

The assembly machines consist of a transfer device with single-purpose workheads and parts feeders at the various workstations. The transfer device can operate on an indexing (synchronous) principle or on a free-transfer (nonsynchronous) principle. These special-purpose machines are expensive and involve considerable engineering development before they can be put in service. The downtime due to defective parts can be a serious problem unless the parts are of relatively high quality. Also, it must be appreciated that these machines are designed to work on a fixed cycle time and are therefore inflexible in their rate of production. If they are underutilized and cannot be used for any other purpose, this will result in an increase in assembly cost.

Programmable Assembly Machines

Programmable assembly machines can allow for more than one assembly operation to be performed at each workstation. This provides for considerable flexibility in production volume, greater adaptability to product design changes, and different product styles. For lower production volumes, a single robot assembly workstation may be preferable. Parts are normally available at the workstations in manually loaded magazines, since the use of parts-feeding devices of the type employed for special-purpose machines is usually not economical. In the case of an automated assembly process, end effectors must be designed and built.

Most automated equipment today will require a software program. This may be a document of programming steps for a programmable controller, or digitized programs on a floppy disk or other electronic media for a computer. In any case, this will become an important part of the "tool documentation" and is often controlled as such.

In recent years, the robot has become a more useful tool for manufacturing. Most of the current assignments of robots are rather simple and repetitive. The robots are employed by mass producers of consumer goods, such as automotive and appliance companies. An area where the robot has had very little impact so far is assembly.

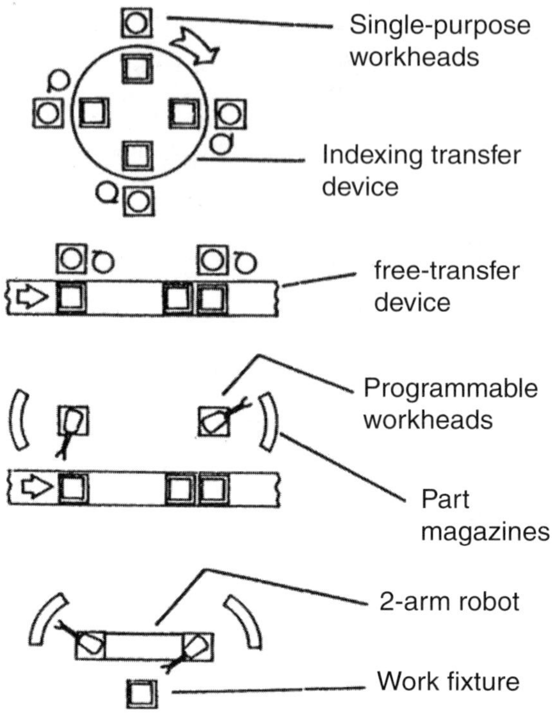

FIGURE 2.14 Automated assembly systems.

Most assembly operations are being conceived for human assemblers who have two dexterous hands, vast assembly experience, and an intricate sensory system. Without such features, robots can perform only very simple, stack-type assemblies.

In order to reach a point in a three-dimensional space, the robot must have three degrees of freedom. If, however, the robot wants to do useful work on an object located at this point, the end effector must have six or more degrees of freedom. The kinematic, dynamic, and control principles for the present industrial robots are amply covered in other literature and are not discussed in this chapter.

With simulation, it is possible to display a pictorial image of the plant on a graphic display, and the manufacturing engineer can observe the creation of a workpiece through its different production stages. Once the layout of the manufacturing floor and equipment has been determined, programming of the machine tools, robots, and other facilities can commence.

Programming

There are explicit and graphical programming tools available for robots. A user-oriented approach for programming is to direct the motions of the robot with task-oriented instructions. This method is of interest for planning and programming of assembly work. For planning and controlling the work of a robot-based manufacturing cell, a centralized data management system must be provided. It is the data repository for modeling and programming the robot and supervising its actions.

Sensors and Grippers

Sensors and grippers constitute the interface between a robot and its environment. The modern intelligent robot systems need to be supported by various advanced gripper and sensor systems. Sensors are, in principle, transducers that transform physical properties (the input signal) into electrical (output) signals. These properties can be electrical, magnetic, optic (surface reflection or transmission), or mechanical (distance, position, velocity, acceleration).

Sensors have a key function to reduce uncertainties concerning the robot itself as well as its environment. Complex tasks for robots (e.g., assembly operations) need a combination of sensors in a manner similar to the use of multiple sensors by human beings for the same tasks. Sensors that simulate the human ability of vision or touch are now coming on the market. For this reason, task-specific sensor configurations are needed for many applications. In manufacturing environments, sensors can be used to measure information about materials, amounts, geometry, place/location, and time.

In all these cases, sensors are employed to exclude uncertainties—that is, to detect them and react appropriately. In the context of assembly by robots, these uncertainties can be traced to two factors: uncertainties from tools (robots, feeders, and fixtures) and uncertainties in the parts themselves (manufacturing tolerances). Two ways to handle these uncertainties in a manufacturing environment are:

1. Avoid all uncertainties in the assembly planning phase
2. Detect uncertainties with the aid of sensors

The first possibility is the traditional approach, which gives rise to inflexible, expensive, and time-consuming systems, especially in the construction of specialized feeding devices. Such a solution makes quick product changes uneconomical because of long changeover times.

The gripper of a robot is the only part that has mechanical contact with the object; its main functions are to grip objects, hold objects, and release objects. Gripping establishes a defined position and orientation relative to the robot during the material transfer route and assembly operation. When releasing the object, the relationship between gripper and object is given up at a specific point.

Generally, a conventional gripping device used for industrial robots is a specialized device that handles only one or a few objects of similar properties (shape, size, weight, etc.). When a single gripper alone cannot cope with the variety of parts to be handled, a multiple gripper, a gripper change system, or a jaw change system can be used.

Grippers can be equipped with sensors for monitoring the gripping functions. An integrated sensor system can monitor the internal state of a gripper (e.g., jaw distance), and the structure of the environment (e.g., object distances). Various gripper systems have been developed that can adapt to the shape of objects. Such grippers are built with flexible fingers with several passive joints, which close around the object to grasp it.

The design of so-called dexterous hands, which imitate the versatility of the human hand, feature multiple fingers with three or more programmable joints each. This kind of hand allows gripping objects of different geometries (cube, cylinder, ball, etc.). Fine manipulations of the object can also be realized by the gripper system itself. This fine manipulation ability results in a task separation similar to that of the human hand-arm system.

A compromise between these highly sophisticated, and therefore expensive, systems and conventional grippers can be achieved by mounting some sort of compliant device between a conventional gripper and the robot. These compliant devices consist of two metal plates connected by elastomeric shear pads and are known as remote center compliance (RCC) devices. RCC devices can compensate positioning faults resulting from reaction forces in the assembly phase.

The design, construction, and prove-out of many of these devices can be very costly and require a great deal of experimentation. As the industry continues to advance, it should become less of a risk to cost and schedule, and therefore more valuable to an "agile manufacturing" plant. Smaller job lots, or continuous manufacturing with variable production rates, should be able to take advantage of a greater degree of automation when the automation is more "flexible." However, for today, a careful study is required in order to justify full automation.

3 Computer-Aided Process Planning

Alexander Houtzeel

3.0 AN OBSERVATION

The majority of the world's most successful competitors in the manufacturing industry have achieved their success by continually improving their existing design and manufacturing processes rather than by using limited funds to develop new ones. These improvement efforts are possible only with systems that manage design and manufacturing data efficiently. *Nobody* needs to reinvent the wheel.

Computer-aided process planning, or CAPP, is a term that has been associated with achieving this manufacturing data management. CAPP has been around for close to 20 years and has recently experienced a revival in the manufacturing industry with the advent of sophisticated workstations, relational databases, and other advanced software technologies. The following is an observation of current manufacturing trends, the evolution of the manufacturing engineering (ME) department, and how process planning, in particular computer-aided process planning, is essential for today's advanced manufacturing engineering. Also included is an overview of the history of CAPP, how it started, how it has evolved, and a brief study of recent CAPP installations at four successful manufacturing companies.

3.1 THE CHALLENGE OF CURRENT MANUFACTURING TRENDS

In mass production, where few if any variations of the same product are allowed, the extra investment in design for manufacturability (including quality control) can be amortized over large lot sizes. An effective interface between design engineering, manufacturing engineering, tool design, and production is required right from the beginning of the product concept if a company is to be competitive on a worldwide scale. However, since today's consumers require greater product variety at competitive prices and high quality, the costs of the design and the manufacturing process can only be amortized on much smaller lot sizes. And, as is typical in a highly competitive environment, this must be accomplished with much shorter lead times. Given this scenario, three key issues become critical:

Low cost and small lot sizes. The efficient management of the design and manufacturing process becomes a difficult challenge where it concerns smaller lot sizes. This is because the changeover from one product variety to the next typically takes a long time in the mass production environment. How does one speed up the changeover for smaller lot sizes? Modern manufacturing and assembly tools must be designed so that changeovers can be accomplished very quickly and efficiently to better use the capital invested in that machinery.

Short lead times. The ability to fill niches in the market with high-quality products requires a tightly interwoven design and manufacturing team and easy access to previously acquired design and manufacturing experience. Reinventing the wheel not only introduces longer lead times but also increases the costs of the design and manufacturing process and puts a strong burden on quality control.

High quality and long-term reliability. The quality of a product is no longer determined simply by its performance at the moment of delivery (the "defect rate"), but also by how well it will work during the promised product life cycle. A product that is delivered without deficiencies but becomes ineffective within 6 months will not entice buyers to purchase that same product again.

Manufacturing trends have placed new, more difficult demands on the manufacturing industry. Now, the only way for manufacturing companies to stay competitive is to create as flexible and advanced a manufacturing environment as possible. This means taking the time to reevaluate the roles of design and manufacturing engineering, and the tools these departments should use to become as flexible and responsive as possible.

3.2 RESPONDING EFFECTIVELY TO THE KEY ISSUES

An effective response to these issues is certainly not a mindless throwing of money onto the problem with hopes that it will go away. Instead, a careful analysis of the design and manufacturing organization is required, along with a clear definition of the company's objectives. Only after this has been accomplished can one begin to consider investment in support systems and, if necessary, in equipment to make the workforce more efficient in preparing for the future.

3.2.1 Sound Engineering

In today's market, no product can expect to be sold effectively and successfully if it is not rooted in sound design and manufacturing engineering. This means not only maintaining a well-educated design and manufacturing staff, but also a set of systems to help these technical personnel to be as effective as possible. This includes providing them with the ability to draw easily on previous experience and to exchange information with others.

3.2.2 Design and Manufacturing Interaction

In far too many companies, the engineering (product design) department seems to be physically and intellectually separate from the manufacturing engineering department. Generally, a company's design group creates products with little or no consideration for how to efficiently manufacture, assemble, and service them. They leave these problems to the manufacturing engineers. More often than not, the concept of designing for manufacturability is still alien to manufacturing companies.

Based on observations made by Dr. Ohno of the Toyota Corporation, it is 10 times more expensive to have a design mistake fixed by the manufacturing department once a part has been released for production than to have the error eliminated by the design team. It costs 100 times more when the mistake has to be fixed by the assembly department, and 1000 times more when the field engineer has to repair the problem at the customer site (George D. Robson, *Continuous Process Improvement*, Free Press, New York, 1991).

3.2.3 Fewer but More Complex Parts

As a general rule, the more parts are involved in a product assembly, the more difficult it is to maintain quality. This is due to the cumulative tolerances involved in making a large number of parts fit together. If two similar assemblies, one consisting of two parts and the other of ten parts, should have the same final tolerance, then the parts in the ten-part assembly have to be machined with five times tighter tolerances.

Obviously, this has very substantial cost effects. Generally speaking, it pays to design a product consisting of few but more complex parts rather than one with many simple parts. Fewer yet more complex parts mean more sound engineering, more knowledgeable designers and manufacturing engineers, and better machine tools that can cope with these parts.

3.2.4 Flexible Manufacturing and Numerical Control

How does a company deal with quick changeovers when manufacturing in smaller lot sizes? The major advantage to flexible manufacturing is that a lot of the preparatory work can be done before a changeover, such as numerical control tapes, tools, etc. The vast majority of work (machining work in particular) can be prepared using numerical control (NC) technology. Not only does numerical control technology allow for the quick implementation of changes, one can also better control the entire manufacturing process, thus ensuring consistent quality. Once a numerical control program has been written by the NC programmer, the machine will continuously make the part in that fashion, provided that the machine tool is always running at its expected standard. By changing the numerical control tapes, other products can be made with a minimum of change time.

3.2.5 Efficient Manufacturing Engineering

No matter how good and complex a new design is, if it is not produced in an optimum way (due to inefficient production or inconsistent quality), then the company is not going to be successful. The driving force in any production environment is the tight control of the manufacturing process. To gain such control, the manufacturing engineering department must generate effective process plans to manufacture the product and also have easy access to existing plans and past experience.

Manufacturing engineering departments should be involved in the product design cycle. They should be included in the discussion of how to best design a part for manufacturing and assembly with the design engineers. Once the design is finalized, the ME department should serve as the control center for determining how the parts are going to be made and assembled. The span of control should include such areas as establishment of the manufacturing process, numerical control methodology, assembly, tooling, etc.

3.3 THE EVOLUTION OF THE MANUFACTURING ENGINEERING DEPARTMENT

The activities of the ME department vary considerably from one facility to another. Some companies still work under the "make fit to print" mentality; others have instituted very complex organizations in charge of the manufacturing process control. It is worth considering how the role of manufacturing engineering has evolved over the last 25 years.

3.3.1 The Traditional ME Function

Traditionally, the ME department looked at the drawings received from the engineering department and essentially wrote on them, "make fit to print." In other words, there was no ME function. Furthermore, in many instances the foreman of the manufacturing or production department would figure out the best way to do things based on his workers' experience. All too often, manufacturing engineering instructions would be totally ignored, and the foreman decided to manufacture the product in an entirely different way. Luckily, that time has passed for most companies, and a more organized approach has evolved. Now at many companies the ME department writes the manufacturing instructions, generates the numerical control tapes, and sends the complete package to the manufacturing department to make the product.

Slowly but surely, a new conception of the ME department is emerging, one that can control the cost of the manufacturing operation. It is ME decisions that determine what the manufacturing process is going to be and, consequently, what capital assets will be used to produce that part.

3.3.2 Management and Dissemination of Manufacturing Information

In an environment where design for manufacturing and design for assembly are increasingly important, it is obvious that the experience either directly available from manufacturing engineering or from a database becomes very important to the design engineering department. Only with knowledge of the past can a design department come up with the best product design, manufactured at the lowest cost, resulting in the highest quality. It is for this reason that the ME department's role should be expanded by including it in the design process. Furthermore, the ME department's traditional area of control, i.e., production activities, should also be expanded to encompass the following: generation of work instructions for manufacturing parts and numerical control tapes, setup of interfaces with the material requirements planning (MRP) and material resources planning (MRPII) systems and production control, ensuring that the proper tools are available from the tool department at the right moment, ordering material, generating time standards, etc. In other words, manufacturing engineering is at the center of a spider web that extends to a great variety of manufacturing functions. This expanded control is the only way to maintain quality and costs in a manufacturing company.

3.3.3 The Tools of the Trade

ME departments, of both the traditional and the modern sort, convey information to the various players in the production process in the form of different documents, such as the process plan, work instructions, tool instructions, and routing forms.

The most inclusive of these documents is the process plan. Ideally, a process plan represents a complete package of information for fabrication or assembly of a detailed part or product. This package may include work instructions for the shop floor, a manufacturing bill of material, a quality control plan, tool planning, effectivity of the process plan for particular part (assembly) numbers, and links to such systems as MRP, time standards, engineering and manufacturing change control, "as-built" recording systems, shop floor control and data collection systems, etc. The result of this creative process is the process plan. The package may also include multimedia information, such as text, graphics, photographs, video, or sound.

Work instructions comprise a set of documents released to the shop floor to produce or assemble the detailed part or product; they are part of the total process planning package. The work instructions are mostly defined as a set of operations with operation numbers. Depending on the type of industry, the work instructions can encompass detailed work procedures for individual steps within an individual assembly (manufacturing) operation, or may contain only summary information for operations. It should be noted that many companies use the expression "process plan" when they really mean work instructions.

The *routing* is a summary of mostly one-liners corresponding to the individual work instructions. The routing normally includes setup and run times. Routing information is most often transferred to an MRP system for scheduling and work-order release purposes and to the shop floor as a traveler with the detailed work instructions.

With ME departments ever expanding their role in the manufacturing (and design) process, more powerful tools are required to provide easy access to in-house manufacturing information.

3.3.4 Computer-Aided Process Planning

The creation of a set of work instructions for the shop floor to produce a detailed part or assemble a product has always been a basic requirement. The "electronic pencil" approach is a great improvement on the old ways of paper-driven process planning, because one has the advantage of word processing on a computer. However, the electronic pencil lacks (1) the sophisticated retrieve/modify capabilities of a CAPP system, thus forgoing existing company manufacturing experience; (2) the potential to connect to other existing systems within a company; and (3) the multimedia capabilities of current state-of-the-art CAPP systems.

With the advent of computers, the generation of a complete process planning package has become easier, but it also requires a substantial integration effort in order to link different software systems together (sometimes on different computer platforms). An effective CAPP system should draw on a variety of different software and database functions to create the process planning package.

CAPP is an interactive software tool used primarily in the manufacturing engineering department to create a coherent set of work instructions that the shop floor (fabrication or assembly) then uses to create a part or product. This tool provides the capability to retrieve and modify existing process plans, or to create a process plan from scratch (using sets of standardized instructions). CAPP also enables the process planner to communicate with a variety of other data resources to complete the set of work instructions.

Although CAPP provides the same functions as the process planner used to perform manually, many new aspects become important to consider:

- Management of engineering and manufacturing changes
- Use of previous experience through retrieval of similar and standard process plans
- Linkage to different systems such as time standards, tool control, MRP, etc.
- Inclusion of computer graphics, images, photographs, or videos to illustrate a manufacturing process
- Identification of procedures to best design a product with the lowest possible manufacturing and assembly cost

3.3.5 Manufacturing Data Management

As a result of product service requirements, product liability fears, government regulations, or plain old good sense, more and more companies are trying to keep better, more retrievable records. "There is gold in them files"—because these records contain the company's best manufacturing experience and ought to be instantly retrievable before process planners go about reinventing the wheel. What product data management (PDM) intends to do for the engineering process, manufacturing data management (MDM) proposes to do for manufacturing—i.e., maintenance and retrieval of the entire process planning package.

Depending on how far a company's ME department has extended its responsibilities over the design and manufacturing process, a level of computer-aided process planning or manufacturing data management should be applied accordingly. In other words, those companies that maintain an ME department that is limited in scope will most likely also maintain a very limited CAPP function. Alternatively, those companies that have worked to integrate the ME function into the design and manufacturing process have installed, or are seeking to install, an elaborate and all-encompassing CAPP/MDM system.

3.4 GETTING TO A PROCESS PLAN: WHAT MUST HAPPEN FIRST?

Several important actions are required before a process plan hits the shop floor. How effectively these actions are taken is crucial in generating the most useful and error-free process plan.

3.4.1 The Review Cycle

In most companies, designs are not automatically released to manufacturing engineering. If a product or part requires a new process or a very difficult process, a substantial review may be needed to examine the manufacturability of the product design both during the design cycle and after the release to manufacturing engineering. In many companies this procedure is formalized in specific committees for value engineering. In others, the chief tool engineer and chief production engineer review and discuss exceptional parts with the design department and come up with a set of changes to the product design to improve manufacturability.

In more advanced environments, this review process (involving both design and manufacturing engineers) is initiated at the very beginning of the product definition (even before one part or product has been designed) and is continued throughout the design cycle. The advanced retrieval capabilities of a CAPP system are very useful in such reviews because such a system provides easy access to previous experience.

3.4.2 Part Analysis

Once a part is released from the review cycle by the chief production engineer or manufacturing engineering manager, the process planner analyzes the new part to determine if a process plan for this type of part already exists. Depending on the data management or CAPP system available to the planner (or lack thereof), he or she will employ certain retrieval methods to check whether information on similar parts is stored in a database or, if operating in a very advanced CAPP environment, whether a set of rules exists to generate the process plan automatically. This analysis determines whether previous experience can be included in the process plan or whether one has to start from scratch. Clearly, this analysis can be greatly expedited by the presence of an efficient retrieval system that provides easy access to past manufacturing experience.

The following short list of common analysis and retrieval methods sums up how efficient (or inefficient) this stage of production can be.

Eyeballing. In this straightforward and most often-used method, the planner visually scans the drawing to determine if the part or assembly is retrievable either by name or number, or if he or she remembers working on a similar part before. Here, the burden is on the planner and his or her memory.

"Black book." The process planner may look at the drawing and go through scribbled notes on past process planning projects in a "black book" (personal notebook) to see if he or she can find some similar experience that can be applied to the part in question.

Classification and coding. A more structured approach to retrieval is by part features that can possibly lead to generative process planning. The process planner may have a computerized classification and coding system available to identify the basic part features and relate those to families of parts already in the database. In this way, he or she can determine if the part is related to a certain part family for which there is already a process plan available. In some cases, a new part may be so similar that a set of rules already in the database will generate the process plan automatically. If the part is not immediately familiar, the system may still recognize enough of its features to retrieve a standardized process plan from the database.

3.4.3 Process Plan Generation

Process plans typically have three main sections: a header, an operation sheet, and call-outs to reference material.

The header contains all the "name plate" information on a particular part (name of the part, part number, lot size, etc.) to meet a process planner's identification and retrieval needs. Other items that should be considered essential to retrieval purposes, and thus should be included in the header, are group technology classification code, engineering or manufacturing revision, plant code, preferred or alternate routing, material, quantities, type of plan, status, name of the process planner, name of the customer, etc. If a company is using a relational database, users can retrieve process plans by using a combination of fields in the header. Fields in the header may also be subject to validation.

The operation sheet provides a detailed description of the machine tools and manufacturing processes required to make a part. Generally, the sheets are divided into columns to organize the information into appropriate categories. The first column typically contains an operation number that identifies the sequence of different operations. The numbers may just be inserted chronologically by the process planner, or the CAPP system being used may provide a resequencing function for different operations.

The second column usually contains a number representing the machine or work center. Essentially this number represents a front-end access code to profile tables that will assist the planner in retrieving more detailed information on machining or assembly operations. Information in these profile tables may include a list of different operations that are possible in a given shop (milling, drilling, etc.), available machine tools and their functionalities, the locations of machine tools, costs per hour, and standard setup and teardown times.

The next column on the operation sheet is usually the operation description column, which provides a detailed outline of how the work is to be preformed. This column can accept text from the machine/work center profile tables, or word processing may be employed to enter text from scratch for entirely new operations. Some advanced systems also have the capability to make small CAD sketches in this column as well as provide photographs, animated computer images, or even videos.

The call-out to reference material enables CAPP system users (both the process planner and the person on the shop floor) to call up standard documents while working with a process plan. Reference material may comprise customer specifications, military specs, welding and assembly descriptions, etc. Some or all of the reference material may have to be made available to the shop.

Most companies have their own process plan layouts that have been in existence for many years. If a company decides to introduce a computer-aided process planning system, it should spend time evaluating whether a new process plan layout is desirable. Often, the "dinosaur" process plan forms of old have a lot of unnecessary information in their layout. A company must decide whether that information will only waste valuable storage space in a database.

3.4.4 Security, Sign-Off, and Change Tracking

Given the proprietary information that resides in a CAPP system, most CAPP users insist on a security system that will prevent errors and tampering with data. In many companies, an elaborate system that includes multiple levels of security codes has been incorporated into the CAPP environment.

Once a process plan has been created, it is customary for several persons in the organization to review and sign off on the plan. These persons may be the chief of the manufacturing engineering department and possibly the foremen in the shop. The process planning system must contain a good sign-off capability including parallel as well as serial sign-off functions.

Finally, CAPP systems should provide a tracking function, such as effectivity and change control, to account for engineering changes and manufacturing changes. This is especially true for those companies that are concerned with product liability and those who typically make frequent changes to a part during the design and manufacturing processes.

3.4.5 Dissemination of Information

Once a process plan is finished and the required sign-offs have taken place, information from the process plan should be available to any of the following organizations within a manufacturing company—although different departments may receive different "info packages":

- MRP or production control, because ultimately this group will release the process plan to the shop floor for production or assembly
- The tool room, to assemble the tool kit and make special tools, jigs, and fixtures
- The shop floor (at the behest of MRP) when the part is ready for manufacturing or assembly
- The materials department, again triggered by the action of MRP, for release of materials
- Quality control
- The NC department, although this is usually part of the ME department

The work measurement and time standards group, if this happens to be a separate group
Support functions such as maintenance and purchasing services
Ad-hoc status reports to management

3.5 VARIANT VS. GENERATIVE PLANNING

CAPP technology has evolved into two methodologies: variant planning and generative planning. A variant CAPP system generally has the ability to retrieve previous process plans by similarity, either directly from a relational database or through a group technology feature retrieval system. After retrieval of process plans for similar parts, the planner can review these and then make changes, if necessary, for the next application. In most situations, companies may have preferred process plans that are stored according to families of parts. The advantages of such a variant CAPP system are substantial over the older, electronic pencil variety of CAPP. The retrieval capabilities inherent in a variant CAPP system prevent a company from reinventing a process plan that already exists in a variant CAPP system's database.

With the advent of artificial intelligence in the early 1980s, it seemed logical to explore the concept that process plans could be generated automatically based on sets of rules that could reside in an advanced database system. These rules would be set up according to a company's manufacturing experience and then installed into a database. With such a system, a process planner would simply identify the features on a part and then automatically call up a corresponding set of rules for the manufacture of each of those features. The objective was to generate a process plan from scratch through this feature identification process. However, like the miraculous folding bicycle, this idyllic product seemed to work only during product demonstrations.

Generative CAPP systems do exist. However, they have been successfully implemented only in those companies where an enormous in-house effort has been made. A vast number of company personnel and company funds must be allocated to the project of analyzing and defining the individual rules for the manufacture of each part feature resident in a company's collection of parts. Rules are determined by shop practices that have evolved over years of experience and are quite different for each manufacturing company and for each of the available machine tools on the shop floor. Although generative systems are fascinating from an academic perspective, their practical implementation is only rarely truly cost-effective.

3.6 THE FIRST 15 YEARS OF CAPP

A brief look at the history of CAPP and how CAPP has advanced during its years of use in the manufacturing world will help to identify what today's industry has come to expect from its process planning function.

The Organization for Industrial Research, Inc. (OIR), appears to have been the first company to develop a commercial CAPP system. The General Electric Light Equipment division in Cleveland became its first customer in 1979. OIR went on to install over 150 systems before it was merged into Computervision in 1984. Although OIR's Multicapp system contained elaborate group technology classification and retrieval capabilities, most customers used it (or are still using it) as an electronic pencil. However, the aerospace and defense industries, both then and now, needed more than just an electronic pencil. Regulations required them to keep elaborate manufacturing records. Consequently, these industries and several other advanced manufacturers developed mainframe-based complex manufacturing data management systems—all in-house. In the late 1980s, several other companies developed commercial CAPP systems. None of these, however, came close to fulfilling the requirements of the aerospace/defense industry, nor did they offer multimedia capabilities. The evolution of CAPP had come to a crossroads. It seemed as though those companies that wanted highly sophisticated CAPP functionality were going to have to rely on their own in-house sources to develop and meet their CAPP needs. Those interested in developing commercial CAPP systems were forced to reassess how they might come up with a commercial CAPP offering that was functional and sophisticated enough to challenge the in-house development option.

3.6.1 Time to Regroup: What Does Industry Really Want?

In 1991, one developer of commercial CAPP systems organized a set of meetings, in both the United States and Europe, with several leaders in the manufacturing industry for the sole purpose of determining what companies envisioned for CAPP in the future. The representatives from these companies all had at least 6 to 10 years of experience in CAPP. These meetings generated an extensive report on the status of CAPP and a detailed list of functional specifications for an advanced CAPP system that would truly meet the needs of the manufacturing industry. The following are highlights from this report on CAPP.

1. Most in-house-developed CAPP systems were much better integrated with other preexisting manufacturing software than their commercially available counterparts. However, the process planners, i.e., the daily users, found the commercial systems much more user-friendly than the in-house-developed systems.
2. All participants initially maintained that their company's requirements were different from anybody else's. After 3 days of discussion, the general consensus was that at least 85% of everyone's requirements were the same. User interfaces and links to other systems, however, were different for most participants. It was determined that an acceptable solution to this problem would be a generic, but tailorable, CAPP system, where user interfaces and links to other systems could easily be set up within a customer-defined macro and then linked to the basic system.
3. Since more and more people are able to read less and less, a multimedia process planning system with the capability to incorporate text, graphics, photographs, and video would be beneficial.

4. To manage the manufacturing data and perform sophisticated retrievals to utilize the best available company experience, a relational database management system was absolutely required.
5. Several participants had invested substantial efforts in the development of artificial intelligence-based generative process planning systems. All had come to the conclusion that such systems are technically feasible but provide unacceptably low returns on investment, especially with regard to general detail parts or assembly operations. However, with *groups* of very similar parts or assembly operations, such as the manufacture of turbine blades for a steam turbine, an AI-based system could be profitable. All participants opted for variant-based process planning systems equipped with standard process plans and sophisticated retrieval capabilities, including retrieval by combinations of part features.
6. Since the maintenance of existing mainframe CAPP and other manufacturing data management systems has become increasingly expensive (and is now starting to be outdated), the UNIX-based client-server environment was thought to be the solution for the near future, with PC networks feasible at a later state—once proprietary manufacturing data were proven to be adequately protected on such PC networks.
7. It was solidly determined that the electronic pencil as a method of process planning was not a promising method for the future, since it only allows a company to create work instructions without any referral to preexisting manufacturing experience. It enables the user to make the same mistakes as in earlier paper-driven systems, only much faster.
8. CAPP systems of the future would have to have interactive links to the design (CAD) database to make the drive toward concurrent engineering a reality. Currently, one or two commercial CAPP vendors are developing links to engineering data management systems (sometimes called PDM or EDM).
9. If CAPP is going to be a spider in the middle of the manufacturing data web, links have to be provided to MRP, time standards, engineering and manufacturing change control, shop floor control and data collection, etc.

3.7 SOME EXAMPLES OF WHAT IS REALLY HAPPENING

More than 10 years after the 1991 meetings, it is worthwhile to take a reality check and compare some recent CAPP implementations with the specifications of the "CAPP of the Future" project.

3.7.1 A Large Subcontractor to a Commercial Aircraft Company Leaves the Mainframe and Goes Server-Based

In 1992, a company that had once been almost exclusively involved in production of military aircraft decided to create a commercial aircraft division geared toward the production of major parts for the Boeing 747. The new aircraft division needed to place itself aggressively in the market. A review of all the costs associated with the 747 product revealed that the data processing costs were very high relative to other costs. This meant a multimillion-dollar reduction in only 18 months, or a two-thirds reduction in data processing costs. It became apparent that this cost reduction could not be accomplished by operating the existing mainframe running their so-called legacy systems. These dated back to the late 1960s and racked up high overhead costs. The company decided to switch from the centralized mainframe processing approach to a server-based distributed processing system.

Use of Off-the-Shelf Products

The company decided to go with UNIX-based off-the-shelf software and was willing to change some of its business processes where necessary, in order to move to a less expensive platform and stay within the vendor's software capabilities. That meant that the tailorability of the off-the-shelf software became a critical issue. Even more important, it meant that the various vendors had to enter into a partnership-like relationship with the company and each other, in order to ensure a successful endeavor and a smooth transition. Above all, this required the integration of a commercial state-of-the-art CAPP system with off-the-shelf MRP, shop floor order control, and manufacturing change control and time standards systems, all using the same relational database system.

The company's new CAPP system was implemented in two phases:

During the first phase, when the mainframe was still in operation, the new server-based system was interfaced with the company's legacy systems so that the mainframe system could still provide detailed part and subassembly plans to the shop floor and support final assembly. This link also provided for a one-time downloading of existing process plans from the legacy system to the new CAPP system. In this way, the new CAPP system maintained the legacy system while enabling planners on the new system to exploit existing text for the creation of new plans.

In the second phase, the CAPP system was integrated with the other new UNIX applications. These included MRP and factory management to produce shop-floor orders for subassembly and fabrication.

The new CAPP system was in place and operating within 60 days after placement of the order, and since then it has been used to perform all process planning at the company's commercial aircraft division. The legacy systems were finally phased out in August 1994, and the mainframe system has been discontinued. The installation and integration of the systems took approximately 18 months, with total costs below budget and a return on investment in less than 2 years.

Ancillary Benefits

Although the main reason for replacing the mainframe system had been to reduce the cost of data processing, there have been considerable ancillary benefits. Some of the changes in processes have already started to pay off enormous dividends.

With the legacy system, it took approximately 57 people 50 days to produce a bill of material (BOM) that could be used for ordering. The BOM resided in two different systems that did not match, and therefore validation was a problem. With the implementation of the new CAPP and MRP systems, the company was able to cut that down to 10 days and do it with about 6 people. The validation and checking groups were taken out of the loop because the system validates itself as entries are made.

In effect, the new approach and particularly the installation of the CAPP system has resulted in the rethinking of the entire flow of manufacturing information. Currently, well over 400 users (50 authors and 350 viewers) employ the CAPP system daily, not only for detail and assembly process planning, but also for tool planning, manufacturing change management, and, in the future, for time standards. Currently the company uses 20 UNIX-based production servers and one server which contains the CAPP database.

This all has added up to a successful conversion from military to commercial operations that leads company employees to speculate about the future: "We want to be a world-class manufacturer of commercial aircraft products. This new systems architecture certainly provides the information management infrastructure to meet our goals."

3.7.2 U.S. Army Arsenal Selects Multiple Vendors for Joint Development of a Very Advanced CAPP System

The arsenal manufactures and assembles heavy arms (such as howitzers) in a vertically integrated manufacturing facility using sophisticated production technology. Among others, it uses a mainframe-based MRPII system, a Tandem-based tool management system, and extensive in-house information on machinability data. In an effort to improve and build upon their current in-house technology, the arsenal recently defined a set of tools to assist process planners, methods, and standards personnel in all phases of their planning activity.

According to their system definition, the new CAPP system was to be client-server based with links to existing systems and with the following capabilities:

Variant process planning that would include expert technology for individual manufacturing steps
Process planning for operations and detailed manufacturing steps within an operation
Online machinability database
Online time standards database
CAPP graphics and interfaces to existing CAD systems

Furthermore, the arsenal defined a wide range of highly sophisticated CAPP requirements that demanded an extension of the basic CAPP system functionality. These specifications included:

User interfaces, including text, graphics, and images
Elaborate checking and sign-off capabilities
Detailed revision history, tracking, and mass update

Vendors Make a Joint Effort to Meet Arsenal's Advanced CAPP Needs

Considering the commercially available CAPP systems and their off-the-shelf capabilities, substantial and expensive customization efforts would be required of each vendor to meet the arsenal's requirements. To reduce these individual customization efforts and take advantage of available off-the-shelf capabilities, two normally competing vendors in the CAPP market decided to make a joint proposal in response to the arsenal's solicitation.

The developers jointly set up the new arsenal CAPP system as follows:

One vendor would supply the general process planning structure, including multimedia user interface, sign-off capabilities, revision history and tracking, relational database installation, links to the other arsenal systems and platforms, and a B-size scanner.

The second vendor would supply the manufacturing technology and time standards software.

The two vendor systems would be seamlessly integrated using one graphical user interface and one relational database.

Flexibility Enables Users to Go from Manufacturing Operations to Manufacturing Steps

This CAPP system is currently being installed at the arsenal and will capitalize on the advanced features of both commercial packages. It is an extremely flexible system, able to call up manufacturing information from a wide variety of sources, to retrieve and generate not only manufacturing operations but manufacturing steps, thus providing feeds, speeds, and time standards for part features.

This joint effort marks the first between two competing vendors and will ultimately produce one of the most sophisticated process planning systems to date, with a projected return on investment for the arsenal in 2 to 3 years.

3.7.3 The Automation Division of a Large Machine Tool Manufacturer Purchases a CAPP System with Group Technology Retrieval Capabilities

Cutting Down on Time to Market

How can we cut down on time to market? This was a challenge that the assembly automation division of the company had to meet. This manufacturer of customized automation assembly systems, known for installations such as the body assembly line at the GM Saturn plant and the chassis line for the new Ford Explorer, was faced with customer demands for completed installations in only 6 months. For a company that was used to completing an installation in a 9- to 18-month time frame, such time-to-market competitiveness was going to require significant changes in their manufacturing procedures.

Repetitive Tasks as a Stumbling Block

The assembly automation division singled out excessive repetition in design and manufacturing tasks as a major stumbling block to improving the rate of product turnaround at their division. Process planners at the plant were re-creating process plans from scratch with each new order. They had no effective retrieval system available that would enable manufacturing engineers to work with existing process plans, simply copy-and-edit, and thus drastically cut down on generation time. The company decided to install a flexible process planning system as well as an up-to-date group technology system.

The Installation of the CAPP System

The company's implementation of a commercially available process planning system was initiated as a two-step process. Step one was the installation of standard process planning software, including screen layouts according to the company's specifications, multimedia capabilities, and links to other company systems (on IBM AS 400). The second step encompassed the installation of the group technology (GT) classification system for metal-forming and metal-cutting parts and analysis of the process planning database. A random sampling of 10% of the company's most recent parts was taken, i.e., parts that had been produced in the last year. The company generates about 12,000 to 13,000 different parts per year; 1,250 of these parts were used for analysis to establish part families and standard routings. With the classification of the sample completed, the process planning database (more specifically, the parts' features) was broken down into part families with similar part features. Parts were sorted according to simple parameters: round/nonround, type of material, similar features, size, and proportion similarities. All of the analysis was performed by the CAPP vendor outside the company facility, enabling the plant to go on with its business with little interference. The company was called upon only at key points in the project to review and approve the classification work.

From Part Families to Standard Routings

With part families established, the final stage of analysis and classification was to set up standard routings for these parts. Again, the company served as an editor in the process, making only minor changes to the standard routings that were produced by the CAPP vendor. A total of only 21 standard routings were set up. These routings managed to cover over 70% of all parts, both lathe and machining center parts, at the division! What did this mean for the company in terms of cutting down time to market? These standard routine templates are now set up in the new CAPP database for planners to call up, edit, and generate with each new order. Reinvention of the wheel has been cut down dramatically. Furthermore, as the company's manager of manufacturing engineering points out, these templates ensure that planners generate routings in a consistent and standardized way. And they can do so with little training and great ease: the CAPP and GT systems all employ a user-friendly point-and-click approach which makes them easy to learn and use. One of the immediate benefits is that parts can be directed to machine tools best suited to the required manufacturing operation.

It is important to note that the company does not produce a "standard" product. Each installation from the integrated automation division is custom-designed. But with each customized installation comes a myriad of similar parts. The division has cut down its similar parts-manufacturing effort significantly (all within 6 months)—thus enabling the group to focus on the heart of the work, the customization of unique parts and the optimization of similar parts.

How CAPP Affected Design Engineering

The new CAPP system helped the company isolate standard routings which enable planners to recycle existing process plans. This same system could also be used by designers to recycle designs, though possibly at a later stage, because the system's flexibility enables it to be fine-tuned to specific design needs. Both existing designs and manufacturing methods can be retrieved from the same database, thereby increasing communication and efficiency between the two departments.

3.7.4 A Large Steam Turbine Manufacturer in England Is Getting Closer to a Nearly Generative Process Planning System for Turbine Blades

Since 1980, this company used an advanced CAPP system for preparing the complete information package necessary to manufacture detailed parts; it is minicomputer-based and runs on a relational database. The use of CAPP had been a very profitable venture for the company for 14 years. However, the time had come for a change to a new-generation CAPP system in the turbine blade department, one that operated in a client-server environment but retained a direct link to the old minicomputer-based system. This link, much like the implementation at the commercial aircraft division, would provide the process planners with the ability to retrieve and modify old process plans without having to download the entire old database.

Turbine blades require a complex machining process, but the process for one blade is very similar to that required for another. Consequently, a turbine blade process planning system that is driven by parameters from CAD that then feed a group technology-type work cell appears to be an excellent solution. On the other hand, this solution does not seem to be even remotely feasible for the manufacturing of other parts of a steam turbine, such as housings, valve assemblies, etc.

3.8 WHERE TO GO FROM HERE?

In the late 1970s, defense and aerospace companies were looking for a solution to meet their advanced process planning needs. Commercial CAPP systems of that time offered an electronic pencil solution, and in some cases, advanced GT techniques, which served the needs of many in the manufacturing industry but did not come close to meeting the needs of aerospace and defense. Thus, the aerospace/defense industry turned inward, developing highly complex in-house systems on the mainframe—at great cost and with very expensive maintenance overhead. Both schools of process planning—the electronic pencil variety and the sophisticated in-house system—have come to a crossroads: both approaches are in need of major change. Now, most industry leaders have decided to move away from both the mainframe systems and the outdated electronic pencil, toward a UNIX-based client-server environment. Furthermore, virtually all have decided to purchase, if possible, commercially available off-the-shelf software, even to the extent that they are willing to slightly modify some of their business practices to fit the off-the-shelf products.

Given their substantial in-house expertise, the companies that are purchasing software are quite selective, preferring to mix and match several vendors to obtain the best technical solution rather than go with a one-stop purchase that may be less optimal. Such a policy does require extensive cooperation between vendors and the buying company. The principal reasons for this move to a client-server environment are:

Cost reduction
Local empowerment of the information service activities
Rethinking of the entire design and manufacturing information structure

Electronic pencil process planning, although still used in many companies, is no longer a wise investment. In the paper age of process planning, the electronic pencil provided a substantial improvement. Its return on investment was very attractive—but that was more than 10 years ago.

As companies start to realize that the intelligent management of product and manufacturing data can be a determining factor in their productivity, the protection of that data also becomes imperative. This an important reason why UNIX-based systems with good data security are the selection of choice for most large companies; however, new PC operating systems that also provide this capability may make PCs equally attractive, especially for smaller companies. A note of caution: It is believed that the new PC operating systems working in a complex client-server network are going to be easier to run than UNIX-based systems, but this may very well be a fallacy.

In the near future, true generative process planning systems (not some sales pitch about it) may not be an attractive solution except where it concerns very narrowly defined groups of parts. On the other hand, feature-driven variant process planning systems with parameter-driven standard process plans should result in large paybacks. This approach also fits well with feature-driven CAD systems. In other words, the gradual linking of product data and manufacturing data management may soon become a realistic venture. At least one or two commercial CAPP system vendors are working along these lines with CAD system suppliers.

Real process planning systems do not come cheap, especially in terms of financial investment and personnel commitment within a company. But there is ample evidence to show that CAPP's return on investment is larger than the ROI on a large machine tool—and the price is about the same!

3.8.1 Getting Started in CAPP

Having reviewed how CAPP implementations have helped four companies better meet the challenges of the manufacturing industry, it is worth summing up some of the key steps involved in preparing to install the right CAPP system for a particular manufacturing site.

To make CAPP part of an integrated design and manufacturing solution, companies will have to put considerable thought into their long-term goals and how a system can be best integrated into existing facilities. Obviously, no one "cookbook" will supply every possible approach for every company's manufacturing environment. In fact, the purchase of a system should come only after a company has had many discussions, both with in-house personnel and outside experts, on what their needs really are. The following steps should be considered.

1. *Form a discussion group.* A philosophy should be established for the company's long-term integrated design and manufacturing solution. On the basis of these discussions, which should involve top management, objectives should be set for fulfilling the philosophy. During this process, the discussion group should consider discussing their ideas and decisions with other companies that are in the same situation. Participation in several seminars, both external and internal, is typically necessary in order to sell the intended objectives to company personnel.
2. *Define information flows.* Once a company has established its objectives for an integrated system, the information flows among different departments should be defined (i.e., between the machine shop and manufacturing engineering, between manufacturing engineering and design, etc.). It makes sense to create an elaborate data flow chart to give a bird's-eye view of the different needs of each department and how they will function in the new environment. One can save a lot of money if these data flows are identified ahead of time, highlighting such details as who needs it, what information is needed when, etc. This information flow plan includes the definition of a software linkage system that will link all the different databases together.
3. *Establish functional specifications.* Once the information flows are defined, the individual departments have to identify which functional capabilities they need to deal with the information flows. Sometimes one system may serve several departments, or it may be designed only for a single department. For example, the CAD needs of the manufacturing engineering

department are usually much less sophisticated than those of the design department. Consequently, the company may have to define two separate systems. At this point the manufacturing engineering department should begin to lay out what it needs from a CAPP system.

4. *Take inventory.* Since most companies have already installed several design and manufacturing systems (some simplistic, others more sophisticated), it is vital to review the functionality of existing systems and analyze how these systems could be used with the new, integrated solution. Several important decisions have to be made, some of which will be politically unpopular. Some systems simply cannot be integrated into the total picture and should therefore be abandoned. One should avoid a scenario where new and old systems exist concurrently without being able to communicate. In some cases, companies may attempt to link old and new systems with "bailing wire." A problem arises when the people who designed the oddball connection leave the company, the entire system breaks down, and the company's integration solution is in jeopardy. A thorough inventory should be taken to determine what systems should stay and what should go.
5. *Decide whether to build or buy.* Once the functional specifications have been set for the required system, the company must decide if the system should be developed in-house or purchased. Companies that have small software departments should not attempt to develop their own systems, but rather purchase the necessary software. For larger companies, economics must be considered. When a capable software department is available (which is often the case with larger companies), the company could easily develop its own systems and probably better gear these new systems toward the functional specifications and data flows required. However, in-house software development is very expensive in both the initial development stages and thereafter in maintenance. It also tends to depend on a few gifted individuals who, when they leave the company, may leave the organization stuck with software that is very difficult to maintain. The alternative is to find software developed by outside parties. These packages tend to be somewhat cheaper. However, third-party software rarely meets all of a company's needs. And there is always the risk that a vendor will go out of business. A financial analysis may be of use when deciding what to do.

3.8.2 Software and Hardware

Given the aforementioned functional specifications for an advanced CAPP system, the software requirements are as follows:

Word processing text and graphics capabilities, and intelligent lookup tables (profit tables)

Linkage to other systems and relational or object-oriented databases

Retrieval capabilities of relational databases using similarities based on group technology

Expert system capability to automatically generate process plans from scratch for certain parts
Rapid response times and user friendliness
Extensive help capabilities

A company that is new to CAPP may decide to start with just text and graphics and then expand to other capabilities later on. It should be recognized, however, that by following this path, one can easily fall into the trap of simply "automating the file cabinet" and believing it is CAPP. So if a company decides to install a system incrementally, it should be sure that it can be expanded without having to rewrite the entire system.

Given the rapid decline in the cost of computer hardware, one should opt for a system with software that is portable across different platforms. The CAPP application should look the same and react the same on different platforms, be it PCs, workstations, mainframes, or a combination of these. This will make training much easier and will also broaden the system's range of application.

With prices of high-end PCs and low-end UNIX workstations converging, prescribing one machine over the other as the platform of choice is difficult. Clearly, PCs, workstations, or a combination of both in a network environment will be far more cost-efficient than running CAPP solely on a mainframe. Not only is this more efficient from an asset-utilization point of view, but also better system response times provide for better usage of personnel.

The option to use X-Windows may improve the process planner's efficiency; however, if the already-installed hardware is unable to support X-Windows without incurring major update costs, that option may be less attractive.

Local area networks are of major importance, but hidden expenses should be considered. A simple local area network function means that if one person requests an existing process plan for modification or review from the file server, then the plan will not be available to another person. This is the standard practice in many manufacturing environments. Supporting more complex data transactions will increase cost substantially.

3.9 AN AFTERWORD ON GROUP TECHNOLOGY

An awareness of the similarities of parts and products has benefited design and manufacturing for many years, usually resulting in substantially higher production throughputs, faster design turnarounds, and better utilization of equipment on the shop floor. The manual search for similarities of parts was very tedious and time consuming until the 1960s, when computers started to be used to search for these similarities. At that time, GT became a practical approach for analyzing a part population for standardization and for machine shop layout.

In the mid-1970s, the grouping of similar parts was enhanced by computerized classification and coding systems. However, despite these pioneering efforts, the results were only marginally successful, in part because the use of similarities for design and manufacturing standardization is a long-term effort. Since then, GT has become an increasingly attractive choice, especially since the availability of relational databases and object-oriented databases has made the retrieval of similar parts much easier.

GT can lead to several applications. In the short term, GT provides information for daily operations, retrieval of product or part design, and retrieval of manufacturing and assembly experience of parts and products before the engineer starts the expensive process of reinventing the wheel. In the long term, GT can provide the analysis of retrieved information, not only for standardization in design, fabrication, or assembly, but also as a tool to successfully introduce design for manufacturing and assembly (DFMA) and the introduction of "concurrent engineering." GT also can be used to better employ the available assets in the machine shop. In others words, standardization in the manufacturing and assembly departments can lead to a rational organization of the production department and provide a tool to analyze what machine tools should be purchased in the future and how machine shops should be laid out.

3.9.1 People and Group Technology

Group technology is a tool box that provides knowledgeable people with the means to come up with cost-efficient solutions. Consequently, its introduction in a company should be accompanied by a clear set of objectives, timetables, and an understanding and commitment by the people involved—including personnel ranging from top management down to the people on the shop floor. Group technology will *not* be successful if it evolves as an edict from the top, nor will it make any inroads if it is simply a backroom activity in some department at the bottom. Departments that *should* be affected by the implementation of group technology include:

The product design department, as a tool to retrieve previous design/manufacturing information to either modify or take as is (GT can also serve as a tool to standardize certain design approaches)

The manufacturing engineering department, as a tool to retrieve, modify, develop, and use manufacturing process information

Both the design and manufacturing engineering departments, as an integration tool to implement the principle of "design for best manufacturing" and "design for best assembly" in order to improve the quality of the product and lower the cost

The materials purchasing department, as a tool to reduce the variety of materials that are used to produce a product

The production, production control, and manufacturing engineering departments, to lay out the machine shop, possibly in so-called group technology work cells, and as a method to purchase machine tools

4 Work Measurement

Kjell Zandin

4.0 WHY WORK MEASUREMENT?

Work measurement in primitive forms has been around for hundreds of years. According to some sources, the Egyptians applied some form of work measurement when they were building the pyramids. It seems fairly obvious to us, since the work of stacking blocks of rock can be considered repetitive. In the Middle Ages, the armor and weapons makers of Catalonia based their prices on the basis of material, quality, and the time it would take to make the product. The time was based on experience. More recently, there have been specific demands for work measurement created by the need to increase productivity (during and after the two world wars) or to establish a basis for incentive payments.

Our society is filled with measurements of all kinds. We measure the distance between two cities or places, and the weight of steaks and other foods in a grocery store. We measure the temperature of the air and of people. We measure the dimensions of parts and components to make sure they will go together. We measure the pressure in automobile tires. We measure how much it rains or snows. We measure the length and weight of a newborn baby and of ourselves throughout life. We even measure how fast someone can drink a beer and how far a golfer can hit a ball, or a football player can carry it. Thousands upon thousands of other more or less useful examples of our measuring spree could easily fill an entire chapter. Practically everything we are, we do, and we own is related to a yardstick.

Measurement is information to us. Without measurements, the world could not function. We base costs, plans, controls, and many decisions on measurements. We need that information to improve the conduct of a business. For instance, we tell a machine operator what to do through blueprints filled with dimensions, tolerances, and other significant facts about a part or product, but seldom how to make it, indicating the process, equipment, tools, and method to use, as well as the time (work standard) for the job. We expect the operator to use the most efficient tools and methods with little or no help. We expect the operator to perform at a high rate. But how can we know what the performance is if we do not measure the work to be done?

Even though we readily accept the science of measurement, as well as being measured in different ways, there is one area where measurements have been either controversial or rejected: *the measurement of work.* However, many industries and organizations have gradually, through improved measurement techniques, education of their people, and commitment to sound management principles, turned the work measurement practice into a logical and natural part of their business conduct. The rewards and benefits have often been both dramatic and lasting.

As with any measurement, the measurement of work brings knowledge. Through this knowledge, factual decisions and improvements can be made and control can be exercised.

The purpose of work measurement is to provide management, as well as the person performing the job, with information about how much time it should take to perform a task according to well-defined work conditions and a specified method. How management uses that information is a different and undoubtedly more significant matter.

Basically, proper measurement of work facilitates and improves management. Using simple or nonscientific forms of work measurement, the time values may, however, be inaccurate or may not match the conditions for the performed work. On the other hand, *scientific* work measurement not only facilitates but provides a sound basis for management. In general terms, work measurement is the basis for planning, evaluation of performance, and estimation of costs. In essence, the manager looks at work measurement as a basis for his or her ability to forecast with confidence. The key word here is "confidence," because if the time for a job is not accurate enough, very often the wrong decisions will be made.

Work measurement is used to establish time standards for individual operations or jobs. Time standards apply primarily to manual operations, even though machine-controlled operations are measured as well, using a different method.

Work measurement is a tool that every manager should use to do a better job—because work measurement brings *knowledge* in the form of definitions of work conditions, method instructions, and reliable, consistent time values. As we shall see, these time values can be very useful in many areas of management.

4.1 METHODS OF MEASURING WORK

4.1.1 Nonengineered Time Standards

The fastest and least expensive way of measuring work is to guess. Very often, one may have little or no information available on which to base a guess. Time standards created by guessing are usually very inaccurate and inconsistent. All they provide is a number, unsupported by the conditions under which the work is being done.

By using historical data in combination with experience with the particular area, time standards may become less inaccurate and less inconsistent compared to pure guesses. In this case, they are called "estimates," because they are estimated by someone who is familiar with the operations or activities to be measured. Estimates are quite often based on a subjective evaluation of the work content supported by available data as the basis for a time standard. Invariably, people overestimate just to make sure that the job can be done in the time allotted. Therefore, one tends to get "loose" standards by using the estimating method. Accuracy and consistency usually do not meet established requirements for good standards.

Another method of establishing estimated time standards is self-reporting. Typically, if a person believes that a job will take 45 minutes to do, the self-reporter will allow 1 hour, just to make sure that the job can be completed within the allotted time and perhaps also to allow time to perform the task at a pace below normal.

All the methods above can be categorized as "nonengineered standards," because no backup documentation supports the time standards. That means that each time a product or part, a job method, or a work condition changes, a new time standard must be determined. Since no backup data are available, it is not possible just to change a time standard based on the change in work conditions.

Stopwatch Time Studies and Work Sampling

Although there are cases when acceptable backup data are produced, very often stopwatch time studies fall into the category of nonengineered standards as well. The same applies for another watch-based system: work sampling. These commonly used methods will likely produce more accurate and consistent standards than the estimating methods. However, the watch-based systems have many drawbacks.

Both the stopwatch time study and work sampling studies require an analyst to observe one or more operators performing the work. In order to obtain a reasonably accurate time standard, a large number of observations must be recorded. This means that the analyst has to spend extensive time in the workshop just observing what other people are doing. By applying these methods, one will only be able to get a picture of the current work content as performed by the operators who are being observed. If anything within this work content changes, a new study (picture) will have to be carried out.

The most controversial drawback is, however, the subjective rating of the performance of the operators being observed. Many arguments, grievances, and arbitrations have resulted from disagreements about a fair judgment of the performance level of the operator. Why then, is the stopwatch time study method so popular? Basically, the method is easy to learn, although the performance rating takes substantial and continuous effort to master. In the manager's eyes, using a watch is the logical way to find out how long a job should take. We all use watches to set standards for ourselves and others.

Work sampling studies are useful to determine utilization and reasons for downtime and delays. However, such studies will only provide data that relate to the period during which the study is made.

Finally, there is a nonengineered method that has been used extensively for setting standards in areas such as maintenance. This method is called "benchmark comparison." Because of the nature of the work, the accuracy requirements are not as stringent as in many other situations. Therefore, a set of scientifically determined time standards called "benchmarks" are developed for typical operations and used as a basis for the determination of time standards for operations that are similar but sometimes very different from the benchmarks. As with the other nonengineered methods, no backup data are available.

Nonengineered methods are frequently used mainly with the purpose of quickly establishing a time value. Some people consider it satisfactory just to have a time value for scheduling, cost estimating, budgeting, etc., rather than develop realistic and consistent time standards that are based on established work conditions and defined work methods and that one can have confidence in. Without backup data, nonengineered standards are low-quality time standards.

4.1.2 Engineered Work Measurement

Prior to starting the manufacturing of a part, component, or product, a detailed engineered drawing is almost always produced. This drawing contains the size and shape of the part, dimensions, tolerances, surface finishes, material, etc., so that the operator will have a very accurate specification of what to manufacture. Engineered work measurement will produce an equally detailed "drawing" of *how* to manufacture or assemble the part, component, or product.

Engineered work measurement uses a predetermined motion-time system (PMTS) as a basis for the measurement. All necessary time values have been predetermined (based on very detailed time studies) and are therefore available to be used by anyone who is trained in the procedures on how to apply these time values. Since all the work measurement time values are predetermined, preproduction standards can be established. This means that the time standard for an operation can be established long before the operation will be performed.

Motion-Based Systems

There are a number of predetermined motion-time systems that will result in documented engineered standards of high quality. These systems can be categorized into "motion-based systems" such as Methods Time Measurement (MTM) and Work Factor (WF). The time elements in these systems have been determined for basic human motions such as reach, grasp, move, position, etc. These systems were introduced in the late 1940s. MTM was published in a book by the late Harold B. Maynard in 1948 which has made MTM the most-used predetermined motion-time system in the world. Because of their detail, these systems are time-consuming and tedious to apply.

Element-Based Systems

Simplified versions of these systems have been developed as "element-based systems." Examples of such systems are MTM-2, MTM-3, General Purpose Data (GPD), Universal Standard Data (USD), Universal Office Controls (UOC), Universal Maintenance Standards (UMS), etc. Usually these so-called standard data systems have been designed for use in specific application areas. These element-based systems are also considerably faster to use than the motion-based systems. However, in the majority of cases, they are being applied only in a manual mode. Further information on MTM is available in the book *Engineered Work Measurement* (Karger and Bayha, 1987).

Activity-Based Systems

The third category of predetermined motion-time systems includes an "activity-based system" called MOST® (Maynard Operation Sequence Technique). The MOST system consists of logical sequence models that cover all the motions included in the activity of moving an object from one location to another. The concept of MOST was developed in 1967, and the complete BasicMOST system was introduced in 1972. The MOST system is considerably faster to use and easier to learn than the motion-based and element-based systems. For these reasons, MOST is a user-friendly work measurement system. More detailed information on MOST systems and the computerized version, MOST Computer Systems, will be presented in Subchapter 4.4.

4.2 METHODS OF DEVELOPING, ADMINISTERING, AND MAINTAINING TIME STANDARDS

All nonengineered systems, and in some cases also engineered systems, are being used to establish a time standard directly for manufacturing and other operations without any intermediate steps. This method is called the "direct measurement method." An operation is defined and the work measurement technique is applied to determine the time standard for it. With large numbers of time standards, e.g., 50,000, 100,000, or 1 million or more, the direct method becomes very time-consuming and costly to apply. Administering and maintaining many direct time standards becomes an unmanageable task, even if a computer is being used. However, when only a few time standards are needed or when time standards for a specific situation or product are required, the direct method is acceptable and even preferable.

Engineered standards are in practically all cases being established by using an indirect measurement method. Instead of applying the work measurement technique to establish a time standard for an operation directly, the work measurement technique is used to develop standard data, or parts of operations, also called "suboperations." Such data units or building blocks can be used in different combinations to determine time standards for operations. It is quite possible to establish 100,000 or more time standards from a base of 200 to 700 suboperations. The task of measuring these suboperations using a work measurement technique and subsequently maintaining a large database of time standards becomes much more manageable.

Therefore, a predetermined time system can be applied either for direct work measurement of defined operations or can be used as a basis for standard data (suboperations). In the case of short-cycle, unique operations such as subassemblies, the direct approach is preferred. On the other hand, if a great variety of the operations are being performed at a work center, the standard data approach is the most efficient and economical method. A worksheet composed of standard data units, each backed up by a work measurement analysis, will provide a fast and simple way to calculate standards. Initially, the desired accuracy level for the resulting time standards should be determined and the worksheet designed accordingly. This means that the tighter the accuracy requirements are, the more data units (suboperations) and the more decisions have to be made in order to set a time standard. A multipage, detailed worksheet will take more time and cost more to use than a single-page worksheet with few elements designed for a lower accuracy level. Consequently, the economics of setting standards is a direct function of the required accuracy of the output.

For instance, if the required accuracy is ±5% with 95% confidence over an 8-hour period, the worksheet may consist of 75 different elements, while a ±10% accuracy with 90% confidence over a 40-hour period may produce a worksheet with only 10 to 15 elements. The difference in application time will be substantial, and since standard setting normally is an ongoing activity, the cost-saving potential is considerable.

In order to further simplify and expedite the standard-setting process, decision models based on expert system technology can be used. In such a case, only simple questions regarding the parts or products to be manufactured need to be answered. The selection of suboperations is made automatically by the computer using the AutoMOST program.

4.3 MOST WORK MEASUREMENT SYSTEMS

4.3.1 Introduction to MOST

Because industrial engineers are trained that with sufficient study any method can be improved, many efforts have been made to simplify the analyst's work measurement task. This has, for instance, led to a variety of work measurement systems now in use. These achievements also led us to examine the whole concept of work measurement to find a better way for analysts to accomplish their mission. This induced the information of a new approach later to be known as MOST—the Maynard Operation Sequence Technique.

4.3.2 The MOST Concept

To most of us, work means exerting energy, but, we should add, to accomplish some task or to perform some useful activity. In the study of physics, we learn that work is defined as the product of force times distance ($W = f \times d$) or, more simply, work is the displacement of a mass or object. This definition applies quite well to a large portion of the work accomplished every day, such as pushing a pencil, lifting a heavy box, or moving the controls on a machine. Thought processes, or thinking time, are an exception to this concept, as no objects are being displaced. For the overwhelming majority of work, however, there is a common denominator from which work can be studied, the displacement of objects. All basic units of work are organized (or should be) for the purpose of accomplishing some useful result by moving objects. That is what work is. *MOST is a system to measure work;* therefore, *MOST concentrates on the movement of objects.*

Work, then, is the movement of objects—maybe, we should add, following a tactical production outline. Efficient, smooth, productive work is performed when the basic motion patterns are tactically arranged and smoothly choreographed (methods engineering). The movement of objects follows certain consistently repeating patterns, such as reach, grasp, move, and positioning of the object. These patterns can be identified and arranged as a sequence of events (or subactivities) manifesting the movement of an object. A model of this sequence is made and acts as a standard guide in analyzing the movement of an object. It should also be noted that the actual motion contents of the subactivities in a sequence vary independently of one another.

This concept provides the basis for the MOST "sequence models." The primary work units are no longer basic motions as in MTM, but fundamental activities (collections of basic motions) dealing with moving objects. These activities are described in terms of subactivities fixed in sequence. In other words, to move an object, a standard sequence of events occurs. Consequently, the basic pattern of an object's movement is described by a universal sequence model instead of an aggregate of detailed basic motions synthesized at random.

Objects can be moved in only one of two ways: Either they are picked up and moved freely through space, or they are moved in contact with another surface. For example, a box can be picked up and carried from one end of a workbench to the other, or it can be pushed across the top of the workbench. For each type of move, a different sequence of events occurs; therefore, a separate MOST activity sequence model applies. The use of tools is analyzed through a separate activity sequence model that allows the analyst the opportunity to follow the movement of a hand tool through a standard sequence of events, which is in fact a combination of the two basic sequence models.

Consequently, only three activity sequences are needed for describing manual work. The BasicMOST work measurement technique, therefore, is comprised of the following sequence models:

The *General Move* sequence, for the spatial movement of an object freely through the air
The *Controlled Move* sequence, for the movement of an object when it remains in contact with a surface or is attached to another object during the movement
The *Tool Use* sequence, for the use of common hand tools
The *Manual Crane* sequence, for the measurement of moving heavy objects by using, for instance, a jib crane (although this is also part of the BasicMOST system, it is used less frequently than the three first sequence models)

4.3.3 Sequence Models

General Move Sequence Model

General Move is defined as moving objects manually from one location to another freely through the air. To account for the various ways in which a General Move can occur, the activity sequence is made up of four subactivities:

A, action distance (mainly horizontal)
B, body motion (mainly vertical)
G, gain control
P, placement

These subactivities are arranged in a sequence model consisting of a series of parameters organized in a logical arrangement. The sequence model defines the events or actions that always take place in a preset order when an object is being moved from one location to another.

The General Move sequence model, which is the most commonly used of all available sequence models, is defined as follows:

A	B	G	A	B	P	A
action distance	body motion	grasp	action distance	body motion	placement	action distance

These subactivities, or "sequence model parameters," as they are called, are then assigned time-related index numbers based on the motion content of the subactivity. This approach provides complete analysis flexibility within the overall control of the sequence model. For each object moved, any combination of motions could occur, and using MOST, any combination can be analyzed. For the General Move sequence, these index values are easily memorized from a brief data card (Figure 4.1). A fully-indexed General Move sequence, for example, might appear as follows:

$$A_6\,B_6\,G_1\,A_1\,B_0\,P_3\,A_0$$

where

A_6 = walk three to four steps to object location
B_6 = bend and arise
G_1 = gain control of one light object
A_1 = move object a distance within reach
B_0 = no body motion
P_3 = place and adjust object
A_0 = no return

This example could, for instance, represent the following activity: "Walk three steps to pick up a bolt from floor level, rise, and place the bolt in a hole."

General Move is by far the most frequently used of the three sequence models. Roughly 50% of all manual work occurs as a General Move, with the percentage running higher for assembly and material-handling work, and lower for machine shop operations.

Controlled Move Sequence Model

The second type of move is described by the Controlled Move sequence. This sequence is used to cover such activities as operating a lever or crank, activating a button or switch, or simply sliding an object over a surface. In addition to the

ABGABPA									
Index x 10	A Action Distance		B Body Motion		G Gain Control		P Placement		Index x 10
	Parameter Variant	Keyword	Parameter Variant	Keyword	Parameter Variant	Keyword	Parameter Variant	Keyword	
0	≤ 2 in. ≤ 5 cm	CLOSE					Hold Toss	THROW TOSS CARRY PICK UP	0
1	Within reach				Light object Light objects simo	GRASP (optional)	Lay aside Loose fit	MOVE PUT	1
3	1–2 steps	1 STEP 2 STEPS	Bend and arise 50% occ.	PBEND	Non simo Obstructed Heavy/Bulky Interlocked Blind Collect Disengage	GET DISENGAGE FREE COLLECT	Adjustments Light pressure Double placement	PLACE REPLACE	3
6	3–4 steps	3 STEPS 4 STEPS	Bend and arise	BEND			Care Precision Blind Obstructed Heavy pressure Intermediate moves	POSITION REPOSITION	6
10	5–7 steps	5 STEPS 6 STEPS 7 STEPS	Sit or stand	SIT STAND					10
16	8–10 steps	8 STEPS 9 STEPS 10 STEPS	Through Door Climb on or off Stand and bend Bend and sit	DOOR CLIMB/DESCEND STAND AND BEND BEND AND SIT					16

FIGURE 4.1 Index values for the parameters used in the General Move sequence model.

A, B, and G parameters from the General Move sequence, the sequence model for Controlled Move contains the following subactivities:

M, move controlled
X, process time
I, align

As many as one third of the activities occurring in machine shop operations may involve Controlled Move sequences. A typical activity covered by the Controlled Move sequence is the engaging of the feed lever on a milling machine. The sequence model for this activity might be indexed as follows:

$$A_1\ B_0\ G_1\ M_1\ X_{10}\ I_0\ A_0$$

where

A_1 = reach to the lever a distance within reach
B_0 = no body motion
G_1 = get hold of lever
M_1 = move lever up to 12 in. (30 cm) to engage feed
X_{10} = process time of approximately 3.5 seconds
I_0 = no alignment
A_0 = no return

Tool Use Sequence Model

The third sequence model comprising the BasicMOST technique is the Tool Use sequence model. This sequence model covers the use of hand tools for such activities as fastening or loosening, cutting, cleaning, gauging, and recording. Also, certain activities requiring the use of the brain for mental processes can be classified as Tool Use, such as reading and thinking. As indicated above, the Tool Use sequence model is a combination of General Move and Controlled Move activities. It was developed as part of the BasicMOST system, in order to simplify the analysis of activities related to the use of hand tools. It will later become obvious to the reader that any hand-tool activity is made up of General and Controlled Moves. The use of a wrench, for example, might be described by the following sequence:

$$A_1\ B_0\ G_1\ A_1\ B_0\ P_3\ F_{10}\ A_1\ B_0\ P_1\ A_0$$

where

A_1 = reach to wrench
B_0 = no body motion
G_1 = get hold of wrench

A_1 = move wrench to fastener a distance within reach
B_0 = no body motion
P_3 = place wrench on fastener
F_{10} = tighten fastener with wrench
A_1 = move wrench a distance within reach
B_0 = no body motion
P_1 = lay wrench aside
A_0 = no return

4.3.4 Elements and Characteristics of MOST

Time Units

The time units used in MOST are identical to those used in the basic MTM system and are based on hours and parts of hours called TMUs (time measurement units). One TMU is equivalent to 0.00001 hours.

The time value in TMU for each sequence model is calculated by adding the index numbers and multiplying the sum by ten. For our previous General Move sequence example, the time is

$$(6 + 6 + 1 + 0 + 0 + 3 + 0) \times 10 = 170 \text{ TMU}$$

corresponding to approximately 0.01 minutes. The time values for the other two examples are computed in the same way. The Controlled Move totals to

$$(1+ 0 + 1 + 1 + 10 + 0 + 0) \times 10 = 130 \text{ TMU}$$

and the Tool Use to

$$(1 + 0 + 1 + 1 + 0 + 3 + 10 + 1 + 0 + 1 + 0) \times 10 = 180 \text{ TMU}$$

All time values established by MOST reflect the pace of an average skilled operator working at an average performance rate. This is often referred to as the 100% performance level that in time study is achieved by using "leveling factors" to adjust time to defined levels of skill and effort. Therefore, when using MOST, it is not necessary to adjust the time values unless they must conform with particular high or low task plans used by some companies. This also means that if a time standard for an operation is properly established by using either MOST, MTM, or a stopwatch time study, the TMU values should be identical or almost identical for the three techniques.

The analysis of an operation consists of a series of sequence models describing the movement of objects to perform the operation. See Figure 4.2 for an example. Total time for the complete MOST analysis is arrived at by adding the computed sequence times. The operation time may be left in TMU or converted to minutes or hours. Again, this time reflects pure work content (normal time without allowances)

MOST® Calculation		CODE	
		DATE	*7/29/91*
PROD/AREA	*ELECTRONIC ASSEMBLY*	SIGN	*A.A.*
OPERATION		PAGE	*11 i*
TITLE	*INSTALL CONNECTOR ON PC-BOARD*	*AT WORKSTATION*	
	ACTIVITY • OBJECT • IN, ON, FOR • PRODUCT EQUIPMENT • TOOL • TO, AT • WORK AREA		
CONDITIONS	*EDGE CONNECTORS ONLY*	PER	*BOARD*

NO.	METHOD DESCRIPTION	NO.	SEQUENCE MODEL	FR	TMU
1	*POSITION EDGE*	1	$A_1\ B_0\ G_1\ {}_1\ B_0\ P\ A_0$		90
	CONNECTOR TO BOARD	3	$A_1\ B_0\ G_1\ A_1\ B_0\ P\ A_0$	2	120
2	*ALIGN CONNECTOR TO*	4	$A_1\ B_0\ G_1\ A_1\ B_0\ P\ A_0$	4	160
	ACCURATE LOCATIONS	7	$A_1\ B_0\ G_1\ A_1\ B_0\ P\ A_0$		60
3	*PLACE SCREW TO HOLE*		A B G A B P A		
	IN CONNECTOR F2		A B G A B P A		
4	*MOVE WASHER TO SCREEN*		A B G A B P A		
	ON BOARD F4		A B G A B P A		
5	*FASTEN NUTS 2 SPINS*		A B G A B P A		
	USING FINGERS F2		A B G A B P A		
6	*FASTEN 2 SCREWS 5 SPINS*		A B G A B P A		
	USING SCREWDRIVER		A B G A B P A		
7	*PLACE BOARD TO RACK*	2	$A_0\ B_0\ G_0\ M_3\ X_0\ I_{16}\ A_0$		190
			A B G M X I A		
			A B G M X I A		
			A B G M X I A		
			A B G M X I A		
			A B G M X I A		
			A B G M X I A		
			A B G M X I A		
		5	$A_1\ B_0\ G_1\ A_1\ B_0\ P_1\ F_3\ A\ B\ P\ A$	2	140
		6	$A_1\ B_0\ G_1\ A_1\ B_0\ P_3(A_1F_{10})\ A\ B\ P\ A$	(2)	330
			A B G A B P A B P A		
			A B G A B P A B P A		
			A B G A B P A B P A		
			A B G A B P A B P A		
			A B G A B P A B P A		
			A B G A B P A B P A		
TIME =	*.65*		minutes (mm.)		*1090*

3/93 [1]H.B. Maynard & Company, Inc.

FIGURE 4.2 Standard calculation form used for analysis work using BasicMOST.

at the 100% performance level. The final time standard will include an allowance factor consisting of P (personal time), R or F (rest or fatigue factor), and D for unavoidable delays (often determined by a work sampling study). Therefore, if the normal time = 1.0 hr. and the allowance factor (PR&D) = 15%, the final time standard = 1.15 hr.

Application Speed

MOST was designed to be considerably faster than other work measurement techniques. Because of its simpler construction, under ideal conditions, BasicMOST requires only 10 applicator hours per measured hour. (MTM-1 requires 300 to 400 applicator hours per measured hour.)

Accuracy

The accuracy principles that apply to MOST are the same as those used in statistical tolerance control. That is, the accuracy to which a part is manufactured depends on its role in the final assembly. Likewise, with MOST, time values are based on calculations that guarantee the overall accuracy of the final time standard. Based on these principles, MOST provides the means for covering a high volume of manual work with an accuracy that can be determined and controlled.

Method Sensitivity

MOST is a method-sensitive technique; that is, it is sensitive to the variations in time required by different methods. This feature is very effective in evaluating alternative methods of performing operations with regard to time and cost. The MOST analysis will clearly indicate the more economical and less fatiguing method.

The fact that the MOST system is method-sensitive greatly increases its worth as a work measurement tool. Not only does it indicate the time needed to perform various activities, it also provides the analyst with an instant clue that a method should be reviewed. The results are clear, concise, easily understood time calculations that indicate opportunities for saving time, money, and energy.

Documentation

One of the most burdensome problems in the standards development process is the volume of paperwork required by the most widely used predetermined work measurement systems. Whereas the more detailed systems require between 40 and 100 pages of documentation, MOST requires as few as 5. The substantially reduced amount of paperwork enables analysts to complete studies faster and to update standards more easily. It is interesting to note that the reduction of paper generated by MOST does not lead to a lack of definition of the method used to perform the task. On the contrary, the method description found with the MOST system is a clear, concise, plain-language description of the activity. These method descriptions can very well be used for operator training and instruction.

Applicability

In what situations can MOST be used? Because manual work normally includes some variation from one cycle to the next, MOST, with its statistically established time ranges and time values, can produce times comparable to more detailed systems for the majority of manual operations. Therefore, MOST is appropriate for any manual work that contains variation from one cycle to another, regardless of cycle length. BasicMOST should not be used in situations in which a short cycle (usually up to 10 sec. or 280 TMU long) is repeated identically over an extended period of time. In these situations (which, by the way, do not occur very often), the more detailed MiniMOST version should be chosen as the proper work measurement tool. In fact, MiniMOST was developed to cover highly repetitive, short-cycle work measurement tasks. At the other end of the spectrum, MaxiMOST was developed to measure long-cycle (2 min. or more), nonrepetitive operations such as heavy assembly, maintenance, and machine setups.

4.3.5 Application of the Sequence Models

The General Move Sequence Model

The General Move sequence deals with the spatial displacement of one or more objects. Under manual control, the object follows an unrestricted path through the air. If the object is in contact with or restrained in any way by another object during the move, the General Move sequence is not applicable. Characteristically, General Move follows a fixed sequence of subactivities identified by the following steps:

1. Reach with one or two hands a distance to the object(s), either directly or in conjunction with body motions
2. Gain manual control of the object(s)
3. Move the object(s) a distance to the point of placement, either directly or in conjunction with body motions
4. Place the object(s) in a temporary or final position
5. Return to workplace

These five subactivities form the basis for the activity sequence describing the manual displacement of the object(s) freely through space. This sequence describes the manual events that can occur when moving an object freely through the air and is therefore known as a "sequence model." The major function of the sequence model is to guide the attention of the analyst through an operation, thereby adding the dimension of having a preprinted and standardized analysis format. The existence of the sequence model provides for increased analyst consistency and reduced subactivity omission.

The sequence model takes the form of a series of letters representing each of the various subactivities (called parameters) of the General Move activity sequence. With the exception of an additional parameter for body motions, the General Move sequence is the same as the above five-step pattern:

A B G A B P A

where:

A = action distance
B = body motion
G = gain control
P = placement

Parameter Definitions

Action distance (A)
This parameter covers all spatial movement or actions of the fingers, hands, and/or feet, either loaded or unloaded. Any control of these actions by the surroundings requires the use of other parameters.

Body motion (B)
This parameter refers to either vertical (up and down) motions of the body or the actions necessary to overcome an obstruction or impairment to body movement.

Gain control (G)
This parameter covers all manual motions (mainly finger, hand, and foot) employed to obtain complete manual control of an object(s) and to subsequently relinquish that control. The G parameter can include one or several short-move motions whose objective is to gain full control of the object(s) before it is to be moved to another location.

Placement (P)
This parameter refers to actions at the final stage of an object's displacement to align, orient, and/or engage the object with another object(s) before control of the object is relinquished.

Parameter Indexing

Index values for the above four parameters included in the General Move sequence model can be found in Figure 4.1. Definitions of all available index values for the four General Move parameters can be found in *MOST Work Measurement Systems* (Zandin, 1980). The definitions for A have been included below as an example.

Action distance (A)
Action distance covers all spatial movement or actions of the fingers, hands, and/or feet, either loaded or unloaded. Any control of these actions by the surroundings requires the use of other parameters.

A_0 < 2 in. (5 cm)
Any displacement of the fingers, hands, and/or feet a distance less than or equal to 2 in. (5 cm) will carry a 0 index value. The time for performing these short distances is included within the G and P parameters. *Example:* Reaching between the number

keys on a pocket calculator or placing nuts or washers on bolts located less than 2 in. (5 cm) apart.

A_1 within reach

Actions are confined to an area described by the arc of the outstretched arm pivoted about the shoulder. With body assistance—a short bending or turning of the body from the waist—this "within reach" area is extended somewhat. However, taking a step for further extension of the area exceeds the limits of an A_1 and must be analyzed with A_3 (one to two steps). *Example:* With the operator seated in front of a well laid out work bench, all parts and tools can be reached without displacing the body by taking a step.

The parameter value A_1 also applies to the actions of the leg or foot reaching to an object, lever, or pedal. If the trunk of the body is shifted, however, the action must be considered a step (A_3).

A_3 one to two steps

The trunk of the body is shifted or displaced by walking, stepping to the side, or turning the body around using one or two steps. Steps refers to the total number of times each foot hits the floor.

Index values for longer-action distances involving walking on flat surfaces as well as up or down ladders can be found in Figure 4.1 for up to ten steps. This will satisfy the need for action distance values for most work areas in a manufacturing plant. Should longer walking distance occur, however, the table can be extended. All index values for walking are based on an average stop length of 2 1/2 ft. (0.75 m).

General Move Examples

1. A man walks four steps to a small suitcase, picks it up from the floor, and without moving, further places it on a table located within reach.

$$A_6 B_6 G_1 A_1 B_0 P_1 A = 150 \text{ TMU}$$

2. An operator standing in front of a lathe walks six steps to a heavy part lying on the floor, picks up the part, walks six steps back to the machine, and places it in a three-jaw chuck with several adjusting actions. The part must be inserted 4 in. (10 cm) into the chuck jaws.

$$A_{10} B_6 G_3 A_{10} B_0 P_3 A_1 = 330 \text{ TMU}$$

3. From a stack located 10 ft. (3 m) away, a heavy object must be picked up and moved 5 ft. (2 m) and then placed on top of a workbench with some adjustments. The height of this stack will vary from waist to floor level. Following the placement of the object on the workbench, the operator returns to the original location, which is 11 ft. (3.5 m) away.

$$A_6 B_3 G_3 A_3 B_0 P_3 A_{10} = 280 \text{ TMU}$$

The Controlled Move Sequence Model

The Controlled Move sequence describes the manual displacement of an object over a controlled path. That is, movement of the object is restricted in at least one direction by contact with or an attachment to another object.

The Sequence Model

The sequence model takes the form of a series of letters representing each of the various subactivities (called parameters) of the Controlled Move activity sequence:

$$A\ B\ G\ M\ X\ I\ A$$

where

A = action distance
B = body motion
G = gain control
M = move controlled
X = process time
I = align

Parameter Definitions

Only three new parameters are introduced, as the A, B, and G parameters were discussed with the General Move sequence and remain unchanged.

Move controlled (M)
This parameter covers all manually guided movements or actions of an object over a controlled path.

Process time (X)
This parameter occurs as that portion of work controlled by processes or machines and not by manual actions.

Align (I)
This parameter refers to manual actions following the controlled move or at the conclusion of process time to achieve the alignment of objects. The index value definitions for the above parameters (M, X, and I) can be found in the textbook *MOST Work Measurement Systems* (Zandin, 1980).

Controlled Move Examples

1. From a position in front of a lathe, the operator takes two steps to the side, turns the crank two revolutions, and sets the machining tool against a scale mark.

$$A_3\ B_0\ G_1\ M_6\ X_0\ I_6\ A_0 = 160\ \text{TMU}$$

2. A milling cutter operator walks four steps to the quick-feeding cross lever and engages the feed. The machine time following the 4-in. (10-cm) lever action is 2.5 sec.

$$A_6\ B_0\ G_1\ M_1\ X_6\ I_0\ A_0 = 140 \text{ TMU}$$

3. A material handler takes hold of a heavy carton with both hands and pushes it 18 in. (45 cm) across conveyor rollers.

$$A_1\ B_0\ G_3\ M_3\ X_0\ I_0\ A_0 = 70 \text{ TMU}$$

4. Using the foot pedal to activate the machine, a sewing machine operator makes a stitch requiring 3.5 sec process time. (The operator must reach the pedal with the foot.)

$$A_1\ B_0\ G_1\ M_1\ X_{10}\ I_0\ A_0 = 130 \text{ TMU}$$

The Tool Use Sequence Model

The Tool Use sequence is composed of subactivities from the General Move sequence, along with specially designed parameters describing the actions performed with hand tools or, in some cases, the use of certain mental processes. Tool Use follows a fixed sequence of subactivities occurring in five main activity phases:

1. Get object or tool.
2. Place object or tool in working position.
3. Use tool.
4. Put aside object or tool.
5. Return to workplace.

The Sequence Model

The five activity phases form the basis for the activity sequence describing the handling and use of hand tools. The sequence model takes the form of a series of letters representing each of the various subactivities of the Tool Use activity sequence:

Get object or tool	Place object or tool	Use tool	Put aside object or tool	Return
A B P	A B P		A B P	A

where

A = action distance
B = body motion

G = gain control
P = place

The space in the sequence model—"use tool"—is provided for the insertion of one of the following Tool Use parameters. These parameters refer to the specifications of using the tool and are

F = fasten
L = loosen
C = cut
S = surface treat
M = measure
R = record
T = think

Tool Use Examples for "Fasten and Loosen"

1. Obtain a nut from a parts bin located within reach, place it on a bolt, and run it down with seven finger actions.

$$A_1\ B_0\ G_1\ A_1\ B_0 P_3 F_{10} A_0 B_0 P_0\ A_0 = 160\ \text{TMU}$$

2. Obtain a power wrench from within reach, run down four 3/8-in. (10-mm) bolts located 6 in. (15 cm) apart, and set aside wrench.

$$A_1\ B_0 G_1 A_0\ B_0 (P_3\ A_1 F_6)\ A_1\ B_0\ P_1\ A_0\ (4) = 440\ \text{TMU}$$

3. From a position in front of an engine lathe, obtain a large T-wrench located five steps away and loosen one bolt on a chuck on the engine lathe with both hands using five arm actions. Set aside the T-wrench from the machine (but within reach).

$$A_{10}\ B_0\ G_1\ A_{10}\ B_0\ P_3 L_{24} A_1\ B_0\ P_1\ A_0 = 500\ \text{TMU}$$

4.3.6 The MOST Systems Family

In addition to the BasicMOST system, several application-oriented versions of MOST are now members of the MOST systems family: MiniMOST, MaxiMOST, and ClericalMOST. A new version, MegaMOST, is under development for future applications.

The MiniMOST System

BasicMOST was not designed to measure short-cycle operations, although the original BasicMOST version can be applied to nonidentical operations of 10 sec. or less and still meet the accuracy criteria. Therefore, the MiniMOST version of MOST was developed to satisfy higher accuracy requirements that apply to very short-cycled, highly repetitive, identical operations. Such operations may be from 2 to 10 sec. long and often are being performed over long periods of time. MiniMOST consists of two sequence models:

General Move:	A B G A B P A
Controlled Move:	A B G M X I A

These sequence models are identical to the two basic sequence models in the BasicMOST version. There is one major difference, however. The multiplier for the index value total is 1 for MiniMOST. Therefore, if the sum of the applied index values is 64, this is also the total TMU value for the sequence model. Another difference compared to BasicMOST is that distances in MiniMOST are measured in inches. The application speed of MiniMOST is about 25:1 under ideal conditions, compared to about 10:1 for BasicMOST.

The definitions and descriptions of the parameters and elements in MiniMOST have been excluded because of space considerations. The second edition of *MOST Work Measurement Systems* (Zandin, 1990) includes a complete review of MiniMOST.

The MaxiMOST System

In order to satisfy the need for a fast, less detailed, but still accurate and consistent system for the measurement of long-cycle, nonrepetitive, nonidentical operations, MaxiMOST was developed. MaxiMOST consists of five sequence models with a multiplier of 100. The sequence models are:

Part Handling
Tool/Equipment Use
Machine Handling
Transport with Powered Crane
Transport with Wheeled Truck

MaxiMOST has a measurement factor of 3–5:1 (analyst hours to measured hours) and is therefore a very cost-effective technique to use in a large number of cases where minute details are unnecessary or even detrimental to proper work instructions. The recommendation is to use MaxiMOST for nonidentical cycles that are 2 min. or longer.

The definitions and descriptions of the parameters and elements in the MaxiMOST system have been excluded because of space considerations. The second edition of *MOST Work Measurement Systems* (Zandin, 1990) includes a complete explanation of MaxiMOST.

ClericalMOST

The ClericalMOST system is based on three sequence models identical to those in BasicMOST:

General Move
Controlled Move
Tool/Equipment Use (two data cards)

MegaMOST

The main purpose of adding a version of MOST on the 1000 multiplier level is to simplify and accelerate the standard setting for long (over 20 min.), nonrepetitive operations in areas such as assembly and maintenance.

While MiniMOST is totally generic and BasicMOST is about 60–80% generic, MaxiMOST is primarily tool-oriented, and MegaMOST is part- and operations-oriented. MegaMOST will be adopted for automated calculation of standards by the computer.

4.3.7 Procedures for Developing MOST Time Standards

A standard MOST Calculation Form should be used for all analysis work using BasicMOST. (Similar forms have been designed for use with MiniMOST and MaxiMOST.) As can be seen from the included example, Figure 4.2, this form consists of four sections:

1. A header identifying the activity to be measured and the work center (area) in which it is being performed
2. A step-by-step method description (left half)
3. Preprinted sequence models in three groups: General Move, Controlled Move, and Tool Use
4. A field for the time value or time standard for the activity (bottom part)

Note: The activity time or standard does not include any allowances at this stage. Prior to applying this time standard, the time value on the form should be multiplied by the appropriate allowance factor (PR&D), thereby constituting the standard time for the operation.

A frequency factor (Fr) for each sequence model can be specified in the column next to the TMU value column for the sequence model. Normally, the space provided on one page of the MOST Calculation Form will allow for analyses up to approximately 1 min.

In all situations where MOST is being used, the "top-down" approach should be followed. A two-step decision model can be put to use, with the first generation being "Is it appropriate and practical to do direct measurement?" If the answer is yes, the work should be measured by using the MOST Calculation Form. If the answer is no, the operations or activities for the plant, department, or work center should be broken down into logical suboperations. Each suboperation is then measured using MOST and placed on a worksheet for the calculation of time standards. In some instances (e.g., for long-cycle operations), suboperations may have to be broken down still one more level and later combined into suboperations before assigning them onto the worksheet.

By following the top-down approach, the database with standard data (suboperation data) will remain compact and more manageable than if the conventional "bottom-up" procedure is applied. In most cases, only 200 to 500 suboperations will be required to calculate any number of time standards.

MOST is an application-oriented or user-friendly system that will require some unlearning and rethinking by its users that are experienced in conventional work measurement. It is a new concept, not only regarding work measurement, but also in the application areas.

4.3.8 MOST Computer Systems

The logical sequence model approach lends itself very well to a computerized application. Therefore, in 1976 the first lines of code were written in an effort to develop a software program that would advance the state of the art of work measurement. While other computerized systems use element symbols or numerical data as input, MOST Computer Systems use method descriptions expressed in plain English. In other words, MOST Computer Systems are language-based systems. Today the computerized MOST program reminds one of an "expert system," although this term was not commonly known when the development started.

Computerized MOST Analysis

The input for a computer MOST analysis consists of (1) work area data and (2) a method description. Based on this information, the computer produces a MOST analysis as output; i.e., the computer actually completes the work measurement task automatically. A simple but representative work area layout sketch is also part of the output. A typical example of a work area description is shown in Figure 4.3, and the MOST analysis for an operation performed in that work area is shown in Figure 4.4.

In designing the program, the basic philosophy of establishing a time standard as a direct function of the work conditions was followed. The computer was therefore programmed to produce a time standard based on well-defined and complete user work conditions. The computer was also programmed not to allow the change of a time value without a change of the underlying work conditions. A change of distance, for instance, or "gain control" or "placement" of an object or a body motion, will result in a different standard. This discipline has proven to increase the uniformity and consistency of the method descriptions and analyses. Equally important is the fact that one does not have to read both the method description and the MOST index values to interpret an analysis. A review of the method is adequate. The index values and the time standard are by-products and direct functions of the method.

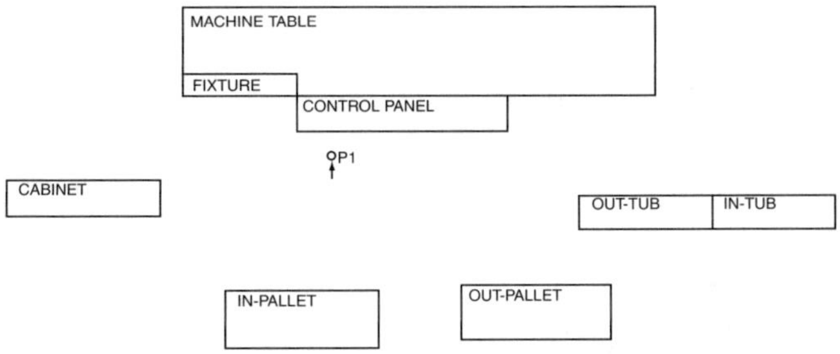

WorkArea Report

Name	Location	Body/Frag/PT
Workplaces :		
MACHINE-TABLE		
CONTROL-PANEL		
CABINET		
IN-PALLET		
OUT-PALLET		
OUT-TUB		
IN-TUB		
FIXTURE		
Tools :		
AIR-HOSE	MACHINE-TABLE	
BEAR-CLAW	MACHINE-TABLE	
PENCIL-GRINDER	MACHINE-TABLE	
BOX-END-WRENCH	FIXTURE	
Objects :		
RAW-PART	MACHINE-TABLE	
MOVE-TICKET	CABINET	
WORK-ORDER-PACKET	CABINET	
PART	IN-PALLET	FRAG
FIN-PARTS	OUT-PALLET	FRAG
Equipments :		
JIB-CRANE	MACHINE-TABLE	
LTD	MACHINE-TABLE	
PIN	MACHINE-TABLE	
UNIVERSAL-VISE	MACHINE-TABLE	
BUTTON	CONTROL-PANEL	
CLAMP	FIXTURE	
Manual Cranes :		
Power Cranes :		
Trucks :		
Operators :		
OPJ	MACHINE-TABLE a	
Carriers :		

WorkArea Report

From	To	Steps
MACHINE-TABLE	CONTROL-PANEL	1
MACHINE-TABLE	CABINET	2
MACHINE-TABLE	IN-PALLET	4
MACHINE-TABLE	OUT-PALLET	4
MACHINE-TABLE	OUT-TUB	1
MACHINE-TABLE	IN-TUB	1
MACHINE-TABLE	FIXTURE	0
CONTROL-PANEL	CABINET	4
CONTROL-PANEL	IN-PALLET	4
CONTROL-PANEL	OUT-PALLET	4
CONTROL-PANEL	OUT-TUB	3
CONTROL-PANEL	IN-TUB	2
CONTROL-PANEL	FIXTURE	2
CABINET	IN-PALLET	3
CABINET	OUT-PALLET	7
CABINET	OUT-TUB	5
CABINET	IN-TUB	10
CABINET	FIXTURE	2
IN-PALLET	OUT-PALLET	5
IN-PALLET	OUT-TUB	6
IN-PALLET	IN-TUB	8
IN-PALLET	FIXTURE	2
OUT-PALLET	OUT-TUB	3
OUT-PALLET	IN-TUB	4
OUT-PALLET	FIXTURE	2
OUT-TUB	IN-TUB	1
OUT-TUB	FIXTURE	4
IN-TUB	FIXTURE	4

FIGURE 4.3 Typical work area description as part of the work area data in ComputerMOST.

Sub-Operation - Method/TMU Report

Sub-Op ID:	547	Status:	Private
Description:	LOAD PART IN FIXTURE AT MULTI SPINDLE VERTICAL DRILL		
Activity:	LOAD/UNLOAD	Obj:	PART
Prod/Equip:	FIXTURE	Tool:	WRENCH
Size:		Orig:	MACHINE
WA No:		other:	
Unit of Measure:	PART	OFG:	2
Workarea ID:	9		
Starting Operator:	Opl		
Starting Location:	MACHINE-TABLE	Total Time:	770 TMU
Applicator:	MCS	Issue:	1
Create Date:	02/16/96	Effect Date:	02/16/96

step	Method Description		TMU
1	PLACE PART FROM IN-PALLET TO FIXTURE A6 B0 G1 A3 B0 P3 A0	1.00	130
2	PUSH CLAMPS AT FIXTURE A1 B0 G1 M1 X0 I0 A0	2.00	60
3	FASTEN 2 NUTS AT FIXTURE 5 SPINS USING FINGERS A1 B0 G1 A0 B0 (P1 A1 F10) A0 B0 P0 A0 (2)	1.00	260
4	FASTEN 2 NUTS AT FIXTURE 4 ARM-TURNS USING BOX-END-WRENCH AND ASIDE A1 B0 G1 A0 B0 (P3 A1 F10) A1 B0 P1 A0 (2)	1.00	320

FIGURE 4.4 ComputerMOST analysis for an operation performed in the Figure 4.3 work area.

How is it possible for the computer to generate a MOST analysis from the input of "only" work area data and a method description? How does the computer select the right sequence model and the correct index values? As explained above, and as can be seen from the example in Figure 4.3, all action distances and body motions are specified as part of the work area data. Therefore, the A and B parameters in the sequence models are assigned an index value from the work area information.

Three additional variables remain to be determined: (1) sequence model selection, (2) the index value for the G parameter, and (3) the index value for the P parameter. This required information has been compounded into one word: a keyword. This keyword, always found in the beginning of each method step, has been chosen from a list of commonly used English activity words such as MOVE, PLACE, and POSITION. For instance, the keyword PLACE means "the General Move sequence model" and a combination of G_1 and P_3 to the computer. MOVE indicates the same sequence model with a G_1 and P_1 combination and POSITION, a G_1 P_6 combination. A GET preceding MOVE, PLACE, and POSITION will render G_3 P_1; G_3 P_3, and G_3 P_6 respectively. (The keywords for General Move are indicated in the table in Figure 4.1.)

Similar keywords are available for all sequence models in MOST Computer Systems. The knowledge of approximately 30 to 50 keywords for BasicMOST will provide the analyst with a sufficient vocabulary to be able to perform most of the analysis work.

Since both the work area data and the method description are entered within a well- structured format, it is possible to dictate this information using a hand-held tape recorder. A person can, in most situations, talk as fast as or faster than an operator can perform an assembly or a machining operation. Therefore, data collection becomes much more efficient. The conventional handwriting of methods is usually cumbersome and inefficient. While the dictation of a method in principle requires the observation of just one cycle, the writing of the same method requires observation of several cycles. The information on the tape is then transcribed by the analyst or a typist on a computer terminal as input to the program. In the future, when a voice-recognition system becomes available for practical applications, this intermediate step can be eliminated. In fact, a TalkMOST system is presently under development that will let the analyst communicate directly with the computer. The input of information will be done by voice and the output either by voice (computer), screen, or printout.

Data Management

The major advantage of a computerized application of MOST lies in the databases (suboperations and standards). These are accumulated as a result of the MOST analysis work and calculation of standards: The filing, searching, retrieving, and updating of the data become extremely efficient and fast compared to a manual system. Some functions requiring manipulations of data, such as mass updating, simulations, and history of standards, are very impractical or impossible to execute manually, while the computer can perform them routinely and quickly.

A complete database system for filing and retrieving suboperations and time standards is the backbone of MOST Computer Systems. The database has, to date, been pushed to handle over 1 million standards.

The filing system for the database also uses the "word" concept. All suboperation data is filed and retrieved under well-defined words in five categories: activity, object/component, equipment/product, tool, and work area origin. The filing system for standards is in all cases customized to fit the user's requirements and includes such conventional header items as part number, operation number, work center number, etc.

The Complete MOST Computer System

MOST Computer Systems is a complete program for measuring work and calculating time standards as well as documenting and updating these standards. It consists of a basic program and a set of supplementary modules. The basic program includes the following features:

1. Work area generator
2. Work measurement (BasicMOST, MiniMOST, and MaxiMOST)
3. Suboperation database
4. Time standard generator
5. Standards database
6. Mass update
7. Documentation of work conditions (work management manual)
8. Auxiliary data
9. Data transfer (electronically to other systems)

Supplementary modules and application programs include (often these programs will be customized to meet the specific requirements of a user):

1. Machining data (feeds, speeds, and process times)
2. Welding data
3. Line balancing
4. Station assignment
5. Process planning
6. Cost estimating
7. AutoMOST
8. ErgoMOST

AutoMOST

AutoMOST is an expert system that can be integrated into MOST Computer Systems to simplify and accelerate several program functions. For instance, the calculation of standards can be either fully automated, by transferring part-related data from the bill of material (BOM) or other system, or semiautomated, in which case a standard is generated after one answers some simple questions regarding the part characteristics. Specific decision models are developed for each user reflecting its product and part configuration as well as other work conditions. The manual selection of suboperations from a worksheet can thus be eliminated. AutoMOST by itself or in combination with voice recognition will raise the level of system automation substantially.

ErgoMOST

The risk of injury in industry is at an unacceptable level in many situations. In order to identify specific problem areas, ErgoMOST has been developed to analyze the workplace design and the work method with regard to force, posture, and repetition, three of the most critical factors causing cumulative trauma disorders (CTDs). After dangerous motions have been identified and improvements have been made to the workplace, ErgoMOST can analyze the effect of the improvements.

The objective with MOST Computer Systems was to adapt the system to cover all possible aspects of establishing time standards in a wide range of situations. Another objective was to make the updating and maintenance of standards efficient and simple. Our intention was also to stimulate industrial engineers in industry, services, and universities and colleges to adopt a positive attitude toward a fundamental and widely used discipline: the measurement of work.

4.4 APPLICATION OF TIME STANDARDS

There are numerous reasons for applying work measurement. Often, however, a company or a manager has a specific purpose for investing in an advanced computerized work measurement system—for instance, productivity improvement or cost control. What everyone seems to have in common is that they want to improve the *accessibility to* and *quality of information,* whether it is related to planning, cost control, budgeting or productivity. *Estimates or historical data are simply not good enough in today's competitive industry!*

Scientific methods have to be applied to provide accurate and consistent information about shop floor operations prior to their performance. Let us briefly review some of the most important justifications for work measurement that managers have to consider.

4.4.1 Productivity Improvement

The first and foremost reason for implementing a work measurement program is (and should be) to improve productivity and/or reduce costs. Simply put, the job of any manufacturing engineer is to improve productivity.

The industrial engineering principles and methods of achieving productivity improvements or cost reductions are many; the industrial engineer has, however, no exclusivity on productivity improvements—everyone in the company should contribute.

A work measurement system becomes an important tool for the industrial engineer and others to accomplish the productivity improvement or cost reduction goals of the company. An overall productivity improvement (or cost reduction) in the workshops of 20–30% or more is often the result of a work measurement project in a U.S. company. The improvement potential is even greater, however, in areas such as industrial engineering, process planning, and cost estimating. It is not uncommon for output (or productivity) to double or triple as a result of the installation of an appropriate computerized work measurement system with the proper application functions attached.

4.4.2 Incentive Plans

The studies showed that about 40% of the companies using work measurement have adopted a wage incentive plan, which obviously requires realistic, accurate, and consistent time standards as a base. Another 40% have measured day-work systems, which need reliable backup standards as well. Without proper work measurement, such wage systems will be either out of control and/or inequitable and unfair to the workers who are financially dependent on them.

On the other hand, a well-designed incentive plan supported by high-quality standards and work instructions is probably the most powerful means and motivator to maintain a high level of productivity. Therefore, if incentives or measured day work are in place, proper work measurement is necessary.

4.4.3 Operations Management

It is almost inconceivable that some managers accept the responsibility for managing a factory operation without having access to accurate and consistent data to accomplish planning and scheduling, determine and evaluate performance, and establish and control costs. However, the majority of managers have learned to appreciate the benefits of work measurement and have committed themselves to the use of proper work standards as part of their management philosophy and system. This is primarily a question of understanding what modern work measurement is and what it can do for an organization.

4.4.4 Cost Estimating

In today's competitive marketplace, it is important not only to exercise control of the costs of producing a component or product, but also to be able to establish an *accurate* cost estimate *quickly*. Quotes and bids must reflect the *real* cost, so that the desired margin can be attained. If the bid is too high, the order may be lost; if it is too low, money may be lost.

Not only will an engineered work measurement system generate a low cost level through the methods engineering effort, it will also provide a realistic cost estimate for the manufacturing of a product or component. If a computerized work measurement system is in use, a cost estimate can be issued promptly. In addition, simulations of costs based on product modifications and variations can be made to adapt a quote rapidly.

The scope of work measurement applications is even broader. Process planning, scheduling, equipment justifications, budgeting, line balancing, manpower or staffing requirements, and product design are other focus areas for work measurement. Each of these areas can become a principal area of application for an individual company, because in order to satisfactorily perform one of these functions, realistic and consistent data (work standards) are a requirement.

In all the application areas indicated above, the use of expert system technology has proven to be of great benefit to the user. Through the development of customized decision trees for the particular application area, the user can obtain the desired output just by answering a few simple questions pertaining to the product or component being processed. A knowledge engineer identifies and collects the best method from domain experts and subsequently develops the necessary tree or trees. This way, expert knowledge is captured, distributed, and preserved for everyone in the company to use. Employees can become experts with a minimum of training by applying the system. Also, the consistency and quality of data improves substantially.

Because of the availability of expert systems and other advanced computer technology, work measurement has become much more accessible both to the user and management. The conclusion is that companies have many good reasons for using work measurement. The fundamental and prevalent reason is, however, *to satisfy the desire and need to know and to predict with confidence.*

4.5 SUMMARY AND FUTURE TRENDS

During the 1950s and 1960s, the work measurement field became inflated with "conventional" derivatives of the original MTM system (MTM-1). That trend has continued to some extent, with one exception: MOST. In the mid-1960s, we believed that a new approach, a more practical and user-friendly method—and, perhaps more important, a faster and simpler technique—was necessary to maintain a reasonably high level of interest in work measurement. MOST seems to have been the answer. Over 20,000 persons representing more than 4,000 organizations have become certified MOST users. MOST has been translated into at least 15 languages and is in use in more than 35 countries around the world. MOST satisfies all the criteria of simplicity, speed, accuracy, consistency, applicability, integrity, and universality that can be put on a modern work measurement technique and system. MOST Computer Systems represents the state of the art in the areas of work measurement and time standards. And users enthusiastically endorse and support MOST.

Traditionally, the manufacturing industry has been and still is the most active user of work measurement. The main reason for this is the emphasis on productivity improvements and incentive systems. Billions of dollars have been saved as a result of the appropriate application of work measurement.

A renaissance in work measurement has been noticeable during the past few years in the defense industry because of a military standard (MIL-STD-1567A) issued by the Department of Defense in 1983. Since then, defense contractors have been obligated to comply with this standard on major contracts. Compliance includes fully documented time standards (80% coverage) based on a recognized predetermined motion-time system. MOST systems have been used successfully by a large number of defense contractors to satisfy the requirements of MIL-STD-1567A.

Also, service industries have shown an increased interest in work standards for staffing, personnel planning, budgeting, and the like.

Despite the efforts by industry to increasingly mechanize manufacturing operations, the measurement of work done by people is here to stay for many more years. The advantages of knowing and being able to plan from realistic and consistent standards are just too great to dismiss. A good example is the well-known NUMMI project, a successful joint venture between General Motors and Toyota. A great deal of its success has been attributed to the back-to-basics standardization that was a cornerstone of the project and was documented in an article by Paul S. Adler (1993).

However, the work measurement and standard-setting disciplines have to become simpler, faster, and more integrated with other functions to attract the attention they deserve.

MOST systems and MOST Computer Systems have met these criteria, but more can be done and more will be done. Today's computer technology has reached a level that cannot be ignored by work measurement specialists. If they take advantage of this technology, time standards can and should become a logical and integral part of any business system, as is the case in many companies already.

The general trend in industry is automation. Therefore, it is obvious that we will see fully automated procedures (using AutoMOST) for calculating and updating time standards based on data developed and maintained by industrial engineers. A direct link to a computer-aided design or material resources planning system with the purpose of producing process plans and cost estimates based on these standards will likely become a reality within the next few years. Finally, it will be necessary to simplify presently used work measurement systems, including MOST, to such a degree that everyone in a company can learn how to use them and benefit from them in their ongoing effort to make continuous improvements.

Work measurement: the science that brings more and better knowledge to people about work and how to improve work.

4.6 BIBLIOGRAPHY

Adler, P. S., Time-and-motion regained, *Harvard Business Review,* 1993.

Cleland, D. I. and Bidanda, B., *The Automated Factory Handbook,* TAB Professional and Reference Books, Blue Ridge Summit, Pa., 1990.

Karger, D. W. and Bayha, F. H., *Engineered Work Measurement,* 4th ed., Industrial Press, New York, 1987.

Maynard's Industrial Engineering Handbook, 4th ed., McGraw-Hill, New York, 1992.

Niebel, B. W., *Motion and Time Study,* 9th ed., Richard D. Irwin, Burr Ridge, Ill., Boston, Mass., and Sydney, Australia, 1993.

Zandin, K. B., *MOST Work Measurement Systems,* Marcel Dekker, New York, 1980.

Zandin, K. B., *MOST Work Measurement Systems: BasicMOST, MiniMOST, MaxiMOST,* 2nd ed., Marcel Dekker, New York, 1990.

5 Control of Production and Materials

Stephen C. Cimorelli

with

Greg Chandler

5.0 INTRODUCTION TO CONTROL OF PRODUCTION AND MATERIALS

This chapter deals with the control of production and material. It is not intended to provide exhaustive coverage of the subject; rather, it seeks to impart a working knowledge of the major elements of the field to manufacturing engineers who must apply them in actual practice, and who must work side by side with production and material control and other professionals.

A sizable number of systems and techniques have been devised to aid those responsible for production and material control. This chapter introduces and discusses the more widely accepted of these. The authors believe that by focusing on the approaches that readers are likely to encounter, and by exploring their applications rather than specific technical details, readers will be better prepared to recognize and apply them to their own situations.

In order to understand and appreciate the importance of these controls, some understanding of the business context in which they are applied must first be gained. Beginning with Section 5.0.1, this introduction seeks to provide that context.

Two of the most prominent organizations dedicated to this chapter's subjects are the American Production and Inventory Control Society (APICS) and the Oliver Wight Companies. Many of the concepts and ideas presented in this chapter were developed by these organizations, both of which offer a wide range of literature and educational programs for those interested in developing a deeper understanding and appreciation of this field. Specific references are found in the bibliography. The remainder of Subchapter 5.0 in particular is based largely on APICS Systems and Technologies literature.

5.0.1 Corporate Strategy

Manufacturing companies, like all companies, exist for the purpose of making money. They do this by designing and producing products that satisfy the wants and needs of their customers in a way that is superior in some form to that of the company's competitors. These competitive advantages are spelled out in the company's marketing strategy, which in turn supports the overall corporate strategy, mission, and objectives. Likewise, a manufacturing strategy is needed to enable the company to achieve these higher-level goals. This is not a one-way street, however. Manufacturing excellence in certain areas can be leveraged to create or gain competitive advantage. See Figure 5.1 for the relationships of the strategies and objectives.

A company's mission statement sets the overall focus for the firm. It defines what business the firm is in and provides a unifying thread tying all products together. For example, a mission statement that says, "We are in the light-rail business" might limit production to small, lightweight trains, while a company "in the mass transportation business" might have a product line as diverse as trains, buses, subways, airplanes, and many other types of vehicles.

Within this focus, objectives for running the business must be set to ensure that the goal of the business—making money—is relentlessly pursued. Typical objectives are (1) financial, related to such things as profit, cash flow, and return on investment; and (2) marketing, related to market share, sales targets, or penetration of new markets. Such objectives may be felt in significant ways in the manufacturing organization as they trigger supporting goals and drive, or limit, new investment in manufacturing systems and technologies.

Corporate strategy defines what resources will be deployed and what actions will be taken to accomplish corporate objectives, ultimately allowing the company to accomplish its mission.

Marketing strategy deals with three primary subjects: market niches, market share, and competitive advantage. Of the three, the most important to the manufacturing organization is competitive advantage—the ability or characteristic of the

FIGURE 5.1 Manufacturing strategy must be part of the overall company mission, objectives, and strategy.

firm that makes customers choose it over other firms offering similar products. This is not to say that market niches and market share are unimportant. It means simply that manufacturing strengths play a large role in providing a competitive advantage in a particular market niche, thereby increasing market share. Typical areas where manufacturing companies seek to gain competitive advantage include price, quality, delivery speed, delivery reliability, flexibility, product design, and after-market service. It is not necessary for a company to gain competitive advantage in all areas, but at least one area must be demonstrably better than the competition for the company to stand out. In order to identify which of these a company should compete upon, they should be evaluated in terms of "order winners" and "qualifiers." *Qualifiers* are those characteristics needed just to play the game, while *order winners* are those characteristics that make customers choose one company over all others.

5.0.2 Manufacturing Concepts

Before moving on to manufacturing strategy, a few additional concepts and definitions bear reviewing: (1) general manufacturing environments, (2) product volume and variety, and (3) product life cycles.

Manufacturing Environments

Manufacturing environments may be grouped into three general categories—job shop, repetitive, and continuous flow. In a *job shop,* products are typically made one at a time or in very small quantities. Parts are routed between machines and work centers that have been organized by the type of work performed, such as sawing, milling, turning, etc. This arrangement offers a high degree of flexibility in that a great variety of products may be produced, but it sacrifices speed and efficiency. Many manufacturers choose to set up certain departments, such as machine shops, around a job shop concept, and seek to improve efficiency by producing parts in batches. This batching allows the time required to set up a machine or work center to be spread across multiple parts, rather than being charged to a single part. A "repetitive manufacturer," by contrast, organizes the plant around product lines, allowing parts to flow through predefined processes. While this arrangement is faster and more efficient in producing selected products, it is less flexible than a job shop. The final category is the "continuous flow" manufacturer. An example of continuous production is an oil refinery, where continuous production of multiple products (gasoline, motor oil, etc.) flows from a primary raw material—crude oil. Regardless of the environment, manufacturing engineers are instrumental in the design and layout of the manufacturing processes.

Product Volume and Variety

Product volumes reflect the quantities of products planned to be produced, usually expressed in terms of product families. The concept of product families and its usefulness to manufacturing professionals, especially those engaged in production and inventory management, will be discussed in greater detail in Section 5.2.2. Product variety deals with the number of end products that the firm produces, or is capable of producing. A general relationship between volume and variety is represented in Figure 5.2, where product variety decreases as volume increases.

To better understand this relationship, consider some parallels between the volume/variety matrix and manufacturing environments. In position 1, exhibiting high variety and low volume, a job shop environment would seem preferable. Continuing down the volume/variety positions, one would likely encounter batch production at position 2, repetitive production at position 3, and continuous production at position 4, where variety is very low but volume is high.

Product Life Cycle

The third concept is product life cycle. Figure 5.3 depicts this cycle. In the introduction phase, when a new product is being test marketed, variety may have to change rapidly to adjust to real versus anticipated customer demands. As a result, low volumes of each product configuration are typical. As we enter the growth phase, design stabilizes, resulting in somewhat lower variety and higher volumes. In the maturity phase, assuming the product is successful and enjoys good customer acceptance, volumes are typically at their peak, with variety being very low. Finally, a product moves into a phase of decline in which customers begin to seek alternatives to the product. In this phase, a company may add a myriad of features and options to satisfy these desires, thereby increasing variety again but reducing the volume of any given model.

Again, some parallels to manufacturing environments may be drawn. The introduction phase often requires the flexibility of a job shop, followed by a move toward batch production in the growth phase, then repetitive production

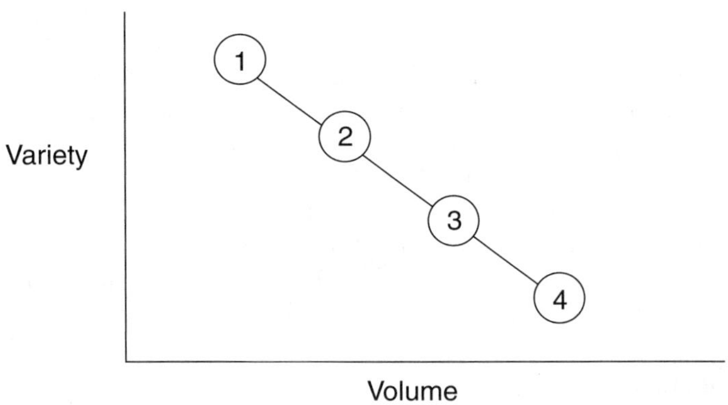

FIGURE 5.2 Product/variety relationship chart. (APICS Systems and Technologies Certification Review Course, American Production and Inventory Control Society, Inc., Falls Church, Va., 1987. With permission.)

during product maturity. Finally, a return to batch production may be required in the decline phase to satisfy the increased variety demanded by customers. While these parallels aid in our understanding of the various concepts, this is not to imply that products move naturally through all three environments. In fact, few product lines compete in more than two. Notice also that the continuous flow environment was never reached. Rarely will this occur, since continuous flow is a special case where production, once begun, proceeds at a constant preset rate, and is either on or off.

5.0.3 Manufacturing Strategy

Manufacturing strategy may be defined as a long-range plan to deploy manufacturing resources in such as way as to support the corporate strategy. In other words, it defines what actions will be taken, and what resources will be deployed to achieve manufacturing's objectives. For example, if the company has defined "delivery speed" as a competitive advantage, manufacturing decisions to lay out machines and work centers in a repetitive process line would be appropriate, while a decision to employ a job shop layout would not.

Such decisions may be categorized as "structural" or "infrastructural." *Structural* or "hard" decisions relate to "brick and mortar," equipment, and technology decisions. Structural decisions tend to support the "resource deployment" element of strategy. Since these decisions typically require significant expense, they are made infrequently, usually by upper management. Manufacturing engineers are often deeply involved in providing technical expertise to these decision makers. *Infrastructural* or "soft" decisions, on the other hand, relate to people, organizations, and systems. These decisions support the "actions" element of manufacturing strategy and are made by middle managers on a more frequent basis.

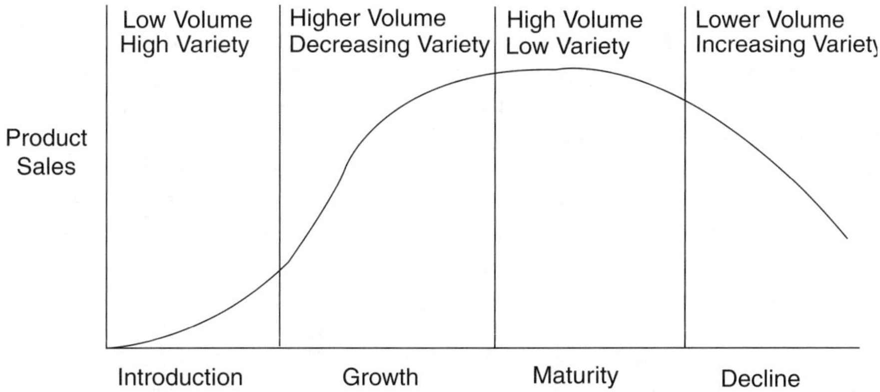

FIGURE 5.3 Product life cycle chart showing the influence on manufacturing requirements. (APICS Systems and Technologies Certification Review Course, American Production and Inventory Control Society, Inc., Falls Church, Va., 1987. With permission.)

A second element of manufacturing strategy deals with capability and the "timing of capacity changes." These capacity strategies are defined as lead, lag, and tracking. In a *lead strategy,* capacity is added before it is needed, enabling a firm to take advantage of a new market opportunity, or to build up inventory prior to a cyclic upturn in demand. This approach is somewhat risky, since anticipated demand may not materialize, leaving the firm with an underutilized asset and a negative effect on the bottom line. By contrast, the *lag strategy* adds capacity only after an increase in demand has been demonstrated. This approach is very conservative and enables a firm to use existing capability to its fullest, but may make entering a new market, or recovering from unanticipated capacity problems, very difficult if not impossible. Finally, the *tracking strategy* is a compromise between the other two. In this approach, the firm attempts to keep market demand and capability closely matched by adding or removing capacity on a much more frequent basis through actions such as overtime, outsourcing, etc. Here again, manufacturing engineers are often key players in developing and executing these capacity strategies.

The final element of manufacturing strategy is "plant focus." A plant may be focused on one of three things: product, process, or order winners. *Product focus* implies that a plant's production is limited to a given product or product line. For example, a plant may produce only one printer, or a small family of similar printers. *Process focus* limits the plant to a process, or a closely related set of processes. Typical examples might be a lumber mill or foundry. Many different products may be produced, but they are all produced from common processes. Finally, there is a focus on *order winners.* In this case, manufacturing resources are organized in a manner that allows a given marketing strategy to be carried out. If delivery reliability is identified as the order winner, our printer manufacturer may choose to carry a high level of finished goods inventory, or may carry excess capacity to ensure that no bottlenecks in production are encountered.

5.0.4 Summary

A company's manufacturing strategy—decisions and actions affecting the deployment of resources—must support its corporate and marketing strategies. The manufacturing strategy may provide competitive advantage in a given market niche, and must consider issues of manufacturing environment (job shop, repetitive, continuous flow), product volume and variety, and product life cycles. Specific issues of manufacturing strategy include structural decisions (plants, equipment), infrastructural decisions (people, organizations, systems), capacity timing issues (lead, lag, tracking), and plant focus (product, process, order winners).

As discussed earlier, this introduction seeks to provide a context for the remainder of the chapter—control of production and materials. It is in no way a comprehensive discussion of the issues presented. Further study of the materials referenced in the bibliography, especially APICS Systems and Technologies Certification Review Course (1987), is recommended for the reader seeking to understand these subjects in greater detail.

5.1 CAPACITY PLANNING AND SIMULATION

5.1.1 Introduction to Capacity Planning and Simulation

The manufacturing sector has spent a great deal of time and money trying to define, calculate, measure, and control "capacity." Many companies have tried to clarify its meaning by prefacing it with adjectives such as rated, nominal, maximum, dedicated, demonstrated, and so on. All of these prefixes attempt to capsulize and connote to the user what that particular company means by the word "capacity." The seventh edition of the *APICS Dictionary* (1992) defines capacity as:

1. The capability of a system to perform its expected function.
2. The capability of a worker, machine, work center, plant, or organization to produce output per time period. Capacity required represents the system capability needed to make a given product mix (assuming technology, product specification, etc.).

Implicit within the definition is the fact that capacity is the *planned* amount that *could* be produced during a given time period. The prefixes presented earlier attempt to narrow the plan down to something less than that presented in the definition. Inherent within any definition of capacity is the need to capture the quantity produced over time.

When a company's products are in high demand, the importance of precision in the calculation of capacity is greatly reduced. The function of capacity calculation is merely ensuring that there is sufficient capacity available. The only way to go wrong is to provide for too little capacity. This situation is further eased if the cost of money is low. The accuracy required, and the difficulty involved in calculation, increases with the increase in competition for the same market.

The 1990s have seen a tremendous increase in the global marketplace, with competition stiffening and more companies, including the blue-chip names, trimming their staffs and moving toward lean, agile manufacturing. The new emphasis is on companies being more flexible, responsive, and proactive to the marketplace and customers' needs. In light of these changes, companies need to have people and systems that behave in a similar fashion.

Enter simulation. Even though it has been around, in one form or another, for over 50 years, it is having a resurgence in popularity due to its flexible nature (and also recent improvements in computer software and hardware). Flexible businesses modify their operational parameters frequently and need ways to predict, understand, and control those changes quickly and with varying degrees of precision. The question of the plant capacity to produce a new product can be an important one. Simulation in its variety certainly fits the bill. This subchapter will attempt to show how simulation can be used to help predict and help control capacity.

5.1.2 Evolution of Capacity Calculation Methods

Since there are as many different definitions of capacity as there are companies defining it, there have been numerous techniques and tools for calculating and predicting it. Some of these methods have proven themselves more useful than others. Different industries seem to have a preference for a given technique or a type of calculation tool.

Initially, a company may choose to predict and/or control capacity by the use of the "intuitive" method. The company will rely on the experience-based, educated guess of a knowledgeable employee (usually in management). This technique is fraught with prejudice and bias, inaccuracies, and guesses not founded in factual data or calculations. Sometimes the company may not realize that the method they are using is the intuitive method. The capacity calculator will get the answer set in his or her mind, discuss his or her ideas with others in management, incorporate their thoughts and insight into his or her position, and fabricate a plan to justify this position. This will usually result in a "buy-in" from management, since they will see some of their thoughts come back to them from the capacity expert.

Another, more sophisticated approach to capacity calculation is the use of some simplistic mathematical techniques. These mathematical techniques try to eliminate the problems associated with the intuitive guesses while incorporating some real system data. By adopting these methods, the company realizes the shortcomings of intuitive guessing and attempts to remove some of these biases. This subchapter will not go into any of the particular techniques but instead list the types of techniques that could fall into this category. These types of techniques are characterized by the use of simple algebraic methods that incorporate time available, time required, scrap factors, equipment/personnel utilization, efficiency, downtime, and so on. An example of such a formula is shown below:

$$\text{Capacity} = \frac{\text{M1}_{\text{avail}} + \text{M2}_{\text{avail}} + \text{M3}_{\text{avail}} + \text{M}x_{\text{avail}}}{(\text{time required on each machine})}$$

where

M1_{avail} = the time machine 1 is available for use, etc.

The calculations tend to be very simple in nature and limited in scope, since they can become lengthy quite quickly. Spreadsheet programs often help in this class of calculations, since they handle large sets of calculations quickly. The formula above can be used to incorporate all items passing across each resource within a department to accumulate the aggregate capacity as well as for individual capacity calculations.

The next step in the evolution of capacity calculations is usually the use of complex mathematical techniques and/or computers. These two ideas are related because they tend to happen together, yet they are distinct since they are unrelated in their view of capacity calculations. They need not coexist within the same organization and often do not.

From the outgrowth of simple mathematical techniques comes the recognition that the capacity calculations need to include more of the overall operations. The company will realize that the interrelations *between* departments play a dramatic role in overall plant capacity. What good is it to have a single department that can produce at a rate of twice any other, since the items flowing to the customer are limited by the slowest flow path? Calculating each individual department's capacity, while useful in a microanalysis, adds nothing to the understanding of plant-level capacity.

This need to include more information within the capacity calculations does not in itself mean that the company must use complex mathematical techniques, but could demand a more effective method of calculation itself—hence the adoption of a computer. The techniques will still be the same as before, simple algebraic relationships, but their magnitude begs for the use of the computer to handle the significantly larger quantity of formulas and relationships. After a few times of calculating these by hand, the company may resort to the intuitive method if it does not migrate to the computer. The computer provides the ideal tool for the organization, calculation, recalculation, and presentation of the larger set of data.

The use of complex mathematical techniques is usually closely related to two things: the adoption of a computer for capacity calculations, or the recognition of the limitations of the simple mathematical techniques. The complex techniques include those types of methods using calculus, statistics, and queuing theory. There are several software programs available that use queuing theory formulas such as the following:

$$L = \text{avg. \# of units in system} = \lambda / (\mu - \lambda)$$
$$W_q = \text{avg. time a unit waits in queue} = \lambda / \mu\, (\mu - \lambda)$$
$$L_q = \text{avg. time unit is in queue} = \lambda^2 / \mu(\mu - \lambda)$$

where

λ = mean number of arrivals per time period

μ = mean number of items served per time period

These techniques offer a vastly improved interdepartmental relationship representation, but themselves offer new and often restrictive limitations. For example, these formulas deal with the system at a "steady state." Have you ever seen your factory in a steady state? These techniques, because of their complex nature, require much more data gathering, emphasis on limiting assumptions, and difficulty in presentation of results. They tend to be less trusted by management, who may not be fully aware of their specific mathematics and calculation techniques. As a result, the company may go back to using the computer and simple mathematical techniques.

The epitome of today's capacity calculation techniques is *simulation*. Its ease of use and flexibility give it the inherent ability to be used for a wide range of capacity calculation situations, from simple to complex. It can be used to formulate a simple model, the creation of which clarifies interrelationships, or to model complex systems beyond the ability of all but the most complex of mathematical computations. The software itself can range from simplistic spreadsheet-like packages to more complex programming languages.

5.1.3 Simulation Basics

This section will define simulation as a time-based, statistical data-driven program that tracks the behavior of the system while entities flow through the system and modify or get modified by it.

That definition implies that only in this type of capacity calculation technique is the activity of an object flowing through the system tracked over time. All the other techniques merely guess or estimate what happens over time by the use of assumptions or averages. It is this ability to track changes over time in conjunction with the use of statistical distributions instead of averages that allows the simulation to be run through several iterations, each one selecting different values from the distributions. This approach allows the model to behave much in the same way as the real world; i.e., each unit flowing through takes a slightly different time to have an operation performed upon it.

Simulation Model Review

To illustrate how simulation software works, lets us use a simple machine shop system of a single machine and a single waiting line (queue) and go through a cognitive example. The flow information is that workpieces (entities) arrive every so often. If the machine is idle, they get right on and their machining begins immediately. If an entity is already on the machine, the parts wait in line for their turn.

Someone has collected data and finds that the time between entity arrivals (called the "interarrival time") at the machine follows a normal distribution with a mean of 6 min. and standard deviation of 1.5 min. Also, the data collected show that the time to process the entity on the machine can be described as a normal distribution with a mean of 5.1 min. and a standard deviation of 2.6 min. From these data one might think that since the machining time is less than the interarrival time, there would not be the need for a large waiting line. However, since the standard deviation, hence variability, for the machining time exceeds that of the interarrival time, there will be occasions where entities (parts flowing through the shop) will pile up in the waiting line.

We will assign a random number stream to both of the distributions. These random numbers allow the model to have the variable behavior similar to the real world. Then, we set the initial conditions for the system to begin with the queue empty and the machine idle. When the simulation run begins, the computer will select a random number between 0 and 1. It will match this number against the arrival distribution, which yields the simulation clock time when the first entity arrives into the system. It will do this for each random number in the arrival stream until the time the next entity due to arrive exceeds the run-time limit we have set. It will store each of these clock times and release an entity into the shop according to this file. The computer will then repeat this for the machine time. When it selects the processing time, it stores it as a duration of time an entity spends on that machine after it gets through with the entity in front of it. Remember, this is all done by the computer, so it happens very quickly!

Now the computer has stored two sets of clock times. One is a simulated time when the entities will arrive into the system, and the other is the duration of time the entity will spend being processed on the machine. I called the interarrival time a "simulated time," by which I mean that the time in the computer model is not linked to the real passage of time. This allows the computer to simulate the passage of time without having to wait for it to happen. It has the effect of compressing time and can simulate years of operation in a matter of seconds. In the following description I will refer to the "clock time," which means the computer's simulated clock time.

Now the computer begins the model tracking. It looks at the first interarrival time and jumps the clock ahead to that time (after all, no sense waiting around—nothing happens until the first entity arrives). Since the first entity finds an idle machine, it hops on and begins to have its machining performed. Meanwhile, the computer model recorded that the time entity 1 spent in the queue was 0. The program then sets the machine to busy, looks at the processing time file, and adds that to the current clock time and stores it. It will then compare that "finish machining" time to the next arrival time to see what happens next, an arrival or a machining finish. It jumps ahead to the next event (we will say it is an arrival time) and updates the system statistics it is tracking, such as:

The average time an entity spends in queue
The maximum time an entity spends in the queue
The total time an entity spends in the system
The maximum queue length
Machine utilization
Maximum and minimum throughput for a given time period (8-hr. shift)
Average throughput

The model repeats these steps until the end of the simulation time (or, alternatively, until a certain number of entities have been machined). It then has tracked one full simulation run. The time it takes to run this model depends on the hardware, software, and model written. For most leading simulation software programs, this simple model can run a year's worth of time (2000 hr.) in under 10 sec.

Even with this simple model, one can experiment to see how a change affects the outcome of the model. If the mean interarrival time increases/decreases or the processing time increases/decreases, what happens to the system measures? At what interarrival time would we have to add a second machine to keep machine utilization between 80% and 90%? We can quickly modify this system, run it for 10 sec. or so, and analyze the output to determine the answers. Remember, though, that the computer model is selecting times based on random numbers. They will be "fast times" or "slow times," depending on how the dice roll! Because of this, it is advisable to run several iterations of the same conditions to approximate the real system.

Steps of a Simulation Study

The general outline shown in Figure 5.4 describes the steps involved in a simulation study. The process is a sound process to investigate the operation and interrelationships of the manufacturing system even if a simulation model is never built. Each of

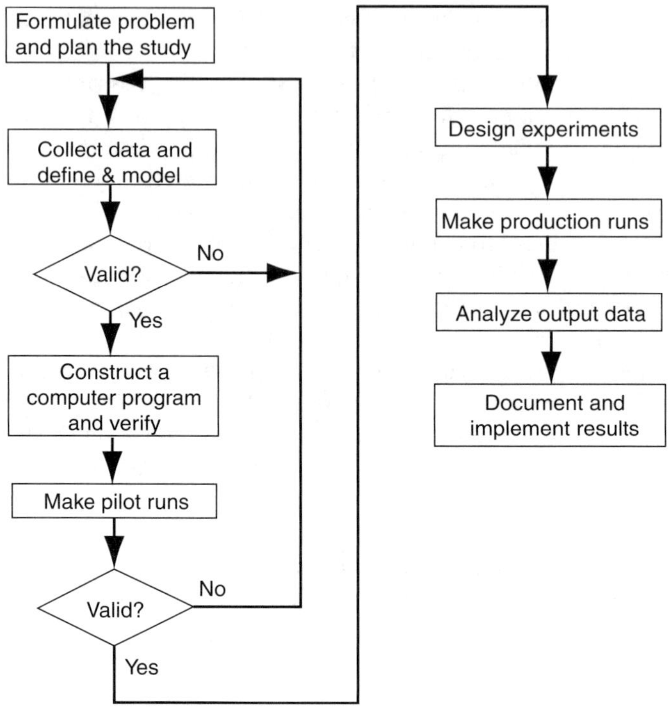

FIGURE 5.4 Iterative steps of a simulation study. (Courtesy of Industrial Engineering and Management Press. With permission.)

the steps will vary in the time needed to perform them based on experience, degree of detail needed to answer the issues the model addresses, and the number of "what ifs" modeled. The largest amount of time is going to be spent during the formulation and data-collection steps, but attention should also be focused on the validation, verification, production runs, and output analysis steps.

The first foundation step in simulation studies is the formulation/planning step. This step is where the objectives of the study are documented. It includes the criteria for determining if one manufacturing system design is better than another. Some of these criteria may include throughput, time of an entity in the system, time of an entity in the queue, time a machine is blocked (unable to work on the next entity until the current entity is removed), time a machine or cell is starved (time spent waiting for an entity), or the size of in-process inventories. This step should also include a schedule (time allotment) and labor projection for the study. This step is not unique to simulation studies. All thorough studies should include some degree of this basic step to limit more detail given than needed, so that more time can be devoted to the unique simulation elements.

The next step is data collection and model definition. The data can be collected on the existing system or estimated for a new system. The data will need to be compiled into statistical distributions for use in the simulation packages. This use of distributions, along with time-tracking elements, is the distinguishing feature of simulation and a key element in the study. Many texts and software programs exist to help one turn raw data into statistical distributions. Several of the simulation software systems allow the use of discrete data as a distribution, as well as the standard distributions, such as the normal, exponential, gamma, and others. Most of the simulation packages will use the time between arrivals as a distribution instead of the number of arrivals per time period.

Next comes the validation step. This frequently overlooked step is an important element in the study, since it is the first chance everyone gets to review the operation of the model. This step allows the people who are working in the actual shop operations to see if the data and model definition will accurately reflect the behavior of the system. If it is accurate enough for the questions defined in the initial step, and credible, management will believe the recommendations that come out of the model building/running process. This step is important because it keeps the people in the system and decision makers involved and understanding the progress and principles of the model. At this stage of the game it is much easier to comprehend the model/data than "walking them through" the simulation code later on. If they have understanding and confidence now, the faith in the construction of the computer model will be easier to establish.

The next step is the creation of the computer simulation model. This step varies in the requirements of the software and time. The important item in this step is that the model mimic the conceptual model in the step above. During this phase, the analyst must decide which kind of simulation system to use (i.e., a general language such as Fortran, a simulation language such as SIMAN or SLAM, or a simulator such as Witness), and if animation of the model is desired. Section 5.1.5 provides more detail about model types and the use of graphical animation in a simulation model.

After the model is created, one will need to perform pilot runs. Notice that "runs" is plural. Since there are statistical distributions for time to perform tasks in the model, the computer will sample from them, sometimes with a fast processing time and sometimes with a slow time. It is this feature, coupled with the need to debug the model, that makes it necessary to run the model several times to see the long-run average operation of the system. These pilot runs need to be validated again, just as the conceptual model was. Tied closely to validation is verification. Verification is the process by which the model is checked to see if it is operating as designed. Both of these elements, validation and verification, help to instill credibility in the model, its output, and any subsequent recommendations.

Next, experiments must be designed for the model. These experiments and their production runs provide the output from which analysis leads to recommendations. The design of these experiments will include some of the following:

The length of time to run the model
The number of runs to perform for each alternative

The initial conditions of the system when the model begins
The point in time when system statistics begin to be collected
What system statistics are to be monitored
Whether the animation will be watched, or the output reports simply reviewed

As with anything, the next task is to analyze the output. This task can be eased by the use of secondary software programs (spreadsheets, statistical analysis/graphing, etc.). Do not shortchange this step, because many times it will lead to more alternative system designs and experiments. Sometimes the analysis can point out subtle bugs in the model that must be corrected before the output is valid to be used. If the system exists in real life, the data collected for model input and the people working in the system can help validate the output.

One last note that should not be overlooked: No matter how accurately the model was created, it will not behave exactly the same way as the real system. For once changes are made to the real system, the random nature of the real world will cause different outcomes. The model approximates the behavior of the real world. Hopefully, this approximation is close to the long-run outcome of the real system in action.

5.1.4 Simulation Model Types

Now that the operation of a simulation program and the steps in a study have been explained, a discussion of the different types of software will be addressed. Since simulation software packages are used to mimic a wide variety of operations, the software packages are themselves diverse. They can be grouped into three categories: continuous, discrete event, and combined. There are other subcategories, such as general-purpose languages, specific-application languages, and simulators. Each of these will be covered in the following paragraphs.

Simulation software for the modeling of continuous operations, such as food processing, chemical plating, etc., involves the use of differential equations as opposed to "static" kinds of calculations. Most of the software manufacturers make products that have the ability to model both continuous and discrete event systems.

Discrete event simulation software can be used to model the vast majority of the manufacturing operations in use today. These systems are characterized by the entities flowing through the system having definite time spans at different operations. The entity will change its state at fixed times (i.e., welding, painting, testing, etc.), as opposed to electroplating, which happens over time. Within this type of software, the modeler defines the steps involved inside the system being modeled and uses the language to perform those operations.

General-purpose languages allow the modeler the freedom to construct specific program code to mimic the system being studied. This class of software is extremely flexible in its ability to model different types of operations and systems. The price paid for this flexibility is the need for skill in the modeler's ability. Since there are no preset routines as there are in the simulators, all the needs (whether general or specific) must be met by the modeler.

Specific-application languages are losing ground in the simulation software field. This class is categorized by its ability to handle specific model items easily, i.e., automated guided vehicles (AGVs), conveyors, or automated storage and retrieval systems (AS/RSs). These systems are being phased out as more of the general languages are adding commands that handle a majority of the common data collection/input requirements of the specific packages.

Simulators as a class represent a vast array of packages. They are characterized by their ease of use, usually through the use of pull-down menus and answering questions generated by the program. They help the modeler build a genetic model by assuming basic information is needed, such as interarrival time, operations times, and the use of common statistical distributions. They will also have preprogrammed routines for some of the specific items as mentioned under specific-application languages. The shortcoming of this class of software occurs when your model has special needs, since rarely will one be able to get into the simulation code to program that function. Some allow the ability to use an outside language such as Fortran to write specific code, but this requires the modeler to add code for the tracking of the statistical data usually handled by the software package.

The line between simulators and general-purpose languages is becoming blurry since the general language programs are trying to incorporate the helpful features of the simulators. Also, the simulators are improving their ability to handle special situations within their simulation language and not make the modeler drop out to another language.

Each of the classes has its pros and cons. It is up to the modeler and the team to analyze each of the packages relative to their needs, resources, and skills. It may well be that more than one package can be used to address different operations, departments, or goals of the model. For example, one could use a rough-cut type of simulator to identify the bottleneck operation, machine, or department and analyze its sensitivity to changes in certain variables, and then use a general-purpose language for detailed analysis of that specific area. No two modelers, models, or studies will be exactly the same. Each needs to be treated as an individual, but usually there are some common elements that should be shared.

5.1.5 Hardware Requirements

Simulation programs are offered in many different styles, languages, approaches, and cost brackets. Likewise, the hardware required to run a simulation program spans the spectrum from minimal personal computers through the powerful PCs and minicomputers (or workstations) and up to mainframes.

As discussed in the previous section, simulation software varies in its approach to modeling, complexity of models handled, flexibility, and ease of use. Similarly, the hardware used must be analyzed in accordance with the intended user of the model, accessibility for the user and modeler, and the availability of secondary analysis software.

The first step in selecting the hardware may have been performed when the software was chosen. If, for example, the software runs only on a particular hardware platform or requires an operating system that can run on only a few platforms, the final choice of hardware is somewhat limited. This section will not deal with particular brand names, but rather with classes of machines and operating systems. The hardware choice may be narrowed down to what currently exists within the organization. This could be due to the lack of funding or merely a misunderstanding of the special requirements of simulation and animation software.

If the person running the model is someone other than the modeler who created it, the user's abilities with a computer will influence the hardware choice. Today's high-end PCs often have operating systems and user interfaces that novice computer users find easier to use than some of the more cumbersome mainframe platforms. If, on the other hand, the end user has a particular preference, expertise, or familiarity with a given platform, there is a distinct advantage in staying with the familiar. No matter how good the model is, no matter how accurately it predicts the behavior of the system, if the user cannot use the hardware on which it resides, it will not be utilized. It stands to reason that if the user is comfortable with the hardware, he or she will be more likely to use, believe, and even defend the model's output.

Also taken into consideration must be the availability of the hardware itself. If the computer will be used frequently by others or is slowed down by running several programs simultaneously (multitasking), the modeler will have a difficult time producing a model and the users will be less likely to use the model simply because it is hard to find an opening on the machine. This situation can be avoided if those individuals involved in the purchase decision are aware of the time requirements of model creation and use. Computers are inexpensive enough today that the benefit generated by the simulation model vastly overshadows the cost of dedicated hardware.

The last, and often overlooked, aspect of selecting simulation hardware is that of the availability of secondary analysis software. Secondary analysis software includes the programs that assist the input, output, analysis, or presentation of the simulation model, i.e., anything used other than specific simulation software. Simulation programs can generate vast quantities of data as output. If the modeler so chooses, he or she can have each action that happens to an entity recorded in a data file. From this data one can analyze various aspects of the system's behavior. In a simulation that lasts a year or more (simulated time), that could generate a tremendous amount of data to sort through and perform statistical manipulations on. A good statistical software program that can receive input from a file created by the simulations program would be worth its weight in gold. Usually the modeler does not realize the usefulness of such a package until hours have been spent in the tedious pursuit of statistical information. The secondary analysis software is not limited to statistics, but also could encompass database management software for the organization of experiments, word processors to help in model code creation or data file creation, graphics programs for creating presentations on results and recommendations, and others depending on specific circumstances and software. Indeed, the simulation computer hardware may need to host an entire suite of software to ease the input and analysis drudgery and free the modeler/user to create effective models that will solve problems and predict system behavior.

Personal Computers

The 1980s saw an explosive growth of PC technology and its market, so the distinction between the high-end PCs and minicomputers has blurred. Some might say that PCs have penetrated well into the mini's territory. In terms of speed, graphics, and multitasking, there has indeed been a overlap in general software areas, but in simulation software there still seems to be a gap (however slight) between PCs and minis.

The first hardware choice may be a PC, since they are well entrenched in the business world. A vast majority of the simulation programs run on this platform. Some programs output ASCII files that can be imported into other software programs for analysis and presentation. Indeed, some simulation software accept two-way passage of information between popular programs that read and write ASCII files. The PC is a familiar object within many organizations and will not compound the resistance that may already exist for a new technology. Training for operations of PCs is minimal, since many people within the company already have experience with other software or at least could assist in training new users. Secondary analysis programs exist for the PC in abundance and may currently exist on those platforms. PCs also enjoy a low price/performance ratio, which should please even the most frugal controller. PCs have the option of running different operating systems (DOS, Windows, OS/2, UNIX, etc.), which could enhance the operation or increase the model size and complexity for greater modeling flexibility.

Minicomputers

Minicomputers have their niche within the simulation field. Some of the simulation language programs can utilize the power, speed, and size provided by these larger machines. Within this class of machines exist the characteristics sought after by the computer-aided design (CAD) engineering crew, so one can be assured that the mathematically demanding simulation software will find a good home here too. Computers are, after all, mathematical manipulators, and simulation programs are a good match for their capabilities. Workstations have the edge over PCs in multitasking, so the modeler can be analyzing one simulation run while the computer is performing the current simulation experiment. This is an underestimated feature that will, if used well, shorten the time required to prepare a recommended course of action.

Mainframe Computers

Mainframes, too, play a role in the complex world of simulation. Their prevalence, speed, multistation capacity, and link with other information within the business allow them to have the first thought in simulation. Indeed, in the 1960s, IBM wrote the software GPSS (General Purpose System Simulation) language and shipped its mainframes with a free copy. For years, it was many people's first look at simulation and what it can do. Since mainframes can multitask at far greater levels than PCs or minis, a business could benefit by several people modeling and using models simultaneously. Mainframes also can process large batches of simulation runs and record vast amounts of data for later, detailed analysis. Often, mainframes are used to run the accounting and material requirements plan (MRP) functions within the business. Therefore, the simulation modelers' access to usage of specific company data can be eased via this link. Many tedious hours of data reentry can be avoided by having the mainframe create files of information structured so that the simulation model can read and use it. This data reentry represents one of the most difficult, time-consuming areas to create and debug. Who, in their right mind, enjoys poring over thousands of records searching for the slipped decimal point or swapped digits?

The drawbacks to mainframe simulation usage are often common with the other hardware platforms: lack of secondary support software, slow response time (not to be confused with execution time), and difficulty in usage of the overall operating system. These aspects should not, in themselves, scare someone away from use of this platform, but rather be a basis of particular simulation environments.

5.1.6 Successful Implementations

This section will present some samples of the diverse and successful uses of simulation in capacity analysis. These few are by no means a complete listing; rather, this is a sampling, showing the variety and magnitude that should encourage readers to open up their imagination and investigate/apply simulation wherever it is possible.

The U.S. Air Force (USAF) contracted for a study to determine the number of air crews to receive training for their strategic bombers. The goal was to have 100% crew readiness based on five crew members per plane on standby for 12 hr. each. The training and readiness pay was very costly. If, for example, it was decided that ten crews per plane was the appropriate number and that was too low, the personnel would cycle through the standby duty too often and possibly suffer fatigue. If the number were too high, we (the taxpayers) would spend millions of tax dollars needlessly. The study showed that a lower number of crews, used on a certain schedule, would achieve the preparedness goal, yet save $20 million per year in training costs.

McDonnell Douglas Missile Systems Company, makers of the Tomahawk Cruise Missile, used simulation to determine when within a 5-year window another ordnance storage bunker was needed. There were over ten other programs sharing the storage bunkers as well. The Department of Defense had specific rules governing what class of items could be stored with other items. There were also federal, state, and local regulations that added to the storage complexity. Also, the quantity of missiles produced varied each year. McDonnell Douglas has an annual capital budget cycle, and it took about 10 months to build a storage bunker, so the company needed to know of the additional requirement 2 years before the required date. The manual calculation methods were difficult, were based on estimates, took several months to complete, and were low confidence builders. The simulation model showed that one of these $400,000 bunkers was not needed within the 5-year time frame. It improved the confidence level in the outcome, and showed how much more product growth could be tolerated before a bunker was needed. An animation of the operations was used to demonstrate the model to both McDonnell Douglas management and the navy program management.

A major city hospital used simulation to determine the quantity, mix, and schedule of doctors, nurses, and other support staff for the entire facility. This included over 450 personnel in many different functions. The model showed the sensitivity to peak demand times and its impact on the manpower schedule. The model predicted when and what type of personnel were needed, which delayed significant hiring surges.

5.1.7 Additional Reading

Law, A. M., Introduction to simulation: A powerful tool for analyzing complex manufacturing systems, *Industrial Engineering,* May 1986.

5.2 SCHEDULING

Having established a clearly articulated manufacturing strategy, we now turn our attention to the subject of scheduling. Recall that the manufacturing strategy describes *actions* involving the *deployment* of manufacturing resources. These actions must be communicated and understood by the entire organization in a way that allows all functions to act in harmony with each other. Activities and events must be properly timed and synchronized in order to be performed most effectively. We call this level of scheduling "major activity planning."

5.2.1 Major Activity Planning

The *American Heritage Dictionary* defines a schedule as "a production plan allotting work to be done and specifying deadlines," while the *APICS Dictionary* (1992) states a schedule is "a timetable for planned occurrences, e.g., shipping schedule, manufacturing schedule, supplier schedule, etc." For our purposes we will need to distinguish between "activities" and "events." An *event* is something that occurs at a point in time, having a duration of zero. An *activity* occurs over a span of time; i.e., it has a duration of greater than zero. For example, the activity "develop engineering drawing for wing flap" must precede the event "release wing flap drawing to manufacturing." In this example, an engineering activity is seen in the larger context of an internal supplier/customer relationship, with design engineering as the supplier of a drawing to their customer, manufacturing. Proper documentation of this relationship on a schedule, extended to all major interactions between internal suppliers and customers, allows all business functions to act in proper synchronization with each other, ultimately delivering goods to the firm's (external) customers in time to satisfy their needs (delivery schedule). For the manufacturing engineer, some important activities/events might include the following: release engineering drawing, create manufacturing bill of material, release planning documents (routings and work instructions) to shop floor, begin first article production, begin rate production, etc. Middle management might be concerned with schedules reflecting routine business activities such as budget preparation, performance reviews, vendor negotiations, etc.

Because of the varying needs of people at different levels in the organization, more than one schedule is required. In fact, many firms employ a multilevel scheduling approach that presents appropriate information to each intended audience, still with the goal of synchronizing activities throughout the business. This "vertical integration" of activities allows top management to communicate the timing of overall business activities to the entire organization, and enables progressively lower levels to carry out those tasks in harmony with each other.

In the sections that follow we will explore this vertical integration in more detail within the context of production and material planning. Specifically, we will introduce a top-down model beginning with a production plan and continuing down through a master production schedule (MPS), MRP, and detailed schedule.

5.2.2 Production Planning

Continuing through Subchapter 5.2.6, we will focus our attention on scheduling activities related specifically to production and material control. An underlying assumption throughout is that a manufacturing resources planning (MRPII) philosophy is employed, and a corresponding MRPII computer system is in place (see Subchapter 5.5). Then, in Subchapters 5.3 and 5.4, brief discussions of two other philosophies will be developed: just in time (JIT) and the theory of constraints (TOC). For now we will deal with scheduling in an MRPII environment.

The top-level production schedule in MRPII is called the "production plan." A production plan defines the overall production rates for each product family (a group of products having similar characteristics). It takes into account all sources of demand, such as customer orders, sales forecasts, spare parts production, interplant production requests, etc. It is developed by senior management as the result of a "sales and operations planning" process, typically on a recurring monthly basis. As such, the production plan becomes management's "steering wheel" on the entire MRPII system.

The *APICS Dictionary* (1992) defines sales and operations planning (previously called production planning) as:

> the function of setting the overall level of manufacturing output (production plan) and other activities to best satisfy the current planned levels of sales (sales plan and/or forecasts), while meeting general business objectives of profitability, productivity, competitive customer lead times, etc., as expressed in the overall business plan. One of its primary purposes is to establish production rates that will achieve management's objective of maintaining, raising, or lowering inventories or backlogs, while usually attempting to keep the work force relatively stable. It must extend through a planning horizon sufficient to plan the labor, equipment, facilities, material, and finances required to accomplish the production plan. As this plan affects many company functions, it is normally prepared with information from marketing, manufacturing, engineering, finance, materials, etc.

This definition introduces another plan, referred to as the "business plan." A business plan specifies, among other things, the amount of money that must be made through the sale of product. The production plan must support these sales objectives, but specifies sales in terms of product rather than dollars.

To summarize, the sales and operations planning function is the process by which senior management becomes regularly and personally involved in MRPII. Decisions made are reflected in the production plan, which in turn steers the master schedule and ultimately drives the entire formal MRPII system.

5.2.3 Master Production Scheduling

The MPS defines what the company intends to produce, expressed in terms of specific configurations, dates, and quantities. These requirements in turn drive MRP. According to the *APICS Dictionary* (1992), the master schedule "must take into account the forecast, the production plan, and other important considerations such as backlog, availability of material, availability of capacity, and management policy and goals." Therefore, any product intended to be produced, whether for sale or other purposes, must be scheduled in the MPS.

At its heart, the MPS is the tool used by the master scheduler to balance supply and demand. *Demand* represents the need for a particular product or component, and may come from a number of sources: customer orders, spare parts, management risk production/procurement, forecasts, etc. (see Figure 5.5).

Once demand has been established, what is needed is *supply* to balance the equation. Each unit of demand must be matched with an MPS order that satisfies it in terms of specific configuration, quantity, and date. These orders fall into three categories:

1. Released orders—approved orders that have already been released to production
2. Firm planned orders—approved orders that have not yet been released to production
3. System planned orders—orders that are not yet approved, but drive lower-level planning activities in MRP

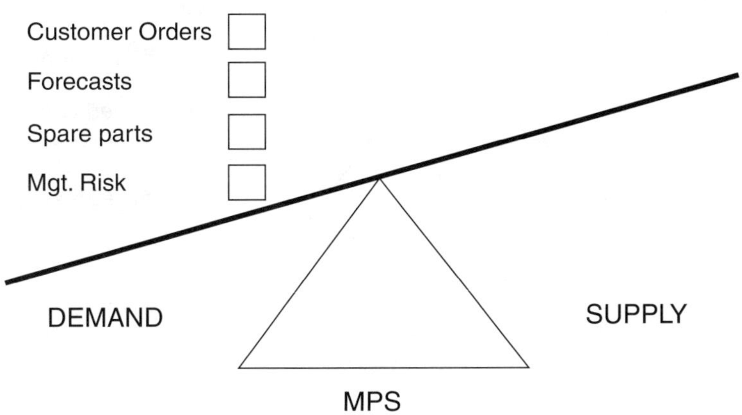

FIGURE 5.5 Chart showing the demand side of the MPS equation.

The resulting supply plan must exactly balance the demand plan—no more, no less. See Figure 5.6 for this principle.

Although this may at first seem unduly restrictive, it is in fact quite reasonable and practical. Remember that the formal MRPII system is the repository for all production activity information. It represents management's decisions for producing products, and the production and purchasing organizations' ability to carry out those decisions. In other words, demand placed on the system with no corresponding supply represents a situation where a management decision to accept demand is not being carried out by production and/or purchasing. Similarly, supply entered into the system with no corresponding demand represents production and/or purchasing's commitment to produce product without management concurrence. The point is to capture and communicate all production decisions in a way that systematically informs the entire organization of all production decisions and permits each department to respond accordingly.

Returning briefly to production planning, we see that senior management participates in a similar process. The sales and operations planning meeting provides the forum for reviewing current performance against the sales plan (high-level demand plan defined in the business plan) and the production plan (high-level supply plan). Resulting decisions, in the form of revised production plans, are entered in the MPS and ultimately communicated to the rest of the organization through the formal system. In this way, senior management steers the entire production and inventory control process. Managing these changes is the subject of the next section.

Master Scheduling Approaches

At the beginning of this chapter, three scheduling environments were presented: job shop, repetitive, and continuous flow. Three other categories are useful for describing

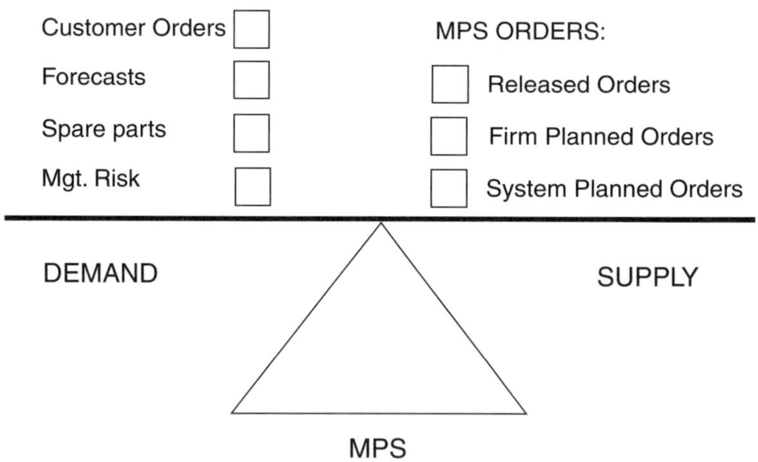

FIGURE 5.6 Chart showing a balanced supply and demand in the MPS equation.

manufacturers, especially in the context of master scheduling: make to order, assemble to order, and make to stock. A *make-to-order* manufacturer produces products only upon receipt of a customer order. Products are generally made from a combination of standard parts and materials, plus custom-designed parts to meet the specific requirements of each customer. An *assemble-to-order* manufacturer is similar, but maintains stocks of completed subassemblies and combines these into custom configurations, also upon receipt of an order. Finally, the *make-to-stock* manufacturer produces standard finished products for off-the-shelf consumption. Figure 5.7 shows typical product structures for each case.

Notice from the figure that in all three, the narrowest section of the product structure is master-scheduled. This is done so as to minimize the manual planning involved. Other planning and schedules systems, such as MRP, can perform the more tedious detailed scheduling from that point down. But what about the levels above these points?

In the assemble-to-order case, a method known as final assembly scheduling (FAS) is often employed, where a customized schedule is created for each custom-ordered product. The FAS (or "finishing schedule," as it is sometimes referred to) schedules the operations needed to finish a product from the point where the MPS left off. The role of the master schedule in this environment is to ensure that sufficient supplies of the needed subassemblies are maintained to be "pulled" by the FAS.

In the make-to-order case, notice that two levels of MPS are needed. At the bottom, raw materials and base components are master-scheduled, usually based upon anticipated consumption rates. This is much easier than attempting to anticipate quantities of a far greater number of possible custom combinations at the top. However, as customer orders are received, a specific customized product structure can be created, and the intermediate levels (subassemblies, options, etc.) may then be planned and scheduled. The scheduling of levels below the MPS falls to MRP.

5.2.4 Materials Requirements Planning

MRP is a process for planning all the parts and materials required to produce a master-scheduled part (refer to Section 5.2.3). The resulting time-phased requirements

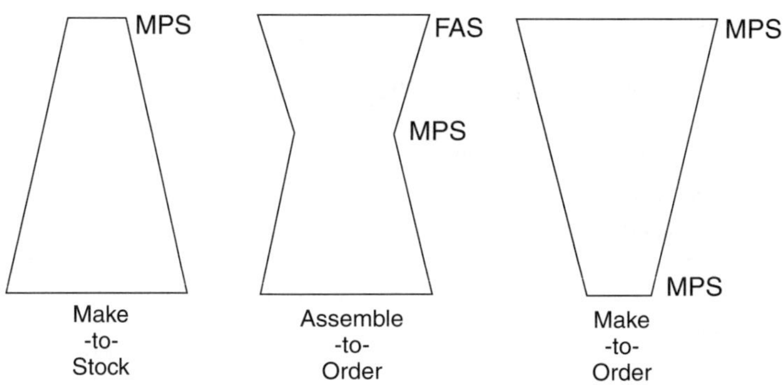

FIGURE 5.7 Different product structures influence the type of scheduling systems required.

take into account parts and materials already on hand or on order, subtracting these from the total requirement for a given time period, to arrive at net requirements. Each resulting MRP-planned order defines, as a minimum, the part number, quantity required, and date.

A detailed description of the MRP process is given in Subchapter 5.5. For now we will limit the discussion to a fairly conceptual level and focus on the role of MRP in an overall scheduling context. Let us begin with the process itself. Figure 5.8 depicts the overall process, beginning with a master schedule. Associated with each requirement in the MPS is a bill of material (BOM), which defines the parts and subassemblies required to produce the MPS part, then the parts required to produce each subassembly, and so on down to purchased parts and raw materials. MRP "explodes" the bill of material, converting the orders at a given level into a set of gross requirements for parts and materials at the next level. Next it subtracts available inventory from the gross to arrive at a net requirement for each item. MRP then creates planned orders to satisfy these requirements, and repeats the process for all remaining BOM levels.

Figure 5.9 shows a simplified BOM for a two-drawer file cabinet. Notice that two drawer assemblies are needed for each cabinet, and that two drawer sides are needed for each drawer assembly. If the MPS contains orders for five cabinets, MRP will calculate a gross requirement of ten drawer assemblies (5 cabinets × 2 drawers/cabinet = 10 drawers). Now let us assume that we already have three drawer assemblies in inventory. MRP will subtract these from the gross requirement to arrive at a net requirement of seven, and create planned orders to build seven drawer assemblies. The process continues down to the next level, exploding the seven drawer assemblies into their component parts, netting at that level, covering the net requirement with planned orders, then continuing down until the entire bill of material has been exploded. But what about the timing element?

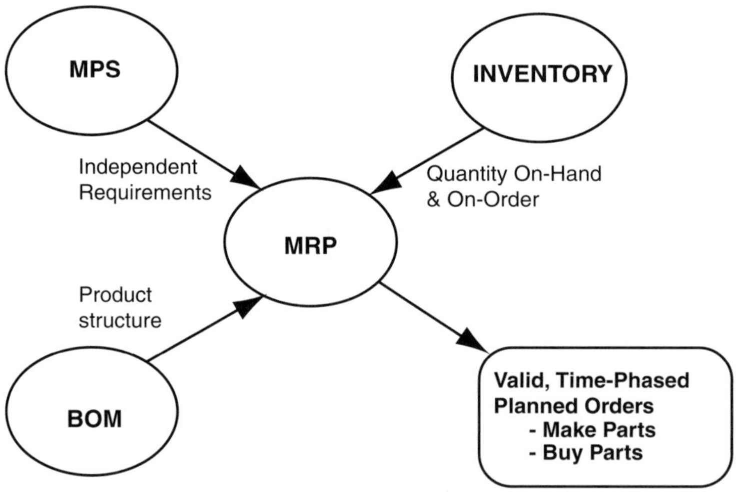

FIGURE 5.8 Material requirements planning.

Let us assume that final cabinet assembly takes 3 days, and each drawer assembly takes 2 days. These "lead times" are used to establish the completion dates that each part must meet in order to satisfy the next-higher-level requirements. In our example, if the MPS order (for five completed cabinets) is due on day 10, then the drawers must be completed 3 days earlier, on day 7. Likewise, the parts required to assemble the drawers must be completed 2 days prior to that, resulting in a due date of day 5.

Two types of demand are evident in this example: "dependent" and "independent." Let us define these terms:

Independent demand—Demand for an item that is unrelated to the demand for other items in the product structure. Demand from customer orders, for replenishing finished goods stock, and for spare parts are all considered independent. Since independent demand cannot be calculated, it must often be forecast.

Dependent demand—Demand that is derived directly from another item or end product defined in the bill of material. Because this demand is driven by some other requirement, it should be calculated rather than forecast.

We would consider the end-item cabinet assembly to be independent demand. Demand for the cabinet frame and drawer assemblies would be considered dependent, because their requirements depend on the demand for the cabinet assembly. Note, however, that a given item may have both dependent and independent demand. For example, if a spare-part order is received for a drawer assembly, this order would represent independent demand, while the demand for drawer assemblies resulting from cabinet assembly orders is dependent.

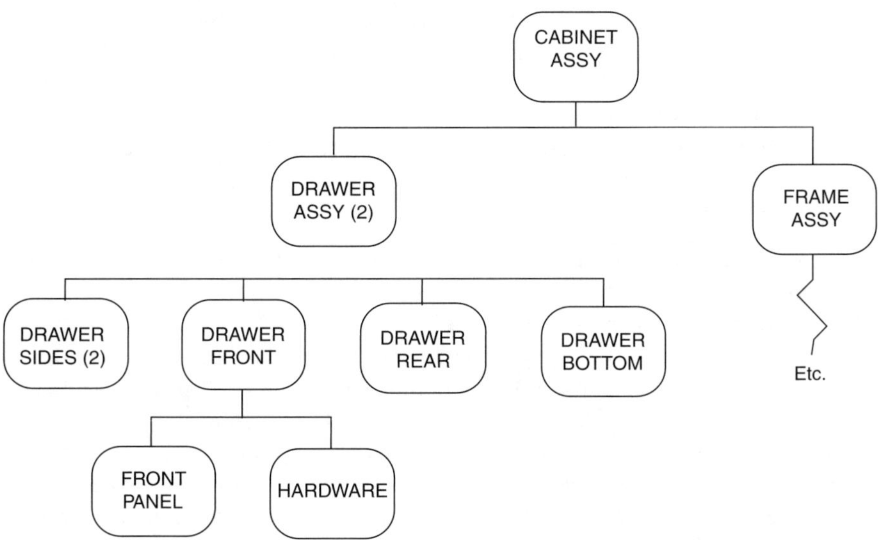

FIGURE 5.9 Simplified bill of material for a two-drawer file cabinet.

Supply and Demand Balancing

In Section 5.2.3 we discussed the need to balance demand from various sources (customer orders, spare parts, etc.) with the company's production plan and MPS. MRP continues this supply/demand balancing process down to the lowest levels of the BOM. Beginning at the top (e.g., a cabinet assembly), the MPS defines a supply plan for finished goods. MRP translates this data into demand for items at the next BOM level. After netting available inventory, it calculates planned orders to supply the remaining requirements. This supply plan in turn becomes the demand placed upon the next lower level, and so on down the structure. In the case of combined dependent and independent demand, MRP must satisfy both sets of demand simultaneously. As changes occur, such as new or canceled orders, inventory adjustments, scrapped parts, etc., MRP dynamically replans to maintain a balance between supply and demand at every level.

There are, however, some self-imposed limitations to this process. In general, the nearer an MRP order is in the future, the more human intervention is required. The further out in the future, the more decisions may be automated. MRP systems generally define three types of orders, distinguished by how far in the future they are planned.

System-planned order—Orders far enough in the future that they are planned entirely by the MRP system. Also referred to as "MRP-planned orders."

Firm-planned order—Orders that are near enough in the future to warrant human intervention, but not close enough to release into production.

Released orders—Orders that have been released to production. Released orders fall outside MRP control.

In the case of near-term orders (firm planned and released), it is probably not wise to allow a computer to make decisions regarding timing and quantity of orders. In all likelihood, resources have already been deployed or are being deployed to execute these orders. The impact of a change at this point in time is severe. People (inventory planners) must therefore make these decisions. In the next section we will discuss how the MRP system assists planners in making these decisions.

Management by Exception

A basic tenet of good management is to handle routine things routinely, and manage the exceptions. Routine decisions are best handled by following tight, clear policies and procedures. A good procedure will define "decision rules" to be followed in specific situations. This approach provides consistency and control; attributes that are vital to the effective management of production and inventory. It also frees managers and workers to apply their skills and intelligence to the exceptions.

MRP employs just such an approach. By automating the routine decision rules for planning parts and materials, MRP frees people's time to deal with exceptions and near-term requirements. MRP communicates these occurrences in the form of "action messages." In each case, the system is recommending an action that it would take if it had control of the order. Let us examine some of the more common ones.

Firm message—Recommends that the planner convert a system-planned order to a firm-planned order. By executing this action, the planner takes control of the order from MRP. If not converted, MRP will retain complete control over the order.

Expedite message—Recommends that the planner accelerate an existing firm-planned order; i.e., reschedule the order to an earlier date. Also referred to as a "reschedule in" message.

De-expedite message—Recommends that the planner slow down an existing firm-planned order, i.e., reschedule the order to a later date. Also referred to as a "reschedule out" or "defer" message.

Cancel message—Recommends that the planner cancel a firm-planned order.

Release message—Recommends that the planner release a firm-planned order to production. By executing this action, the planner removes the order from further scrutiny by MRP.

In all but the first and last messages, MRP is communicating an imbalance in supply and demand. In the case of reschedules (expedite, de-expedite), the total planned quantity is balanced but the timing is off, resulting in future periods of either too much or too little supply. In the case of a cancel message, the total planned quantity (supply) exceeds the total requirements (demand), i.e., an imbalance of supply over demand.

The final message is a recommendation to release an order to production, removing it from MRP's view. At this point, control of the order falls to the shop floor control system, which is the subject of the next section.

5.2.5 Detailed Scheduling (Shop Floor Control System)

Detailed scheduling as defined in an MRPII context is the assignment of start and/or finish dates to each operation on a routing. Within MRPII this level of scheduling falls in the domain of the shop floor control (SFC) system. In this section we will review the detailed scheduling process itself, as well as introduce some basic elements of production activity control (PAC). Section 5.2.6 will discuss PAC in greater detail.

Every shop order (released MRP order) has an accompanying document known as a "routing," which describes how a given part is manufactured. Each step of the manufacturing process is given an operation number. Accompanying each operation is a brief description of the work to be performed, tooling to be used, and standard setup and run times. These are the basic elements, although other data such as queue times, reference documents, and required operator skills are often included as well. Routings also call out nonmanufacturing steps, such as picking of parts from stores, movement of parts between operations, inspection operations, and test requirements.

In addition to the routing, a set of detailed work instructions is provided. These documents are often referred to by such names as planning documents, process sheets, and assembly manuals, and are produced by manufacturing engineers.

Forward Scheduling

The detailed scheduling process differs significantly from that performed by MRP. MRP uses a "backward scheduling" algorithm, which defines due dates by which orders must be completed in order to satisfy the requirements at the next higher BOM level. SFC, on the other hand, often employs a forward scheduling algorithm. Starting at today's date and calculating forward, each operation is assigned a start and/or finish date. Depending on the precision required, some systems schedule operations by date and time (hour, minute, etc.). Figure 5.10 depicts a sample calculation.

A typical routing will begin with instructions to pick parts from stores. These parts are then moved to a staging area where they sit in queue until assigned to a worker. Some preparation is generally required (setup) before actual work on the part begins. Once begun, work normally continues (runs) until it is completed for all parts in the order. Therefore the amount of time required to process the order through run operations is the run time multiplied by the order quantity. Parts are then moved to the next staging area and the queue–setup–run sequence is repeated. Finally, the order is inspected and returned to stock. An actual routing typically has many more operations, as well as intermediate inspection points, test operations, etc. However, this simplified routing will suffice for our example.

Assuming that today is considered day 1, operation number 1 (pick), which has a duration of 2 days, will start at the beginning of day 1 and complete at the end of day 2.

Sample Detailed Scheduling Calculation

Pick	Move	Queue	S/U	Run	Move	Queue	S/U	Run	Inspect	Stock

Pick = 2 days
Queue = 5 days
Move = 1 day
S/U = 4 hours
Run = 0.2 hours/part
Inspect = 4 Hours
Stock = 2 days

Oper. No.	Description	Hours	Ord. Qty.	Dur (Hrs)	No. Days
1	Pick	16.0		16.0	2.0
2	Move	8.0		8.0	1.0
3	Queue	40.0		40.0	5.0
4	Setup	4.0		4.0	0.5
5	Run	0.2	100	20.0	2.5
6	Move	8.0		8.0	1.0
7	Queue	40.0		40.0	5.0
8	Setup	4.0		4.0	0.5
9	Run	0.2	100	20.0	2.5
10	Inspect	4.0		4.0	0.5
11	Stock	16.0		16.0	2.0
			Totals	180.0	22.5

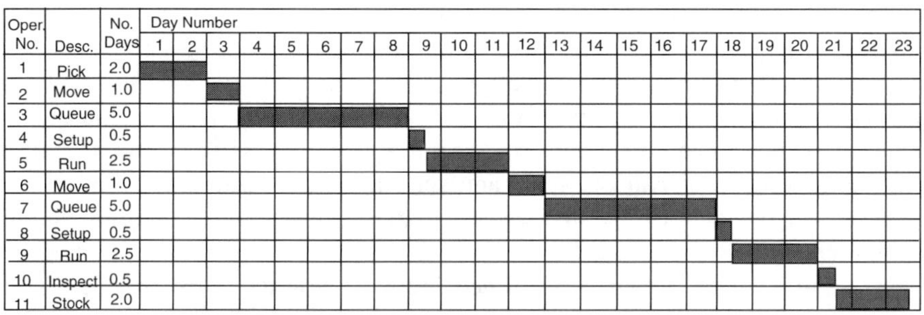

FIGURE 5.10 Sample calculation of forward scheduling.

Operation number 2 (move) will start at the beginning of day 3 and complete at the end of the same day. The order will then sit in queue until the end of day 8, when it is ready to be worked. The work area (machine, work bench, etc.) is set up, and by the middle of day 9 is ready to begin processing the order. The total order requires 20 hr. (2.5 days) to complete, finishing at the end of day 4. Continuing this calculation, we see that the order is completed and parts returned to stock by the middle of day 23.

For the order in our example to be completed on time, the MRP lead time must be at least 23 days. In practice, lead times are often inflated slightly to allow for unanticipated delays in their processing. Our sample order might have a lead time of 25 days. If the MRP planner had released this order just in time to complete on day 25, then according to our calculations, the order would be completed 2 days early, resulting in 2 days of "slack time." We will see in the following section how this slack time can be used as one means of prioritizing orders. Before continuing, some observations about the above example will help to illustrate two general principles.

First, notice the relationship between setup and run time. Since the setup time is very large compared to run time, it is necessary to produce a sizable quantity in order to hold down the cost per part. For example, if only one part was produced, the cost of labor (or machining) for that single part would be 4.2 hr. (4 hr. setup + 0.2 hr. run). If 10 parts were produced, the total time required would be 6 hr. (4 hr. setup + 10 × 0.2 hr. run), or 0.6 hr. per part. Manufacturers must use caution however, not to produce too many parts, since parts must then be stored. Storage and related costs must be considered when determining order lot sizes. This subject is discussed in more detail in Section 5.6.3. (A better alternative is to reduce the setup time, thereby avoiding the need to produce large batches. This approach is advocated by "just in time," and is discussed in Subchapter 5.3.).

Secondly, notice that a large portion of the time that this order spends in the shop is tied up sitting in work queues. In our example, 44% (80 ÷ 180) of the order's time is spent in queue. This is not unusual. In fact, in many companies, especially job shops, orders spend more time in queue than in all other operations combined, making the management of work queues one of the most important elements of production activity control. One method of controlling queues is called input/output control and is discussed in Section 5.2.7. The relationship between queue time and lead time follows.

5.2.6 Production Activity Control

In this section we will deal with five related subjects: control of queue size, priority sequencing rules, dispatching, shop floor data collection, and performance measurement. A sixth major area of production activity control—capacity management—is covered in Section 5.2.7, as well as in Subchapter 5.1. We begin the discussion with control of queue size.

In the previous section it was pointed out that queue time makes up a large percentage of an order's lead time, especially in a job shop environment. In order to manage queue size effectively, an essential ingredient in overall queue management, an understanding of this relationship is vital.

Let us begin by looking at the relationship between order lead time and work-in-process (WIP) levels (the number of parts/orders in various stages of completion in the shop). Figure 5.11 illustrates the point. Assume that one order per day is scheduled to be completed, beginning on day 5. If the lead time is 5 days as depicted in the top chart, the first order must begin on day 1. By day 5, when the fifth order has been released, there will be a total of five orders in the shop. This level will remain as long as work is released and completes at the same rate. The bottom chart shows what would happen to WIP levels if the lead time could be reduced to 3 days. To satisfy the first completion requirement on day 5, the first order would need to begin on day 3. By day 5, and for all subsequent days, a total of three orders will be in work, a reduction in WIP of 40%.

So how is this reduction in lead time accomplished? Recall our earlier discussion on the length of time that orders spend in queue. The more orders that are in WIP, the longer an order entering a queue will have to wait its turn to be processed.

By arbitrarily reducing the total lead time from 5 to 3 days, we will succeed in reducing the number of orders waiting their turn, thus reducing queue time. Notice that no reduction in actual operation times was required. By simply recognizing the circular effect of queue size on lead times, of lead time on WIP levels, and of WIP levels on queue size (Figure 5.12), it is possible to reduce the number of orders open at any given time, thereby simplifying the management of the shop (and producing considerable savings to the company) while maintaining the required completion dates.

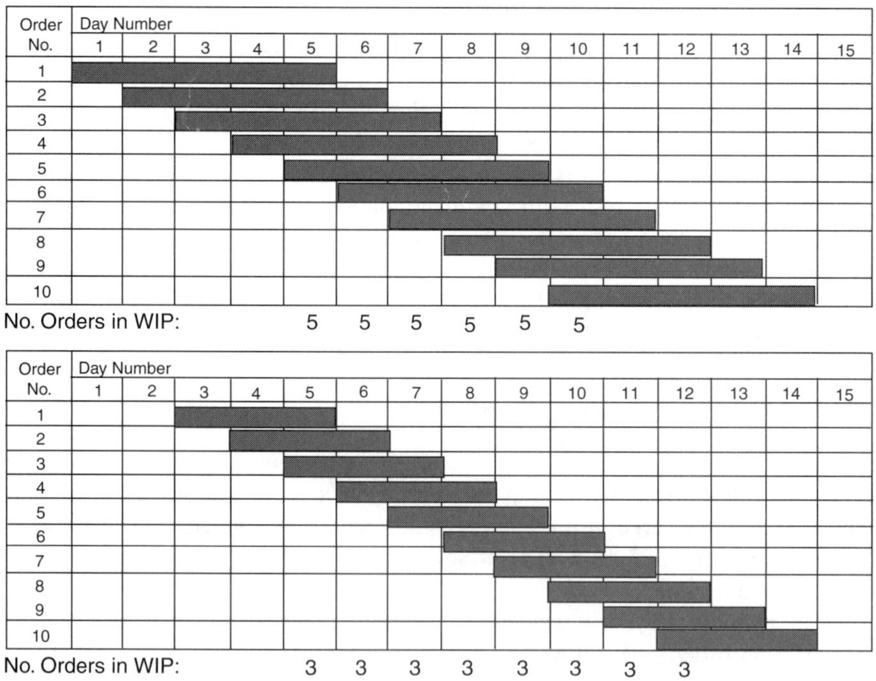

FIGURE 5.11 Effect of lead time on work-in-process levels.

Production control managers who fail to recognize the effects described above tend to react to poor delivery performance in just the opposite way. If deliveries are not being met, the rationale is that they do not have sufficient time to complete the orders they are given. To compensate, they *increase* lead times. This of course has the opposite effect than they anticipated. By increasing lead times, WIP increases, queue times lengthen, and the burden of managing these increased queues only makes delivering on time more difficult. Those responsible for production scheduling, especially at the MRP level, must diligently guard against this practice. Since manufacturing engineers are often responsible for establishing lead times, they play a key role in this process.

Priority Sequencing Rules

In the previous section we saw how proper control of lead times influences queue size. But how are the queues themselves managed? What tools and techniques allow production control management to determine relative priority of all the orders in a given queue? This section briefly lists some of the more common approaches.

Operation due date—Each operation is assigned a due date, typically using the forward scheduling technique discussed in the previous section. Orders with operations having the earliest due date are selected first.

First-in, first-out (FIFO)—Orders are worked in the sequence in which they entered the queue, i.e., the oldest order (first in) is worked first (first out).

Critical ratio—Orders are worked based on a calculated ratio dividing time remaining by work remaining, where time remaining is the difference between today's date and the order's due date, and work remaining is the sum total of all remaining operation times. A ratio of less than 1.0 indicates that

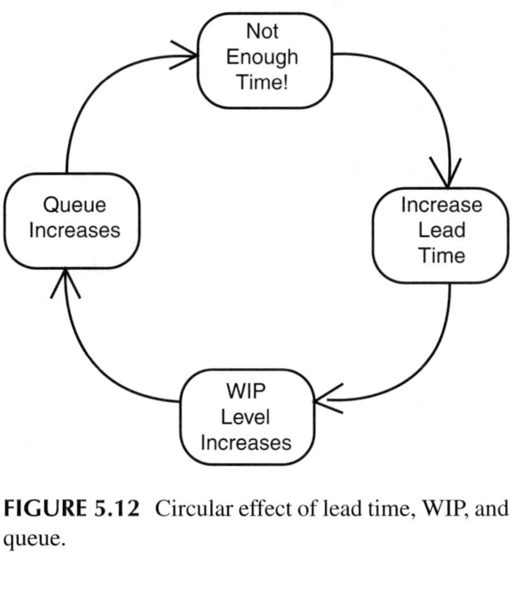

FIGURE 5.12 Circular effect of lead time, WIP, and queue.

the order is behind schedule (i.e., time remaining is less than work remaining); a ratio equal to 1.0 indicates that the order is on schedule; a ratio greater than 1.0 indicates that the order is ahead of schedule (i.e., time remaining is greater than work remaining). A negative critical ratio indicates that the order is already past due. Orders with the smallest critical ratio are selected first.

Slack time—Orders are worked based on the amount of time between the order completion date (calculated by forward scheduling) and the order due date. Positive slack indicates ahead of schedule, zero is on schedule, and negative is behind schedule. Orders with the least slack (or most negative) are selected first.

Shortest operation next—Orders whose next operation is shortest are selected first. The rationale is to process the greatest possible number of orders, thereby minimizing the number of orders in queue. This rule must be used with caution, and is probably best used as a tie breaker in combination with some other rule(s).

Most operations next—Orders with the greatest number of remaining operations are selected first. The rationale is that orders with fewer operations are easier to schedule into available work centers, and therefore the more difficult orders (i.e., those with more operations) should be scheduled first.

Informal—Numerous informal priority sequencing rules exist, including hot lists (maintained manually to override the formal system), colored tags (e.g., assign orange-tagged orders top priority), and "he-who-screams-the-loudest."

Dispatching

Dispatching is "the selecting and sequencing of available jobs to be run at individual workstations and the assignment of those jobs to workers." A *dispatch list* is "a listing of manufacturing orders in priority sequence. The dispatch list is usually communicated to the manufacturing floor via hard copy or CRT display, and contains detailed information on priority, location, quantity, and the capacity requirements of the manufacturing order by operation. Dispatch lists are normally generated daily and oriented by work center" (*APICS Dictionary,* 1992).

In Section 5.2.5 we discussed the process of detailed scheduling. The resulting operation-level schedules are prioritized using a variety of sequencing rules, and these priorities are communicated to manufacturing via dispatch lists. In order for schedules and priorities to be maintained in a timely manner, some way of communicating the status of individual orders must be provided. This is the role of shop floor data collection.

Shop Floor Data Collection

The shop floor data collection system provides the means of collecting information on a wide variety of shop floor activities, including the following:

Employee time and attendance
Order status

Scrap and rework
Labor accumulated against orders
Work center capacity output
Machine downtime

Data collection can be done manually, but as bar coding and other data collection technologies advance, more and more manufacturers are turning to automated systems for this purpose. Computer terminals located throughout the factory may be used to query the status of a particular order online at any time. Other information, such as work center queues, capacity data, etc., might also be available.

Within the context of this chapter, we will focus primarily on data related directly to shop orders (status and labor). We will deal with the subject of order status first. A typical order "life cycle" is depicted in Figure 5.13.

A typical scenario might employ a bar-coded shop order traveler, usually a heavy card stock paper document identifying the order number, part number, and operation numbers matching the information contained on the order's routing. These data elements are usually printed using standard text, accompanied by their bar-code equivalents. When an order is released, it may be assigned a status of "released." As the order progresses through its life cycle, its status will change repeatedly and might reflect any of the following:

In work
Held for … (various reason codes)
Complete
Closed
Etc.

In addition, operation status information might also be reported, typically indicating the last operation completed. This information, however, does not appear by magic. It comes only through the disciplined efforts of many people—machinists, parts movers, inspectors, etc. Each worker has a responsibility to report his or her progress on orders as they are processed, thereby providing this information through the system. These data are then available to support production management decisions such as priority sequencing and dispatching.

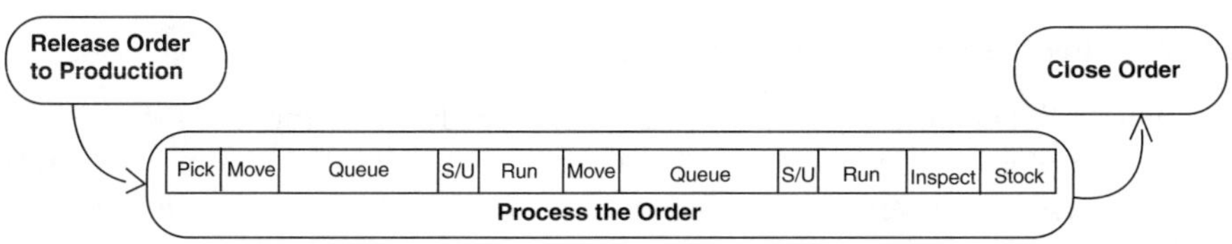

FIGURE 5.13 Life cycle of a typical order.

As orders are completed, returned to stock, and closed, inventory balances are updated, and MRP takes these new balances into account in its next planning cycle. This closed-loop process is discussed further in Subchapter 5.5.

In addition to status information, most data collection systems record the amount of time charged to each operation. While labor is almost universally collected, some systems also permit both labor and machine (or other process) time to be collected. This information is used for various purposes such as capacity planning (refer to Section 5.2.7 and Subchapter 5.1), inventory costing, and performance reporting, which is the next subject.

Performance Reporting

Three primary performance indices are of interest to manufacturing engineers—efficiency, utilization, and productivity. Definitions and a brief example of each follows:

Efficiency—A measure of the actual output versus the standard output. If the standard time to produce a part is 4 hr. and the part is actually produced in 5 hr., the efficiency is $4 \div 5 = 0.8$, or 80%.

Utilization —A measure of the percentage of a resource's available time that is actually used. If a resource, say a machine, is available 8 hr./day, 5 days/week, and is actually used for 35 hr. in a given week, its utilization for that week is $35 \div 40 = 0.875$, or 87.5%.

Productivity—An overall measure of work center efficiency comparing the total standard hour output of a work center to the "clock time" expended. If a work center completes a total of 38 standard hours in a 40-hr. week, its productivity for that week is $38 \div 40 = 0.95$, or 95%.

While these measures have much validity in measuring the capabilities of a given work center, they must be applied with caution. All three indices measure localized performance, with little or no regard for the overall effectiveness of the total manufacturing operation. In fact, it may be desirable for some work centers to operate well below 100% in any of these measures, especially if the work center is a "nonbottleneck." "Bottleneck" resources, on the other hand, must strive for 100% in all three measures to maximize the overall productivity of the operation. These principles are central to the theory of constraints discussed in Subchapter 5.4.

5.2.7 Levels of Capacity Planning

Throughout Subchapter 5.2 we have dealt with scheduling without regard to capacity. In other words, we have implicitly assumed that unlimited capacity was available to produce whatever was scheduled. In reality, however, capacity is a very real constraint in manufacturing companies, and must be considered when developing schedules.

In this section, four levels of capacity planning will be addressed:

1. Resource requirements planning
2. Rough-cut capacity planning
3. Capacity requirements planning
4. Input/output control and finite loading

Figure 5.14 identifies these levels, the purpose of each, and their relative position in the scheduling hierarchy. A few definitions must be introduced before continuing.

Infinite loading—Assigning work to a work center without regard to capacity. The resulting calculated load may show periods of overload and/or underload as compared to available capacity; i.e., infinite loading reflects how much capacity is needed to perform the schedule.

Theoretical capacity—The calculated maximum capacity, usually based on the standard hours a work center is able to produce in a given time period.

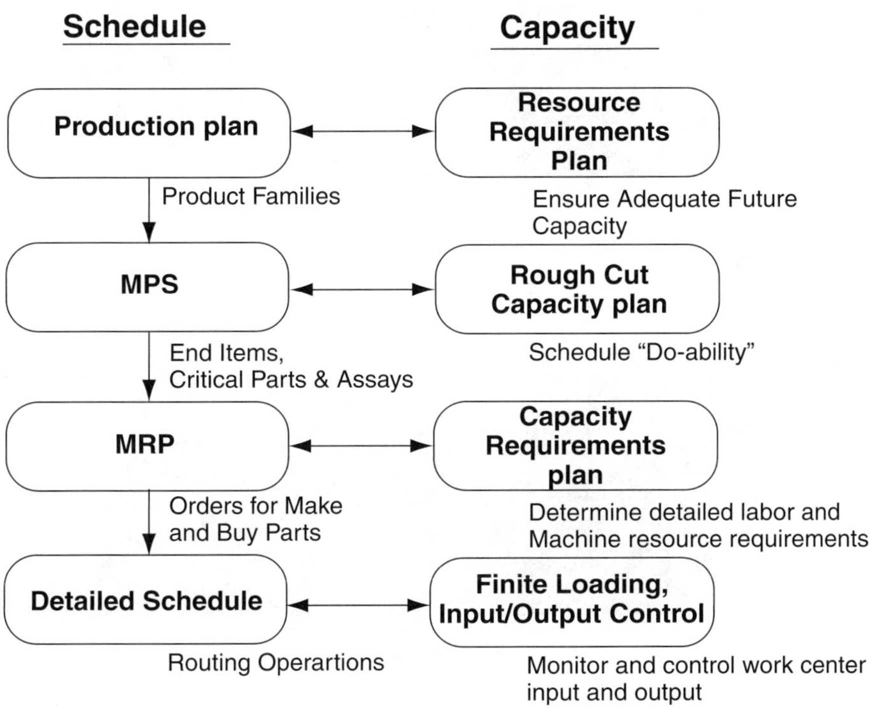

FIGURE 5.14 Levels of schedule and capacity planning/control.

Theoretical capacity makes no adjustments for nonproductive time such as routine maintenance, repair, shutdown, inefficiency, etc.

Demonstrated capacity—The proven capacity of a work center, demonstrated over a period of time. Usually based on actual standard hour output per day, week, etc.

Figure 5.15a and Figure 5.15b provide an overall context for these definitions. A typical model used to represent capacity and load is the "bathtub diagram" shown in Figure 5.16. Let us use this diagram to examine a microview of a work center first. As orders are released and moved through the shop, production control assigns them to the proper work center, as depicted by the spigot. If the work center is a "gateway" (the first work center on a routing), production control may exercise some discretion over when to release the order, thereby controlling the "input rate." Upon entering a work queue, orders become part of WIP and increase the work center's load, as shown by the water level in the bathtub. The rate at which WIP is processed (i.e., orders are

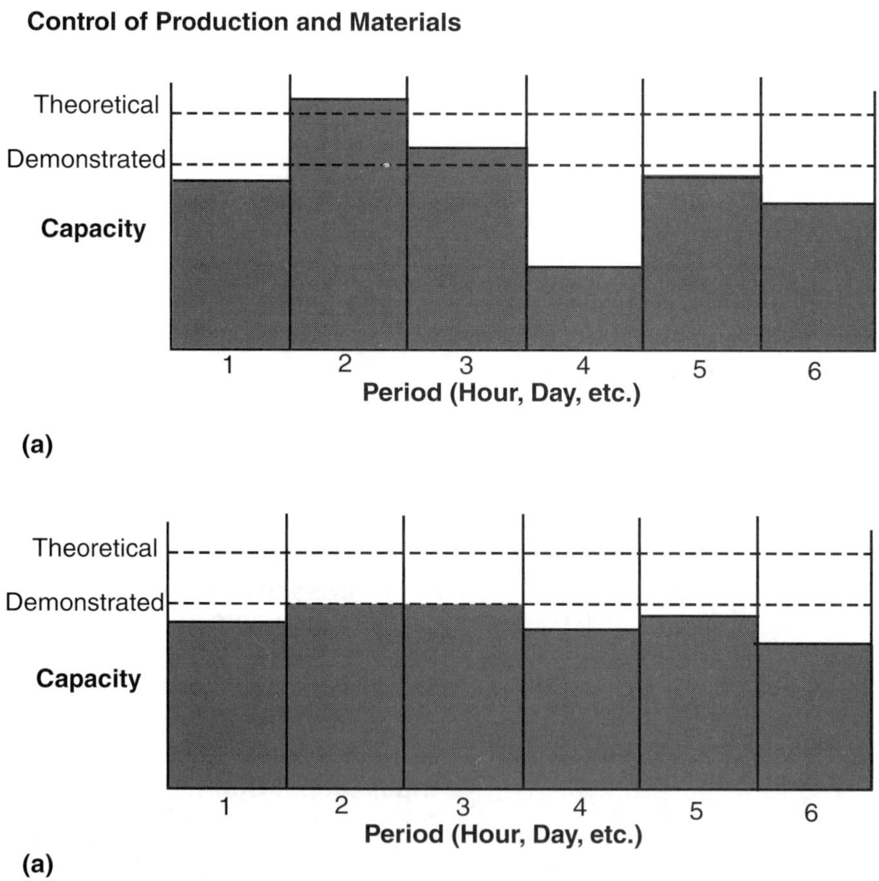

FIGURE 5.15 (a) Infinite shop capacity loading. (b) Finite shop capacity loading.

completed) is a function of the work center's capacity. This capacity is also somewhat adjustable through actions such as overtime, worker reassignments, etc., as depicted by the stopper in the tub's drain. The elapsed time between when an order enters a work queue and when it leaves is the manufacturing lead time for that operation.

Now let us take a macroview of Figure 5.16 as representative of an entire plant. If work is released to the plant at a higher rate than it is completed, work will begin to pile up on the shop floor. In addition, the overall load will increase, thereby increasing the actual lead time of each order. (Recall the discussion related to Figure 5.12 in Section 5.2.6 regarding the circular effect of WIP, queue time, and lead time.) Conversely, if the output rate is greater than the input rate over a period of time, work will dry up, leaving manufacturing resources idle. Maintaining a balance between the input and output rates is largely a production control responsibility. However, as we will see in the following paragraphs, concern for this balance must be demonstrated at all scheduling and capacity planning levels.

Resource Requirements Planning

As discussed in Section 5.2.2, a production plan defines the overall production rates for each product family (a group of products having similar characteristics). The associated level of capacity planning is known as resource requirements planning (RRP) (Figure 5.14). At this level, the primary concern is to ensure that adequate resources will exist to support the production plan, with consideration given to long-term, "brick

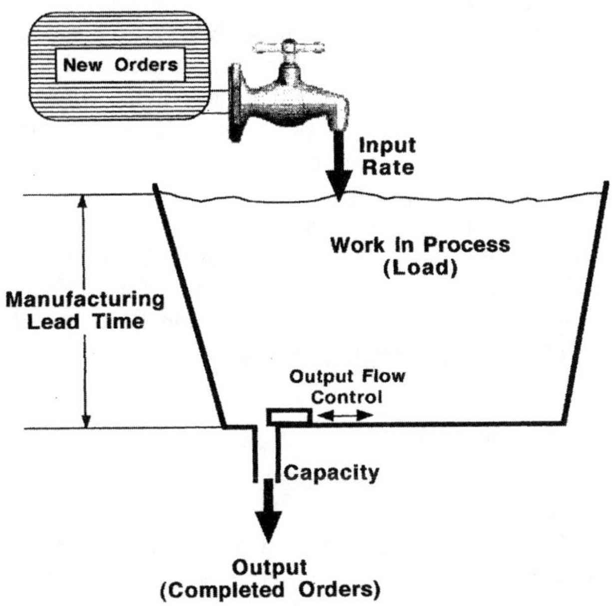

FIGURE 5.16 Capacity and shop load model.

and mortar," and overall manufacturing capacity. Units of capacity might include the following:

Total plant square feet
Warehouse volume
Total machining hours

The horizon for these decisions is considered long-term, typically 2 to 5 years with quarterly or longer time periods.

Figure 5.17 depicts an overview of the RRP process. For each product family, a "bill of resource" must be defined that identifies the resources the company wishes to plan and the amount of each resource required to produce a typical product in that family. In our example, "sawing," "painting," and "assembly" have been identified. The units of capacity (U/C) for sawing and assembly are in hours, while painting is in square feet. This will allow the company to calculate future requirements for sawing and assembly hours, perhaps to identify the need for a new saw, additional assembly workers, added shifts, etc. Calculating painting requirements in square feet might indicate that the company is concerned that existing paint booths are not large enough to accommodate the projected production rates, and management wishes to know when an additional booth needs to be added.

The "Bill of Resource Summary" table summarizes all three bills of resource. This table is then multiplied by the "Production Plan" table to calculate the resource requirements. The "Resource Requirements Plan for Sawing" table shows the results of this calculation on the resource "sawing." This process is repeated for the remaining two resources (not shown in the figure).

Notice that no lead times were defined in the bill of resource. While most RRP systems allow for lead-time offsetting of resources, it is fairly common to make a simplifying assumption that the resources will be used in the same time periods that end products are produced. In our example, which uses yearly time periods, this assumption is quite valid. We will see in the next section how lead time offsetting affects capacity planning.

Once resource requirements are known, the next step is to compare them to available resources. If the current sawing capacity is 2000 hr. per year, the RRP plan indicates an overcapability situation beginning in 1997. The company might decide to add a saw in that period, perhaps adding an additional 1000 hr. of capacity. The resulting resource plan is depicted in Figure 5.18.

Rough-Cut Capacity Planning

The process of rough-cut capacity planning (RCCP) is identical to that described above, except that it almost always takes lead-time offsetting into account and uses smaller time periods. Recall from Section 5.2.3 that the MPS defines what the company intends to produce, expressed in terms of specific configurations, dates, and quantities. The purpose of RCCP is to validate the MPS, i.e., to ensure that it is attainable, or doable. While RRP is used to identify the need for major resource

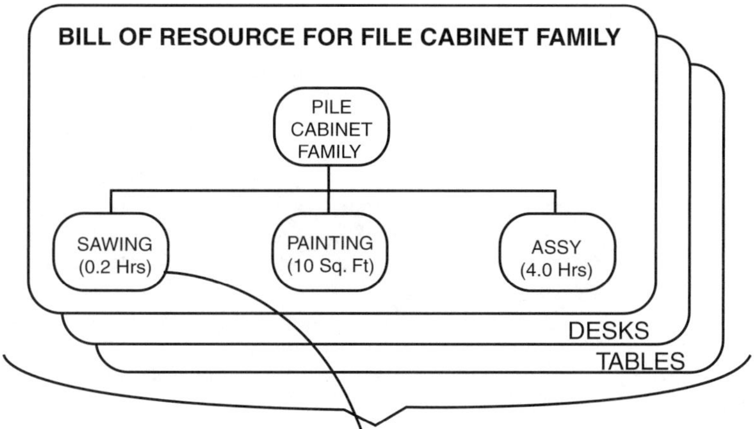

Bill of Resource Summary	U/C	File Cabinet	Desks	Tables
Sawing	Hrs.	0.2	0.5	0.1
Painting	Sq. Ft.	10.0	20.0	20.0
Assy	Hrs.	4.0	6.0	2.0

Production Year Plan	1995	1996	1997	1998	1999
File Cabinets	500	550	600	600	600
Desks	2000	2500	3000	3500	3500
Tables	3000	3750	4500	5250	5250

Resource Requirements Plan for Sawing

Sawing (Hours)	Year 1995	1996	1997	1998	1999
File Cabinets	100	110	120	120	120
Desks	1000	1250	1500	1750	1750
Tables	300	375	450	525	525
Total Hours	1400	1735	2070	2395	2395

FIGURE 5.17 Resource requirements planning process.

adjustments, RCCP is used primarily to ensure that *existing* resources are used effectively. However, not all resources are planned at this level. Only those resources considered critical are planned by RCCP. A critical resource is generally thought of as a known or suspected bottleneck that is likely to be a limiting factor in attaining the MPS. The bills of resources used at this level generally come from an analysis of the BOM structure for each MPS item, and the routings associated with every part in the BOM. Extracting the load and time-phasing information for critical resources allows the bill of resources to be developed. Other resources might be

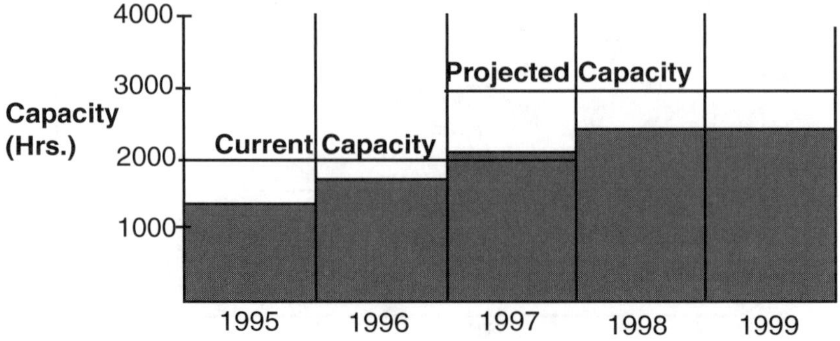

FIGURE 5.18 Resource graph for sawing, showing current and planned capacity versus projected workload.

added to this basic structure at management's discretion. Examples might include the following:

Hours on a lathe that is used by multiple products
Workers with a critical skill, such as precision welding
Cash
Hours required to pick parts from stores
Packaging operation(s)

The horizon for these decisions is considered medium-term, typically 6 months to 2 years. Figure 5.19 and Figure 5.20 show the RCCP process, adding the effect of lead-time offsetting of resources. For the sake of simplicity, only cabinets are shown. An actual case would include all products in the MPS.

Beginning with Figure 5.19, let us walk through a calculation of sawing requirements for all cabinets. As indicated in the figure, 3 weeks prior to completing a cabinet A assembly, 0.2 hr. of sawing is required. The MPS table in Figure 5.20 shows a requirement for eight such cabinets in period 4. Therefore, 3 weeks earlier (period 1), the requirement for sawing is 8×0.2 hr. $= 1.6$ hr. Repeating this process for all products in the MPS and summing the totals results in the total number of hours indicated at the bottom table of Figure 5.20.

As in the RRP example above, the next step is to compare these totals against the sawing capacity. The graph in Figure 5.20 shows the resulting capacity plan, assuming two capacity levels, demonstrated and theoretical. Any overload conditions should be corrected before running MRP against the MPS. This may require minor adjustments in capacity through overtime, off-loading work, etc., or may require the MPS to be modified. Loads falling between the two capacity lines are generally managed by adding capacity, while loads above the theoretical (maximum) capacity almost always require revision of the MPS. In our example, no overloads exist, indicating the MPS is attainable.

Once an attainable MPS is established, MRP may be run. Its associated level of capacity planning is discussed next.

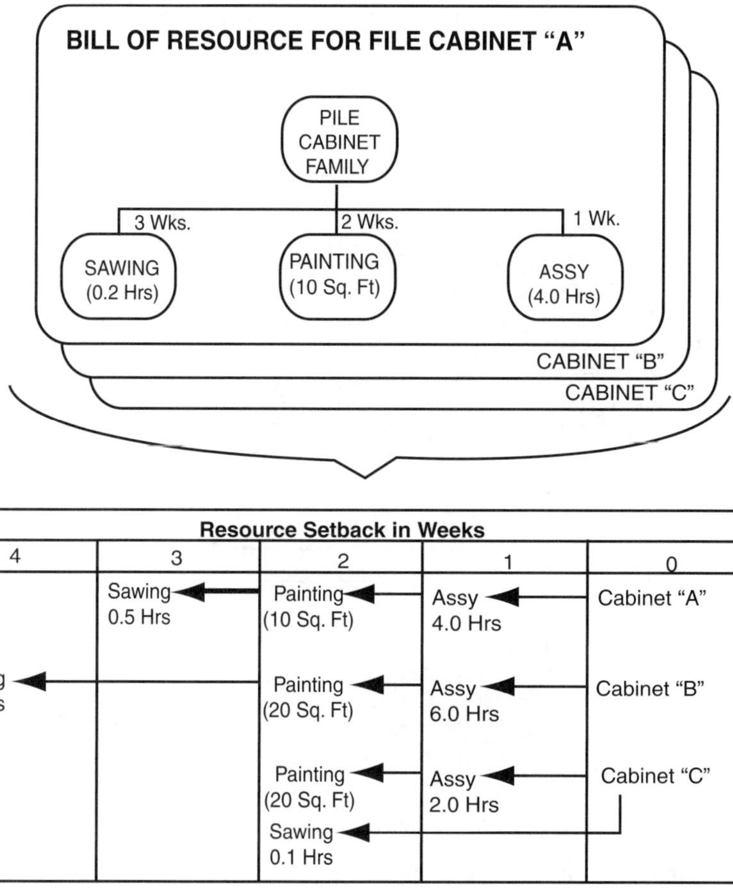

FIGURE 5.19 Rough-cut capacity planning (RCCP) of critical or bottleneck resources resulting in a time-phased bill of resource.

Capacity Requirements Planning

Once MRP has run, and produced a new set of planned orders, capacity requirements planning (CRP) may begin its task. In addition to MRP-planned orders, CRP takes into account all released and open orders, considering the status of each to determine present location and next operation (see Section 5.2.6, on shop floor data collection). The first step in CRP is to determine the start and/or finish dates and times for each order's remaining operations. Recall the detailed scheduling approach discussed in Section 5.2.5, where detailed operation-level schedules were developed by the shop floor control system using a forward scheduling calculation. CRP backward scheduling is identical to this process, only in reverse.

The purpose of CRP is to provide a forecast of load at each work center, enabling production control to manage the current and anticipated workload at each. A sample work center load report is shown in Figure 5.21. Notice that, in addition to the load

Master Production Schedule (MPS)										
	Week Number									
	1	2	3	4	5	6	7	8	9	10
Cabinet "A"	8	8	8	8	8	8	10	10	10	10
Cabinet "B"	32	32	32	32	32	32	40	40	40	40
Cabinet "C"	30	30	30	30	30	30	30	30	30	30

8 x 0.2 Hrs. = 1.6 Hrs
3 Week Offset

40 x 0.5 Hrs. = 20 Hrs
4 Week Offset

Rough Cut Capacity Plan for Sawing										
	Week Number									
	1	2	3	4	5	6	7	8	9	10
Cabinet "A"	1.6	1.6	1.6	2	2	2	2			
Cabinet "B"	16	16	20	20	20	20				
Cabinet "C"	3	3	3	3	3	3	3	3		
Total Hrs.	20.6	20.6	24.6	25	25	25	5	3	0	0

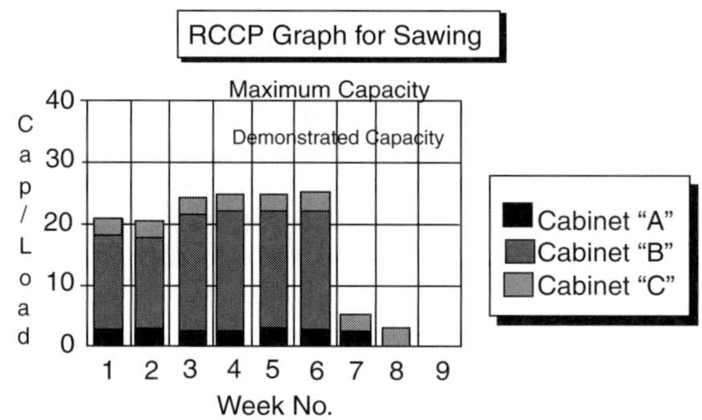

FIGURE 5.20 Example of the RCCP process.

graph, detailed information on orders is provided to show the sources of each period's load. This information is critical when decisions such as moving work between periods must be made. Notice also that an "infinite loading" approach was used—i.e., no attempt was made to limit the load in each period to the available capacity. This is fairly typical in CRP. Finite loading is generally reserved for the shop floor control system, as discussed in the next subchapter.

Notice also that dates are shown as numeric values representing manufacturing days. While this is fairly common in older systems, most newer software converts these values to their corresponding calendar dates and displays them using normal date conventions, e.g., 940304, 3/4/94, 4-Mar-94, etc.

Finite Loading and Input/Output Control

In Figure 5.15 we saw a comparison of infinite versus finite loading. The finite load graph is repeated in Figure 5.22 with further detail to show the process involved. The

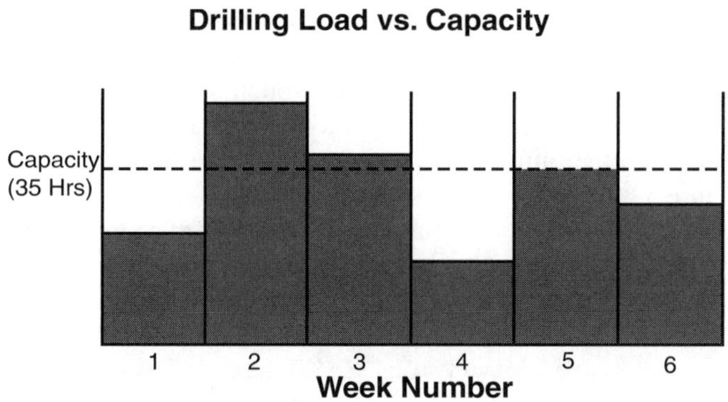

Work Center: Drilling						
Capacity: 35 Hours/Week						
Week	Load Hours	Part Number	Start Date	Finish Date	Order Number	Order Status
1	10	ABC	230	231	123	In Work
	5	DEF	232	232	234	In Work
	8	GHI	233	234	345	Released
Total	23					
2	4	JKL	235	235	456	Released
	12	MNO	236	237	567	Planned
Total						

FIGURE 5.21 Capacity requirements planning (CRP) example taken from the work center report, showing drilling requirements and capacity.

loads moved from periods 2 and 3 represent operations containing sufficient load hours to bring the total down to the capacity level. Notice that work was moved forward into an underloaded period. If the loads depicted by this graph were developed by a CRP backward scheduling calculation, this would likely mean that the operations moved will finish behind schedule, since the process of backward scheduling results in just-in-time due dates and operation dates. If, on the other hand, the loads were developed by SFC using a forward scheduling calculation, moving work forward would more likely consume some of the order's slack time, without jeopardizing the order due date.

Assuming that we now have a viable capacity-balanced schedule, the next step is to ensure that each work center's load is properly managed. One method for accomplishing this is called "input/output control."

The input/output report is used to monitor the actual versus planned load entering a work center and the actual versus planned load leaving it. Its purpose, therefore, is to monitor planned versus actual work flow through the work center. Its horizon is considered short-term, typically covering 4 to 8 weeks in the past (i.e., actual flow) and 6 to 8 weeks in the future (i.e., planned flow). A sample input/output report is shown in Figure 5.23.

A few observations will aid in our understanding of this report. First notice that, over the previous 5 weeks, total actual input exactly matches total planned input,

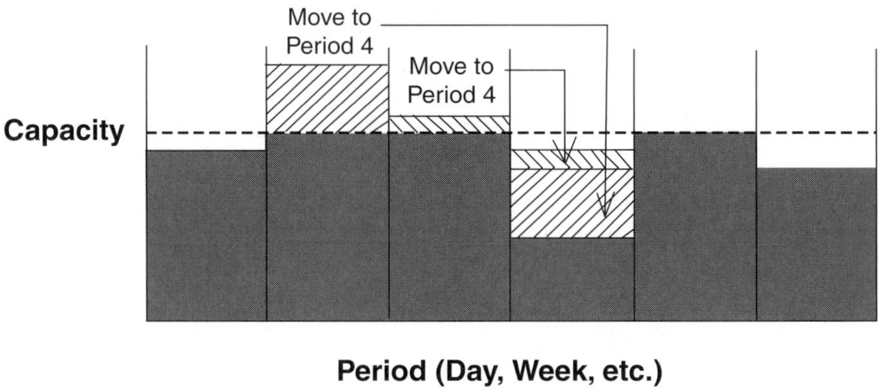

FIGURE 5.22 Example of the shop floor control system adjustment of the infinite loading of work orders, to balance capacity by moving orders earlier in the queue and achieve a more balanced workload.

		Week													
		Actual					Planned								
		–5	–4	–3	–2	–1	0	1	2	3	4	5	6	7	8
Planned Input		25	25	25	25	25	25	25	25	30	30	30	30	30	30
Actual Input		20	22	30	30	23									
Cum. Delta (Plnd-Actl)		5	8	3	–2	0									
Planned Output		20	20	20	25	25	25	25	25	30	30	30	30	30	30
Actual Output		18	20	15	25	28									
Cum Delta (Plnd-Actl)															
Actual Backlog	25	27	29	44	49	44									
Planned Backlog		30	35	40	40	40	40	40	40	40	40	40	40	40	40

Desired Backlog: 40 Hours

FIGURE 5.23 Sample input/output report showing the planned and actual receipt of orders in a work center, versus the planned and actual output.

resulting in a cumulative delta of zero hours. While week-to-week fluctuations occurred, overall the flow into the work center matched the plan. Likewise, the actual output is reasonably close to its plan, adding only 4 hr. to the cumulative delta.

Second, notice that at the beginning of our tracking period a backlog of 25 hr. existed, versus a goal of 40 hr. By planning more input than output over the next 5 weeks, we sought to increase the backlog to the desired level. However, because less output was produced than was planned (i.e., cumulative delta of 4 hr.), the actual backlog is 4 hr. greater than desired. If a cumulative input delta had also existed, the difference between planned and actual backlog would be offset by that amount as well.

5.2.8 Summary

Throughout this subchapter we have examined the basic elements of scheduling and capacity planning employed in an MRPII environment. A production plan sets the overall production rates for families of products over a relatively long time frame and provides the input to the resource requirements planning (RRP) system. The resulting resource plan is used by upper management to support long-range decisions regarding brick and mortar and other major capacity adjustments. The master production schedule (MPS) translates the production plan into a set of requirements, expressed in specific configurations, quantities, and dates, over a medium time frame. Its associated level of capacity planning, known as rough-cut capacity planning (RCCP), tests the MPS against available capacity on critical resources. Once an attainable MPS is established, material requirements planning (MRP) explodes these requirements down a bill of materials (BOM) product structure, netting against available inventory, to create a set of planned orders for both manufactured and purchased parts. Capacity requirements planning (CRP) calculates the resulting work center loads, taking into account planned, open, and released orders. Finite scheduling may be used to further adjust schedules at a routing operation level, allowing only as much work to be scheduled as each work center can produce. Input/output control can be used to monitor work flow through a work center and identify the need for capacity adjustments in the immediate time frame.

5.3 JUST IN TIME

APICS defines *just in time* (JIT) as:

> A philosophy of manufacturing based on planned elimination of all waste and continuous improvement of productivity. It encompasses the successful execution of all manufacturing activities required to produce a final product, from engineering to delivery and including all stages of conversion from raw material onward. The primary elements of zero inventories (synonym for JIT) are to have only the required inventory when needed; to improve quality to zero defects; to reduce lead times by reducing setup times, queue lengths, and lot sizes; to incrementally revise the operations themselves; and to accomplish these things at minimum cost. In the broad sense it applies to all forms of manufacturing job shop and process as well as repetitive. (*APICS Dictionary*, 1992)

Because a just-in-time philosophy encompasses all a manufacturing company's activities, it is wise to start small and build on success rather than attempt an all-out conversion. In *Just-in-Time: Surviving by Breaking Tradition* (1986), Walt Goddard recommends approaching a JIT implementation in three steps: Begin with people issues, then move on to physical plant issues, and finally deal with changes that the earlier improvements will require in the way of computer systems.

It is important to recognize that the just-in-time journey is not linear, but circular. It is a philosophy of continuous improvement, not a destination. No matter how good a company becomes, how excellent its quality, how short its lead times, how good its prices and delivery performance, it can always improve. In fact, it must improve to stay ahead of the competition, which is always in hot pursuit.

In this subchapter we will briefly examine each of the key elements introduced in the definition given above. Readers should note that this is a thumbnail sketch of JIT only. Additional sources of information on JIT are found in the bibliography.

5.3.1 People Issues

When dealing with people issues, it is advisable to start with education. Put the users in charge, and organize them in teams. Provide professional facilitation training and detailed subject matter training to the team leaders to ensure they are prepared to handle team meetings, and can address both the people and technical issues that are sure to arise. Cross-train all workers so they can perform multiple tasks, not just in specific manufacturing jobs, but in teamwork skills, process simplification, and other improvement techniques. Such a "whole-person" approach utilizes all the knowledge and capabilities of employees, provides improved flexibility for the company, and benefits employees at the same time.

5.3.2 Elimination of Waste

At its heart, JIT is the relentless pursuit and elimination of waste. Waste is defined in this context as anything that does not add value to the product, such as poor quality, double handling and storage of materials, large inventories, long setup times, and paperwork. The goals of JIT are to produce quality products, on time, at the best price, and with the shortest possible lead time. Waste in any form interferes with a company's ability to meet these goals. Beginning with lead times, we will discuss each of these elements and introduce methods for reducing or eliminating them.

5.3.3 Reduced Lead Times

In Section 5.2.6 (Figure 5.12), we examined the circular effect of lead times, WIP, and queue times. If lead times are reduced, lower work in process levels are required to produce at a given rate, in turn reducing queue times. Because queue times (prior to JIT) account for the largest single element of a part's lead time, reducing queue time further reduces lead time. We noted, however, that no impact on actual processing time—made up of setup and run times—was involved in this cycle. Here we turn our attention to setup reduction as a key element in reducing lead times.

To better understand setup time and ways in which this time may be reduced, we must start with a clean definition. We will define setup time as follows, further breaking it down into two elements:

Setup time—The time from making the last good part A to the first good part B.
Internal setup—Activities that can only be performed when a machine is not running.
External setup—Activities that can be performed while a machine is running.

The traditional approach to dealing with setups has been to assume that they are fixed. To compensate, the setup time (and cost) must be amortized, or spread, among many parts by producing in batches. A time-honored technique for determining batch size is the "economic order quantity" (EOQ). Figure 5.24 shows this approach graphically. Two costs are considered. First is the cost per part (descending line), indicating the effect of amortizing setup on an increasingly large batch. Second is the carrying cost (ascending line), which recognizes that storage of parts, once produced, is not free. The EOQ is the point at which the combined total cost (sum of the two lines) is minimized.

Because the bottom of the total cost curve is relatively flat, the calculated EOQ is frequently adjusted by management decree to "make at least 3 month's worth," or similar policy decisions. Such approaches ignore other important considerations, however. First, large batches increase the amount of time orders sit in queue, increasing lead times and decreasing flexibility. Second, quality problems are typically not found until parts from the batch are used in a later operation. By then, so much time has elapsed since they were produced that determining the root cause is far more difficult than if it had been identified at the source. A third factor is the cost of loss, damage, or obsolescence. The longer a part is held in inventory, the greater is the chance that it will be lost or damaged, or that it will be superseded by a new design. Even if design changes can be delayed until on-hand inventory is used up, the company pays a price in customer satisfaction by not being able to bring design improvements to the marketplace quickly.

A far better approach is to limit batch sizes, ideally to one, but to at least begin down the path. The place to start is with setup reduction. In the above example, if we reduce the setup time to 30 min. (0.5 hr.), we see that the EOQ becomes 2, and the additional cost to build just one is only 50 cents (Figure 5.25). It is now feasible to build only the quantity needed, with less lead time than was previously required to build the entire batch, thereby avoiding all the waste previously attributed to batch production.

The question remains, though, of how we achieve the setup reduction. A key approach is to perform as much of the setup as possible in parallel with the current job's run time. In other words, convert as many of the setup steps as possible to "external" setup and perform them all ahead of time. Then, minimize the remaining "internal" setup steps.

Setup Time (Hrs.)	10
Run Time (Hrs.)	0.1
$/Hour Rate	$12.00
Carrying Cost Rate	$2.50

Cost Per Part =
((S/U + BS x Run) x $/Hr) / BS

Carrying Cost =
BS x Carrying Cost Rate

Batch Size	Cost Per Part	Carrying Cost	Total Cost
1	$121.20	$2.50	$123.70
2	$61.20	$5.00	$66.20
3	$41.20	$7.50	$48.70
4	$31.20	$10.00	$41.20
5	$25.20	$12.50	$37.70
6	$21.20	$15.00	$36.20
7	$18.34	$17.50	$35.84
8	$16.20	$20.00	$36.20
9	$14.53	$22.50	$37.03
10	$13.20	$25.00	$38.20
11	$12.11	$27.50	$39.61
12	$11.20	$30.00	$41.20
13	$10.43	$32.50	$42.93
14	$9.77	$35.00	$44.77
15	$9.20	$37.50	$46.70
16	$8.70	$40.00	$48.70
17	$8.26	$42.50	$50.76
18	$7.87	$45.00	$52.87
19	$7.52	$47.50	$55.02
20	$7.20	$50.00	$57.20

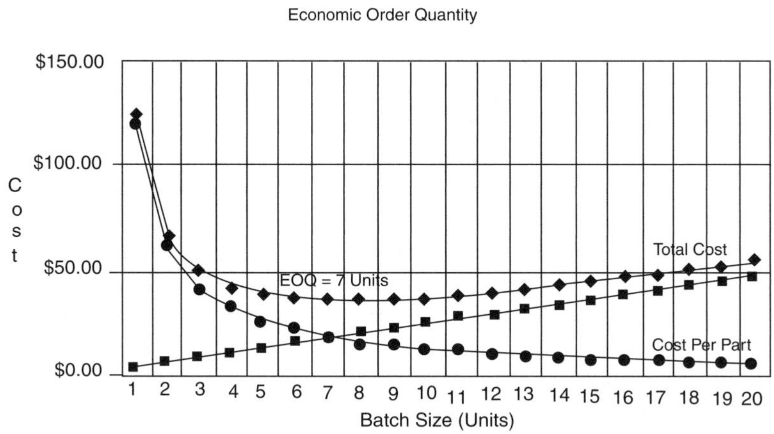

FIGURE 5.24 Example of the time-honored system of economic order quantities—prior to considering JIT or some other system.

Some specific techniques for reducing setup times are outlined by Walt Goddard in *Just-in-Time: Surviving by Breaking Tradition* (1986):

1. Standardize the external setup actions; replace adjustable gauges with permanent ones.
2. Put all probe and blow-off hoses on one side of die.
3. Put a bench at the side of the press at the same level as the press opening, to hold the next die.

Setup Time (Hrs.)	0.5
Run Time (Hrs.)	0.1
$/Hour Rate	$12.00
Carrying Cost Rate	$2.50

Batch Size	Cost Per Part	Carrying Cost	Total Cost
1	$7.20	$2.50	$9.70
2	$4.20	$5.00	$9.20
3	$3.20	$7.50	$10.70
4	$2.70	$10.00	$12.70
5	$2.40	$12.50	$14.90
6	$2.20	$15.00	$17.20
7	$2.06	$17.50	$19.56
8	$1.95	$20.00	$21.95
9	$1.87	$22.50	$24.37
10	$1.80	$25.00	$26.80
11	$1.75	$27.50	$29.25
12	$1.70	$30.00	$31.70
13	$1.66	$32.50	$34.16
14	$1.63	$35.00	$36.63
15	$1.60	$37.50	$39.10
16	$1.57	$40.00	$41.58
17	$1.55	$42.50	$44.05
18	$1.53	$45.00	$46.53
19	$1.52	$47.50	$49.02
20	$1.50	$50.00	$51.50

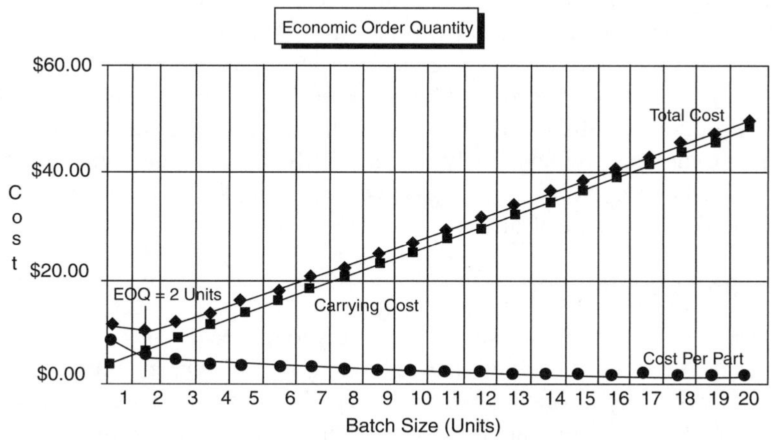

FIGURE 5.25 Reducing setup time for a batch reduces cost per part and total cost—showing that EOQ is now 2 units (rather than previous example of EOQ = 7).

4. Color code all hose connections: air, hydraulic, water, etc.
5. Use parallel operations—deliver all components to support die setup (use a check-off sheet to ensure all are present prior to setup).
6. Design a quick locating system—positioning pins and holes, with quick fasteners.
7. Standardize all die receptacles.
8. Add a tonnage monitor on the press to detect two pieces in die before damaging dies.

9. Involve tool-and-die designers in setup reduction programs so all new designs incorporate quick change-over concepts.
10. Use two-way radio between the setup man and lift-truck operator who removes and delivers dies.
11. Photograph completed operation as a guide for the setup man: location of tables, wrenches, baskets, etc.
12. Review material flow charts with reduced movements in mind.
13. Make as many of the setup activities as possible internal to the run time. That is, do as much of the setup as possible without shutting down the machine.
14. Standardize all bolt sizes.
15. Code parts on the dispatch list for major or minor setups to aid scheduling.
16. Standardization and use of common parts in the product will reduce the number of different parts required. If design engineering does not design a new part, no setup is required for it.

Remember to start small. The idea is to strive for continuous improvement, not to reach the ultimate solution in one quantum leap. Setup reduction goals, however, should be aggressive. When dealing with machine setups, for example, the goal should be for "single-minute exchange of die" (SMED). Any set time less than 10 min. (i.e., measured as single minutes, versus tens of minutes, hours, etc.) qualifies as SMED. A good analogy for those who consider this an impossible goal is that of a racetrack pit crew. While it might take the average person 15 min. to change a tire and gas up, the pit crew takes only 15 sec.! This comes from teamwork, detailed study of the process, proper tools, practice, and attitude. It can be done. And the payback is well worth the effort.

5.3.4 Reduced Inventory

A direct result of reduced lead times is the ability to reduce inventory. At the finished-goods level, if a product can be produced within the delivery lead time required by customers, why carry finished goods inventory? The same holds true of internal customers. If parts can be built one-for-one as the next process calls for them, why build them ahead of time only to pay for storage, additional handling, and potential costs of obsolescence, loss, or damage?

Many companies use inventories to hide problems associated with poor scheduling, poor quality, and large setups. A good analogy is a river with a rocky bottom. The traditional approach is to raise the water level in the river, thus hiding the problems. This of course is expensive and wasteful, but far easier than dealing with the problems head-on. The JIT alternative is to slowly lower the water level (i.e., reduce inventory), exposing the rocks (problems). Deal with each problem as it occurs, and not just the immediate symptom. Address and correct the root cause. Then lower the water some more. Another approach is to dive below the surface to identify problems before they surface. In either case, the intent is to solve root-cause problems, so they are gone forever.

5.3.5 Zero Defects

As the name implies, a just-in-time system requires that the right part be delivered to the right place just as it is needed. But if the part is of poor quality, on-time delivery is worthless. Quality has been defined by Phil Crosby as "conformance to requirements," and by J. M. Juran as "fitness for use" (Goddard, 1986). The former, "conformance to requirements," is of primary interest to internal customers who must install or use the parts previously produced in a subsequent operation. The latter, "fitness for use," is often the final customers' ultimate yardstick. Unless a product does what the customer wants it to do in a particular application, it fails this criterion, even though it may conform to a set of technical requirements.

The goal of a quality improvement program should be to strive for zero defects. Along the path, manufacturers must first seek to replace traditional definitions of "acceptable quality," measured in percentage points, to parts per million. Where 98% quality might have been acceptable yesterday, today's target is 6 sigma, or just over 3 parts per million. One method used to great benefit by JIT practitioners to ensure high quality is "statistical process control" (SPC). What SPC does is to shift the focus from inspection of completed parts to monitoring the process used to produce parts. If a process is capable of producing quality parts, and the process is kept in control, then all the parts produced by the process will satisfy the "conformance to requirements." Figure 5.26 shows three sample SPC control charts. The basic elements include upper and lower control limits (UCL/LCL), within which the process must remain to be considered "in control." (*Note:* These limits identify the natural variability of the process, not the design tolerances of the parts being produced. The UCL and LCL must be within these design tolerances. What is being measured is "process control," so that even if a process has gone out of control, it may not have produced bad parts yet.) Also included is a centerline and observations occurring over a span of time. Notice that two out-of-control conditions exist. The first is depicted by the middle chart, showing a single point out of the specified control limits. The second, depicted by the bottom chart, shows a situation where seven sequential points have occurred between the center line and the upper control limit. Such a situation is not statistically random and should be interpreted as the process being out of control.

A natural result of SPC is to allow workers to monitor their own processes and the quality of the parts they produce. This is referred to as "operator verification," "self-inspection," and "quality at the source," among other names. Such responsibility is often coupled with authority to stop the production line when a quality problem is found. While this may sound extreme, it serves to focus everyone's efforts on solving the immediate problem, preferably at the root-cause level so the problem is unlikely to occur again.

If SPC is to ensure the production of quality parts, the natural variation in the processes must be held within the specified control limits. Perhaps the best way to accomplish this is with routine preventive maintenance (PM). Virtually all machine manufacturers specify PM programs for their machines. Follow them. Schedule PM routinely and do it religiously. This is key to ensuring that processes stay in control. An excellent practice is to make machine operators responsible for at least some

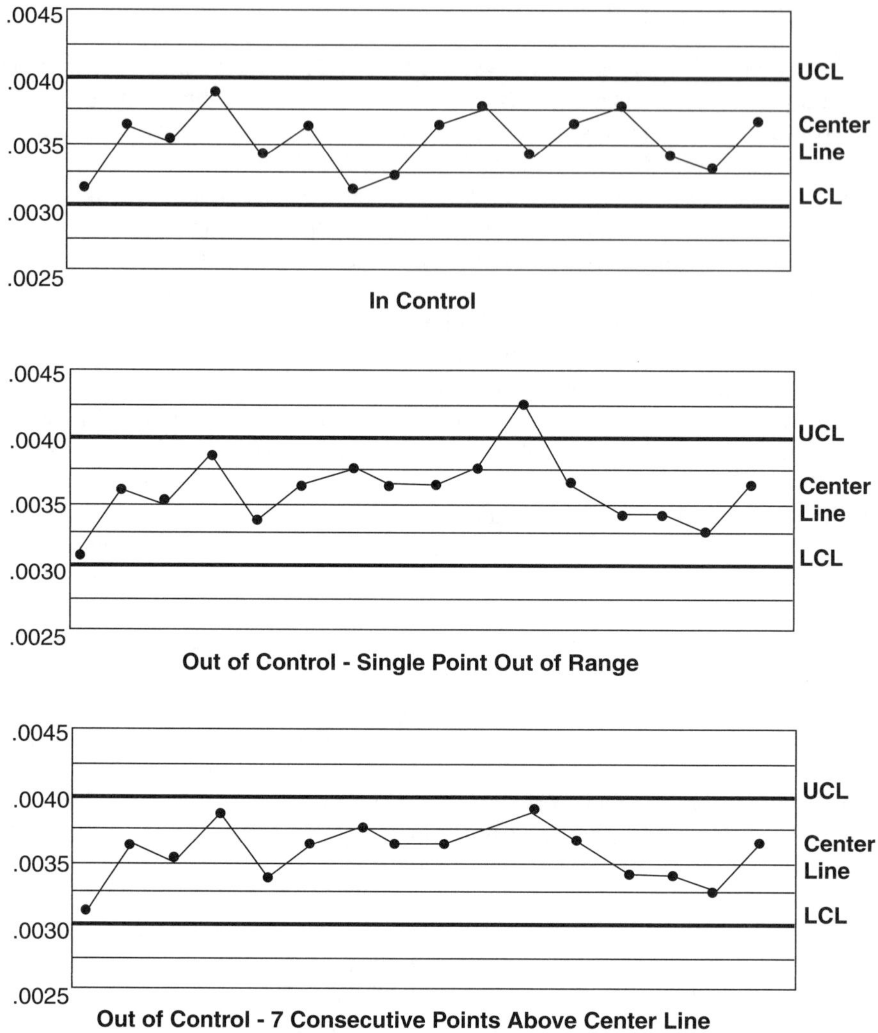

FIGURE 5.26 Sample SPC control charts.

PM activities. They know from experience how the machine should look, feel, and sound. Who is better, then, to keep them running at peak performance?

Yet another way to improve quality is to use foolproofing, or fail-safe techniques. The intent is to ensure that a process can be performed only one way—the right way! Some examples include (Goddard, 1986):

Use checklists or monitoring devices to detect parts missing on assembly. Do not allow them to pass to the next operation until corrected.

Use locator pins to ensure parts are aligned correctly, or dies are installed properly during machine setup.
Perform weighing operations to detect missing or extraneous parts or materials.

One final subject dealing with quality involves design engineering. No amount of effort, skill, or desire on the part of manufacturing can overcome the effects of poor product design. For this reason, design engineering must develop designs with manufacturing limitations in mind. This "concurrent engineering" (or "design for manufacturability," as it is often termed) is critical to the success of a just-in-time effort. Some examples to clarify the possibilities follow.

Replace multiple sheet metal components with a single molded part.
Replace highly complex parts with simpler parts, taking advantage of "compensating tolerances."
Move options higher in the bill. Avoids adding high-percentage options as a standard and reconfiguring to customer requirements later.
Move difficult-to-install components lower in the bill, where access may be easier.
Move design engineering offices close to production floor.

While some of these may sound contradictory, the message is to find what works best in a given situation. There is no cure-all. Each case is different, and must be addressed by the responsible functions and teams.

5.3.6 Valid Schedules

If an objective of JIT is to have the right part at the right time, then it stands to reason that schedules, which specify time and place, must be valid. In Subchapters 5.2 and 5.5 the subject of scheduling is addressed from an MRPII perspective. It has been suggested, erroneously, that MRPII and JIT are incompatible. Nothing could be further from the truth. In fact, most successful practitioners of JIT use MRPII very successfully, and consider the resulting schedule validity that MRPII brings as critical to their success with JIT. Where the difference often lies is at the material planning level (i.e., "little MRP"). While many companies practicing JIT use MRP to schedule orders, an alternative scheduling technique that others have found beneficial is based on the "Toyota production system." The Toyota system is identical to MRPII from the business plan level down to MPS. It is not until the MRP level that the systems diverge.

Where MRP plans orders based on future demand, the Toyota system employs a pull-scheduling approach using *kanbans,* a Japanese term meaning "visual signal." *Kanbans* can take the form of cards, golf balls, squares painted on the floor, or any other visual signal indicating the need for a particular part. *Kanban* is similar to a two-bin, reorder point system (see Section 5.6.4). An empty bin (or square, card, etc.) signals the need to replenish what has been used. *Kanbans* are most effective in stable, repetitive environments, where lead times are fixed and demand is regular.

The primary limitation of *kanban* is the lack of future visibility. For example, if an engineering change causes a part to be superseded by another, continuing to replenish the old part could be disastrous. To compensate, Toyota explodes the BOM against the master schedule to provide a 90-day summarized forecast of all components. When a particular part is no longer needed, a special *kanban* is inserted into the system to signal when its production is to cease.

The bottom line, regardless of the scheduling approach employed, is to make only what is needed by the immediate consumer. If nothing is needed, *do not produce.* Doing so only builds excess inventory, which is of course waste.

A useful technique employed by both systems (MRP and *kanban*) for stabilizing production rates, a critical element of JIT, is known as "mixed-model scheduling." The goal of mixed-model scheduling is to make some of every product every day. For example, if the ratios of products A, B, and C are, respectively, 60%, 30%, and 10%, a mixed-model schedule for a total of ten items per day might look like this:

Day 1	Day 2	Day 3
(6 A's, 3 B's, 1 C)	(6 A's, 3 B's, 1 C)	(6 A's, 3 B's, 1 C)

Mixed-model scheduling is usually employed at the MPS level. Prerequisites to such scheduling are the ability to produce in small order quantities and to switch production quickly from one product to another. As discussed earlier, this is the result of having short or nonexistent setups.

5.3.7 Vendors

The key point to be made when bringing vendors on board in a JIT program is to lead by example. A company must begin by getting its own house in order, with active programs aimed at quality, setup reduction, dependable scheduling, and order quantity reduction. The best indication to a vendor that you are prepared to lead the way is to provide stable, valid schedules over a long enough period of time to establish your own credibility, *before you even approach them.* Next, provide them with education. Tell them what it's all about and prepare them for what they are about to see. Then bring them in and show them what you have done internally that has enabled you to provide the schedule stability they have already witnessed. However, don't wait too long. You don't have to be expert at JIT, just far enough along to show concrete improvement and the commitment to continue.

The goal in bringing vendors on board is to replace the traditional adversarial role with vendors, and between competing vendors, with partnerships between the company and a greatly reduced vendor base. The benefit for the company is in having vendors who are able to deliver quality parts, just in time, at a better price. The benefits to the vendors who survive are the same, providing a competitive edge over their competition, which their own marketing people can use to good advantage. In addition, the remaining vendors will enjoy a larger share of your business, since they will no longer compete with others for the same parts and materials.

Not all vendors will be willing to come along, and not all should be considered. First weed out those with chronic quality or delivery problems. Stack the deck in your favor and take only those with a high likelihood of success through the process. This process in itself may take a year or more. Use this time for your own in-house efforts, then begin the process of bringing the survivors on board. Because the vendor base is reduced, buyers can take on a new role, becoming "vendor managers," who will assist the companies they work with in their own JIT efforts, ranging from SPC to setup reduction to valid scheduling. Rather than searching for additional "just-in-case" suppliers, and expediting when problems occur, they can spend their time developing the partnering relationships so critical to success with JIT.

Let us look ahead now to a few problems (opportunities!) that will have to be addressed once vendors are successfully delivering just in time. First is the issue of transportation. When vendors are delivering more frequently (i.e., daily or more frequently versus monthly), managing the traffic in receiving can become a problem. A good way to overcome this is to use public carriers with either a central delivery point and vendors delivering to the carrier, or a milk-run approach where the carrier makes frequent pickups from each of the vendors' sites and then delivers to the company at a prearranged time. Carriers must also be educated in JIT philosophy, so they understand the importance of on-time delivery. The consolidation and partnering arrangements made with vendors apply to carriers as well.

Second is the potential impact on the accounting department. The goal should be to not overwhelm them with paper. While it may be acceptable with infrequent deliveries to process all shipments through an existing accounts payable process, daily or hourly deliveries might quickly bury accounting in paper. The best approach is to involve the accountants early on in the implementation, and devise ways to resolve such problems ahead of time. One approach is to do away with individual orders and invoices, replacing them with a once-a-month invoice. Some companies go so far as to eliminate this process entirely. Since they know how many of a vendor's parts are needed to complete each end product, they simply total the number of products sold and in work once a month and pay the vendor via electronic funds transfer for that amount of product.

Similarly, if receiving inspection is to keep pace, simplifying or eliminating traditional processes is a must. Applying the same logic as worker self-inspection, once a vendor has demonstrated the ability to produce with consistently high quality, they might be allowed to bypass receiving inspection altogether. Combined with JIT delivery of small lots, the logical extension of this is to allow them to deliver directly to the shop point of use, bypassing not only receiving inspection but the warehouse as well.

Finally, consider electronic links between the company and its vendors. The goal here is the elimination of paper (waste, since it adds no value). In its simplest form this might involve giving vendors inquiry-only access to your scheduling system, enabling them to view online the current and future needs for their products. More advanced capabilities might include electronic PO placement, invoicing, and even electronic funds transfer to pay for parts received.

5.3.8 Plant and Processes

An objective of JIT is to move a company's manufacturing as much toward a process-flow, or repetitive, environment as possible. Another is to enhance visibility between operations so that communication between work centers is enhanced, and problems, once identified, can be more readily dealt with.

One method of accomplishing these objectives is the use of manufacturing cells. Cellular manufacturing brings machines and workstations together to make similar parts or products. (Cellular manufacturing is related to group technology [GT], in that GT classifies parts in such a way as to identify candidate parts for a given cell.) In a traditional job-shop layout, parts are routed between machines and workstations that have been arranged by function. Many companies' managers are astounded to learn just how far a part travels and how many non-value-added steps are involved in production. By bringing machines together in cells, these problems are avoided. Work enters a cell and is processed sequentially across the appropriate machines, leaving the cell as a completed part. Reductions in distance traveled from thousands of feet, or even miles, to tens of feet, and in lead times from weeks to hours are not uncommon.

A resulting benefit also comes in the form of simplified scheduling and tracking. Without cellular manufacturing, each step or group of steps in the process may have required a new part number with its own level in the bill of material. Individual shop orders are certainly required in this situation. With cellular manufacturing, none of this is needed. A single part number moving through the cell is all that is required. And because the part moves through so quickly, the need to track intermediate steps in the process is eliminated. Cellular manufacturing, then, has the effect of "flattening" bills of material, simplifying scheduling, reducing paperwork, and reducing waste.

A second change in plant layout frequently employed is the use of U-shaped assembly lines. Where assembly lines exist, arranging them in a U-shaped configuration has the dual effect of improved visibility and reduced space, both of which improve communication between workers on the line, enhancing teamwork and product quality.

5.3.9 Computer Systems

The previous discussions lead to some conclusions about a company's operating software, such as MRPII (Goddard, 1986). First, to accommodate reduced lead times, time periods must be daily or smaller. The ability to replan daily or more frequently is also a must, preferably in a manner similar to net-change MRP (see Section 5.5.4). To accommodate point-of-use inventory, the ability to identify multiple inventory locations, not just bin locations in a common stockroom, will be needed. If paperwork, such as shop orders, is to be reduced or eliminated, electronic or other means of maintaining inventory balances, labor collection, and other performance-related data must be provided. One such approach is known as "inventory back flush" or "post-deduct" logic. Rather than a traditional order-based approach, where inventory is pulled from stock, the inventory matched to an order, the parts produced, and finally the inventory returned to stock to be received, back flushing automatically performs these transactions when an order is completed and moved to its "consuming" location.

5.3.10 Summary

JIT manufacturing is the relentless pursuit and elimination of waste. Waste comes in many forms, including poor quality, invalid schedules, excess inventory, unnecessary material handling, excess floor space, and anything else that adds cost but does not add value to a company's products. A number of tried-and-true approaches for reducing or eliminating each of these elements of waste have been demonstrated by JIT pioneers, and are available for companies wishing to begin their own journey.

The benefits are tremendous. Combined, these approaches, and the attitudes they foster, improve every competitive element of a manufacturing company. Quality is improved through statistical process control and shortened feedback; delivery performance is improved through reduced lead times, smaller order quantities, and valid schedules; and cost is lowered through the elimination of waste.

Workers contribute to a process of continuous improvement through a whole-person concept that treats them as the experts they are, not just as cogs in a division-of-labor wheel. They are given the education and tools required to do the job right, and the authority to stop the line when something goes wrong. Problems are addressed at the root-cause level, not hidden with excess inventory.

Successful JIT implementations begin by addressing the people issues, then move to physical changes in the plant's processes and finally to ensuring that the company's computer systems support the requirements of the new processes. All functions participate, including the company's vendors, who are dealt with as partners, not as adversaries.

JIT is a journey, not a destination. No matter how good a company becomes, it can always improve. JIT provides the banner under which to proceed.

One final word of caution: Since a successful JIT program hinges on people, don't sabotage your efforts by rewarding initial gains with a layoff. That is the surest way to guarantee the program's failure. Instead, take the opportunity to cross-train workers, provide new skills, or work on the next JIT improvement. When the benefits begin to take hold, customers will notice. Higher quality at less cost and superior delivery performance in less time than the competition will bring new business, avoiding the need for layoffs.

5.4 THEORY OF CONSTRAINTS

In his book *The Goal* (1984), Eliyahu M. Goldratt introduces, in novel form, the principles of his "theory of constraints" as applied to a manufacturing environment. Also known as optimized production technology (OPT), the application of the theory's principles allows the novel's main character to turn his failing plant into the corporation's most profitable. Following is a brief discussion of the major elements of *The Goal*. Readers are encouraged to read the book, which brings these principles to life in a way not otherwise possible.

5.4.1 The Goal

The goal of a manufacturing company is to make money. However, this goal is sometimes achieved in the short term, to the detriment of the company's long-term viability, by selling off assets, for example. Therefore, the goal should be "to make money now as well is in the future." Goldratt introduces three measurements to monitor progress toward this goal: throughput, inventory, and operational expenses. He defines these as follows (Goldratt and Fox, 1986):

Throughput—The rate at which the system generates money through sales.
Inventory—All the money the system invests in purchasing things the system intends to sell.
Operating expense—All the money the system spends in turning inventory into throughput.

To achieve the goal, throughput must be increased while simultaneously reducing both inventory and operating expense.

Further study reveals some interesting differences between these definitions and their more standard business definitions. For example, throughput is traditionally measured at the point where finished goods are completed. By contrast, applying the above definition, unsold finished goods are considered inventory; increasing this inventory works counter to the goal and should be avoided. A second example might be inventory carrying costs. Carrying costs typically include such expenses as the cost of capital invested in inventory, taxes, insurance, obsolescence, warehouse space, etc. These costs are traditionally prorated and applied to each part in inventory. In the definition above, these costs are all considered operating expense.

5.4.2 Troop Analogy

An excellent example used in the book to illustrate these points is the analogy of a Scout troop on a hike. In this example, the following definitions apply:

Throughput—The rate at which the entire troop (the system) progresses; i.e., the distance covered by the slowest Scout.
Inventory—The distance between the first and last Scout. As the first Scout passes a given spot, this spot is added to inventory. The system retains this inventory until it is converted to throughput (passed by the last Scout).
Operating expense—All the energy the system (Scout troop) spends in turning inventory (distance between first and last Scout) into throughput (distance covered by the last Scout).

Let us add two more definitions before continuing the analogy:

Dependent event—An event that cannot occur until some prior event occurs.
Statistical fluctuation—Random variations in the time to perform an operation.

Returning to the example, imagine that the Scouts are allowed to hike at their own pace. Before long they will have spread out, with the faster hikers in front and the slower in the rear. Using our definitions, the system will have created throughput only at the rate of the slowest hiker (named Herbie in the book), and will have consumed a great deal of energy and inventory in the process.

To correct this situation, we need to make each hiker dependent on another, and limit the distance between hikers by controlling the rate at which they walk (i.e., limit statistical fluctuations). This is best accomplished by moving Herbie to the front of the line, and instructing the remaining Scouts to line up from slowest to fastest, and not to pass each other (i.e., make them dependent on each other).

Again, imagine the troop hiking under these rules. With Herbie at the front, any gaps that open between Scouts can be readily closed, since all other hikers are able to increase their pace to catch up. Since we have done nothing to change Herbie's pace, the system will continue to generate throughput at the same rate, but with two significant differences. First, inventory (distance between the first and last hiker) will be considerably less, and second, less energy (operating expense) will be expended since the faster hikers will not be allowed to charge ahead.

We have now succeeded in accomplishing two of our objectives: decrease inventory and operating expense. But what about the third? In the story, Herbie carries a large backpack full of goodies. To increase his speed, and thereby the entire system's throughput, his load is distributed to other Scouts. The result is easy to imagine. All three measurements (throughput, inventory, operating expense) move in the desired direction, in keeping with the goal.

Let us look at some additional definitions and apply them to our troop analogy.

Bottleneck—Any resource whose capacity is equal to or less than the demand placed on it.
Nonbottleneck—Any resource whose capacity is greater than the demand placed on it.

In our example, Herbie is the bottleneck. All the other Scouts are by definition nonbottlenecks, because each contains some amount of excess capacity.

5.4.3 Balance Flow with Demand

A traditional approach to managing capacity is to balance capacity with demand. In theory, by not paying for unnecessary capacity, manufacturing costs can be reduced. This theory, however, ignores the problem of statistical fluctuation described above. Without this extra capacity, the gaps between dependent operations cannot be closed, leading to higher than necessary work-in-process inventory. Instead, what should be sought is to balance flow with demand. The goal should be to make the flow through the bottleneck equal to demand from the market. Two principles must be kept in mind to accomplish this: (1) make sure the bottleneck's time is not wasted, and (2) make the bottleneck work only on what will contribute to throughput today, not 9 months from now.

How is the time of a bottleneck wasted? One way is for it to sit idle during breaks, shift changes, meals, etc. Another is for the bottleneck to produce parts that could be made at a nonbottleneck resource. Yet a third is for it to process parts that are already defective, or which will become defective in a subsequent operation. In other words, don't waste the bottleneck's time on parts unless you are sure they are good and will stay that way. All three time wasters must be relentlessly eliminated. The point to remember is this: *A minute wasted at a bottleneck is gone forever. It can never be replaced!*

To further emphasize this point, Goldratt suggests that the per-hour cost of a bottleneck is the total expense of the system divided by the number of hours the bottleneck produces. This approach provides a financial perspective on the reality that the *entire system is controlled by the bottleneck.* It should be managed as though it cost as much as the total system!

But what about nonbottlenecks? How should they be managed? First, we must challenge the notion that all machines must be kept busy all the time. When a nonbottleneck is producing a part that is not immediately needed, it is not increasing productivity, it is creating excess inventory, which is against the goal. Instead, we should *expect* nonbottlenecks to be idle part of the time, allowing their excess capacity to be used to close gaps created by statistical fluctuation. This applies to all resources, including people!

The *APICS Dictionary* (1992) identifies a five-step approach to applying the theory of constraints:

1. Identify the constraint (bottleneck) of the system.
2. Exploit the constraint (i.e., tie it to market demand).
3. Subordinate all nonconstraints (i.e., make sure they support the needs of the constraint and never hold it up).
4. Elevate the constraint (i.e., increase its capacity).
5. If the constraint is broken in step 4 (i.e., if by increasing its capacity it is no longer the constraint), go back to step 1 to identify the next constraint.

5.4.4 Summary

The goal of a manufacturing company is to make money now and in the future. To do so, manufacturers must seek to increase throughput while simultaneously reducing inventory and operating expense. To accomplish this, identify the bottleneck in the manufacturing system, then balance the flow through the bottleneck with demand from the market. Subordinate all other resources (nonbottlenecks) to the bottleneck by allowing them to produce only as much as is needed to keep the bottleneck supplied.

Doing anything more only increases inventory and/or operating expense, in conflict with the goal.

5.5 MATERIAL REQUIREMENTS PLANNING SYSTEMS

In Subchapter 5.2 we presented a top-to-bottom scheduling approach with the assumption of an MRPII system being in place. In this subchapter we will provide some historical context on the evolution of MRP systems and further describe the processing of the basic MRP engine. Two basic replanning techniques—regenerative and net change MRP—will be presented, and we will end with a brief discussion of a technique for measuring the effectiveness of MRPII, the ABCD checklist.

5.5.1 Evolution of MRP

Prior to the 1960s, manufacturing companies depended primarily on reorder-point techniques to plan parts and material requirements. While some manufacturers sought an alternative approach, most directed their energies toward finding the most economical order quantities and on developing machine loading techniques to optimize the utilization and efficiency of factory resources. Those who did seek an alternative pioneered the way for today's MRP systems. When MRP was first developed, computers were just being introduced into manufacturing companies. The development of material plans was therefore done manually, requiring a month or more from top to bottom of the product structure. It was not until the 1970s, when computers became commonplace, that computerized MRP came into its own.

The basic MRP system was nothing more than a tool for breaking down an MPS into its component parts by exploding a bill of material. After accounting for available material, planned orders were generated, creating a time-phased plan for ordering manufactured and purchased parts. As the order release dates arrived, orders were issued to the factory and to vendors, to be managed by the execution functions—production and purchasing. This "order launch and expedite" approach led to the creation of hot lists, shortage reports, and expediting as the major elements of a production control system. What was missing was some way to report actual status of the released orders, both in the shop and with vendors.

As orders are processed, material is issued and received, master schedules are revised, and other real-life changes occur, the material plan needs to be revised. By capturing this information and feeding it back to MRP, a new plan may be generated that takes these changing conditions into account. The MRP system then issues action messages to production and inventory control workers recommending appropriate intervention.

The advent of shop floor control systems, capacity requirements planning, and purchasing systems allowed for the basic MRP system to be upgraded. When all these elements were combined, the term "material requirements planning" was no longer sufficient to describe the new system, and the term "closed-loop MRP" was introduced. Further development brought capabilities such as "what-if" simulations, various levels of capacity planning and control, and the incorporation of financial data requiring yet another term. Oliver Wight coined the term "MRPII," changing the meaning of the acronym from material requirements planning (MRP) to manufacturing resources planning (MRPII). Figure 5.27 depicts a fairly typical MRPII system overview.

5.5.2 MRP Processing

The basic planning engine in MRP systems is the material requirements planning subsystem itself. In this section we will examine the process by which MRP develops its time-phased material plans. Figure 5.28 provides an example.

Beginning with the top table, we will proceed through each row and review the processing steps involved. Recall from Section 5.2.4 that our goal is to translate a master production schedule into a time-phased material plan (i.e., planned orders) for component parts and materials. We begin with an assumed master schedule for part A as shown in the "Gross Requirements" row of the top table. The second row,

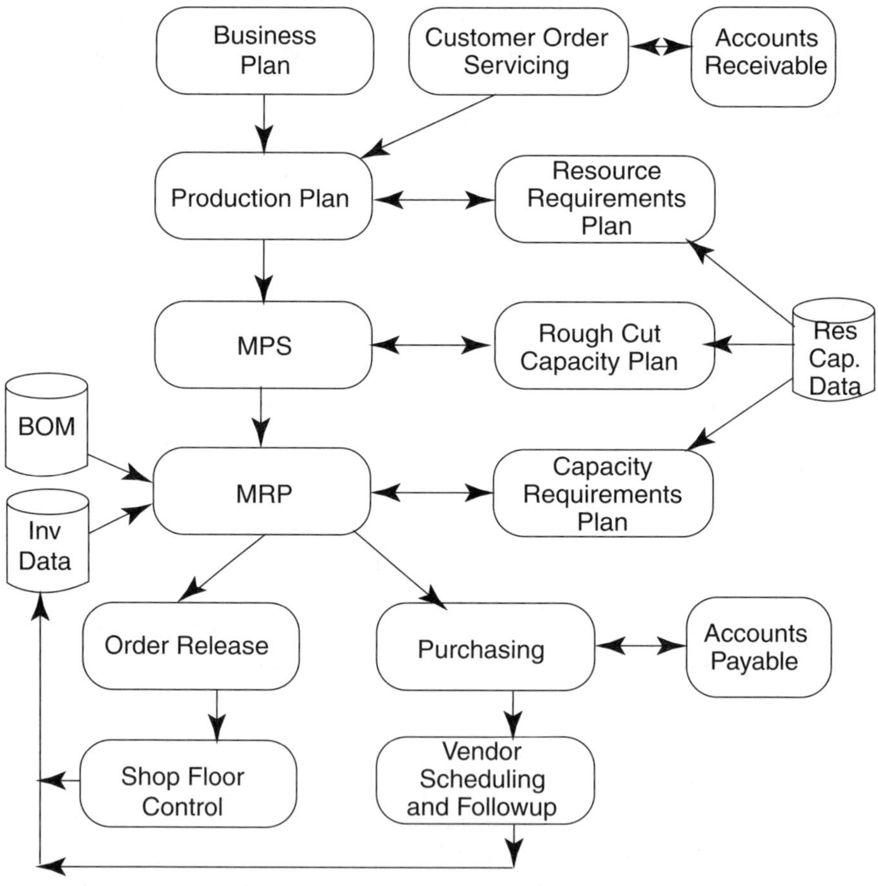

FIGURE 5.27 Typical MRPII system overview.

"Scheduled Receipts," indicates that an order is already in production for a quantity of 50 parts and is scheduled to complete in period 2. The next two rows calculate future inventory balances, beginning with a starting quantity of 30 parts. The "Projected on Hand" row ignores future additions to inventory resulting from planned orders, while "Projected Available" includes these quantities. The respective formulas are:

$$\text{POH in period 1: } (\text{POH})_1 = \text{SI} - (\text{GR})_1 + (\text{SR})_1$$
$$\text{POH in period } n\text{: } (\text{POH})_n = (\text{POH})_n - 1 - (\text{GR})_n + (\text{SR})_n$$

Lot Size = 50
Lead Time = 2
Safety Stock = 0

Part Number "A" — Preliminary Record

		Periods 1	2	3	4	5	6	7	8
Gross Requirements		20	25	25	10		30	20	10
Scheduled Receipts			50						
Projected On Hand		10	35	10	0	0	–30	–50	–60
Projected Available	30								
Net Requirements									
Planned Order Receipts									
Planned Order Releases									

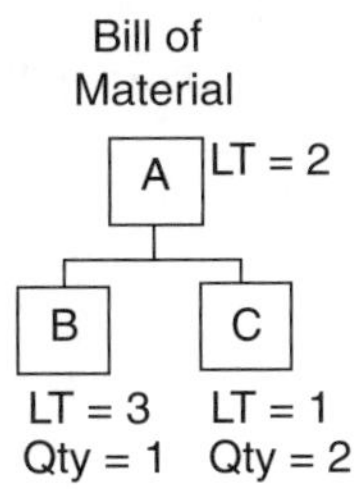

Lot Size = 50
Lead Time = 2
Safety Stock = 0

Part Number "A" — Completed Record

		Periods 1	2	3	4	5	6	7	8	Balance Check
Gross Requirements		20	25	25	10		30	20	10	
Scheduled Receipts			50							
Projected On Hand		10	35	10	0	0	–30	–50	–60	
Projected Available	30	10	35	10	0	0	–30 / 20	0 / 0	–10 / 10	
Net Requirements		0	0	0	0	0	30	0	10	
Planned Order Receipts							50		50	100
Planned Order Releases					50		50			–60
										40

Lot Size = 30
Lead Time = 3
Safety Stock = 25

Part Number "B" — Completed Record

		Periods 1	2	3	4	5	6	7	8	Balance Check
Gross Requirements					50		50			
Scheduled Receipts				30						
Projected On Hand		35	35	65	15	15	–35	–35	–35	
Projected Available	35	35	35	65	15 / 45	45 / 45	–5 / 25	25 / 25	25 / 25	
Net Requirements		0	0	0	10	0	30	0	0	
Planned Order Receipts					30		30			60
Planned Order Releases		30		30						–35
										25

FIGURE 5.28 Material requirements planning MRP subsystem showing details of the part records.

$$\text{PA in period 1: } (\text{PA})_1 = \text{SI} - (\text{GR})_1 + (\text{SR})_1 + (\text{POR})_1$$
$$\text{PA in period } n\text{: } (\text{PA})_n = (\text{PA})_{n-1} - (\text{GR})_n + (\text{SR})_n + (\text{POR})_n$$

where

POH = projected on hand
PA = projected available
SL = starting inventory (initial projected available)
GR = gross requirements
SR = scheduled receipts
POR = planned order receipts
n = period number 1, 2, 3, 4 …

Working through a few periods for "Projected on Hand" results in the following:

$$(\text{POH})_1 = \text{SI} - (\text{GR})_1 + (\text{SR})_1 = 30 - 20 + 0 = 10$$
$$(\text{POH})_2 = (\text{POH})_1 - (\text{GR})_2 + (\text{SR})_2 = 10 - 25 + 50 = 35$$
$$(\text{POH})_3 = (\text{POH})_2 - (\text{GR})_3 + (\text{SR})_3 = 35 - 25 + 0 = 10$$

"Projected Available" is similar, except that it is used as the "trigger" for planning a new order, and then takes the planned order receipt quantity into account when projecting future inventory quantities. If the resulting quantity falls below the specified "safety stock," MRP plans a new order, then recalculates the projected-available balance. Working through a few rows, we see the following:

$$(\text{PA})_1 = \text{SI} - (\text{GR})_1 + (\text{SR})_1 + (\text{POR})_1 = 30 - 20 + 0 + 0 = 10$$
$$(\text{PA})_2 = (\text{PA})_1 - (\text{GR})_2 + (\text{SR})_2 + (\text{POR})_2 = 10 - 25 + 50 + 0 = 35$$
$$(\text{PA})_3 = (\text{PA})_2 - (\text{GR})_3 + (\text{SR})_3 + (\text{POR})_3 = 35 - 25 + 0 + 0 = 10$$
$$(\text{PA})_6 = (\text{PA})_5 - (\text{GR})_6 + (\text{SR})_6 + (\text{POR})_6 = 0 - 30 + 0 + 0 = -30 \text{ (first pass)}$$

Since this calculation is less than the safety stock of zero, MRP must plan an order. The quantity and timing of the order are determined by the lot size (50) and lead time (2). If the order is to be received in time to satisfy the requirement it must be released 2 periods earlier, i.e., in period 4. Recalculating with the planned order quantity results in the following:

$$(\text{PA})_6 = (\text{PA})_5 - (\text{GR})_6 + (\text{SR})_6 + (\text{POR})_6 = 0 - 30 + 0 + 50 = 20$$

We now continue through the remaining periods:

$$(\text{PA})_7 = (\text{PA})_6 - (\text{GR})_7 + (\text{SR})_7 + (\text{POR})_7 = 20 - 20 + 0 + 0 = 0$$
$$(\text{PA})_8 = (\text{PA})_7 - (\text{GR})_8 + (\text{SR})_8 + (\text{POR})_8 = 0 - 10 + 0 + 0 = -10 \text{ (first pass)}$$

MRP plans a second order for 50 units, released in period 6 for receipt in period 8, then recalculates the projected available:

$$(\text{PA})_8 = (\text{PA})_7 - (\text{GR})_8 + (\text{SR})_8 + (\text{POR})_8 = 0 - 10 + 0 + 50 = 40 \text{ (second pass)}$$

The remaining row, "Net Requirements," is shown only when the projected-available balance drops below the specified safety stock. Its value is the difference between the safety stock and the first-pass calculation of projected available. In other words, it is the quantity needed to bring the projected-available balance back up to the safety stock level. In our example, only periods 6 and 8 show net requirements. Their values reflect the quantity required to bring negative projected-available balances up to zero.

Once MRP completes planning a "parent" part number, its next step is to explode the resulting requirements down to the next level in the bill of material. The gross requirements of each "child" are equal to the planned order release of the parent multiplied by the child's quantity. In our example, since only one B is required for each A, the gross requirements for B equal the planned order releases of A. The gross requirements for C, on the other hand, will be double these quantities (i.e., 100 in periods 4 and 6 each), since two C's are required for each A.

MRP then performs its calculations on the next-level parts, resulting in planned order releases; explodes to the next level; and continues down the bill of material until all levels have been calculated.

One final check is useful to further our understanding of this process, and to ensure that the calculations performed at each level are correct. The "Balance Check" shown at the right of the two completed records show the total planned order quantity minus the projected inventory shortfall (i.e., negative on hand). The difference represents how much inventory should remain after satisfying the shortfall, and should therefore match the final projected-available balance. This is the case in both our tables.

5.5.3 Lot for Lot versus Fixed Order Quantity

In the above example we illustrated how MRP operates using a fixed order quantity. (See Subchapter 5.6 for a variety of lot sizing rules which arrive at fixed order quantities.) This is only one of the ways in which material can be planned. MRP also has the ability to determine its own lot sizes based upon a lot-for-lot calculation. In this approach MRP processing is identical to that described above, except only the net requirement is planned, as opposed to some predetermined fixed quantity. Figure 5.29 shows the results of lot-for-lot planning as applied to part A.

Notice that no "remnant" material is planned. In other words, only the immediate need is satisfied, with no material remaining at the end of a period to cover demand in the next. Because of this, inventory is minimized, making lot-for-lot production the favored technique in a JIT environment (see Subchapter 5.3).

5.5.4 Measuring System Effectiveness: The ABCD Checklist

In 1977, Oliver Wight (*The Oliver Wight ABCD Checklist for Operational Excellence*, 1993) created the first ABCD checklist to aid companies in determining how effectively they were using MRPII. The original list consisted of 20 items, grouped into three categories: technical, focusing on design; data accuracy, to determine the reliability of key data; and operational, designed to gauge employees' understanding and

Lot Size = Lot-for-Lot
Lead Time = 2
Safety Stock = 0

Part Number "A" - Completed Record

		Periods								Balance Check
		1	2	3	4	5	6	7	8	
Gross Requirements		20	25	25	10		30	20	10	
Scheduled Receipts			50							
Projected On Hand		10	35	10	0	0	–30	–50	–60	
Projected Available	30	10	35	10	0	0	-30 / 20	0 / 0	-10 / 10	
Net Requirements		0	0	0	0	0	30	20	10	
Planned Order Receipts							30	20	10	60
Planned Order Releases					30	20	10			–60
										40

FIGURE 5.29 Detail records for lot-for-lot planning in MRP.

use of the system. The checklist has evolved over the years and is now accepted as the industry standard for measuring MRPII. In its present form, the checklist is organized into five basic business functions:

1. Strategic planning processes
2. People/team processes
3. Total quality and continuous improvement processes
4. New product development processes
5. Planning and control processes

Each function is scored individually, resulting in a rating of A, B, C, or D. As an example of how ratings should be interpreted, the following is excerpted from the first and last items listed above:

Strategic Planning Process

Class A—Strategic planning is an ongoing process and carries an intense customer focus. The strategic plan drives decisions and actions. Employees at all levels can articulate the company's mission, its vision for the future, and its overall strategic direction.

Class B—A formal process, performed by line executives and managers at least once per year. Major decisions are tested first against the strategic plan. The mission and/or vision statements are widely shared.

Class C—Done infrequently, but providing some direction to how the business is run.

Class D—Nonexistent, or totally removed from the ongoing operation of the business.

Planning and Control Processes

Class A—Planning and control processes are effectively used company wide, from top to bottom. Their use generates significant improvements in customer service, productivity, inventory and costs.

Class B—These processes are supported by top management and used by middle management to achieve measurable company improvements.

Class C—Planning and control system is operated primarily as a better method for ordering materials, contributing to better inventory management.

Class D—Information provided by the planning and control system is inaccurate and poorly understood by users, providing little help in running the business.

5.6 INVENTORY MANAGEMENT

Inventory management is the business function that deals with the planning and control of inventory (APICS Inventory Management Certification Review Course, 1987). Inventory is commonly thought of as materials, parts, and finished goods, which allow a company to satisfy their customers while enabling manufacturing to produce these products in an efficient manner. Typical classifications of inventory include raw materials, work in process, and finished goods, each of which serves a particular purpose or function, such as:

Covering random or unexpected fluctuations in demand—normally handled with safety stock.
Covering seasonal fluctuations—off-season production to cover demand during high-sales periods.
Providing a "buffer" between factory and distribution centers, i.e., "transportation" inventory. Usually the amount of inventory equals the demand rate multiplied by the transportation time.
Providing a "buffer" between manufacturing operations—cushions variations in production between sequential work centers.

The primary goals of inventory management include:

Minimize inventory (dollars) in order to increase return on investment—ROI = net income ÷ total assets (includes inventory).
Provide high customer service levels—cover off-the-shelf demand for goods, or provide faster delivery than would be possible by producing from scratch upon receipt of an order.
Stabilize production rates—allow for stabilization of workforce and equipment utilization.
Generate profits through the sale of finished goods.

Some typical measures of how effectively the inventory management function is performed are:

Total value of inventory—goal is to minimize.
Inventory turns (the rate at which products are sold versus the average level of inventory maintained)—goal is to maximize. Inventory turns = cost of

goods sold ÷ average value of inventory on hand. *Example:* If 2500 items with a standard cost of \$100 were sold in a year, then the cost of goods sold = \$250,000. If the average value of inventory on hand = \$50,000, then inventory is "turned over" five times per year (250,000 ÷ 50,000).

Periods on hand (inventory "coverage" measured in periods)—the goal is to maintain a desired level to cover projected sales. *Example:* If \$50,000 of inventory is on hand, and 750 units with a standard cost of \$25 are projected to be sold each month (cost of sales = 750 × \$25/month = \$18,750), then POH = \$50,000 ÷ \$18,750/month = 2.67 months.

In the sections that follow we will describe some general concepts and approaches that allow inventory to be managed in such a way as to contribute to the goals and functions presented above. We will also present some techniques to answer the three fundamental questions that must be answered by inventory management (APICS Inventory Management Certification Review Course, 1987):

1. What to order
2. How much to order
3. When to order

5.6.1 General Concepts

ABC Analysis

ABC analysis of inventory, also know as inventory stratification, is the process of categorizing inventory into classes. ABC classifications are determined based on the Pareto principle of the 80/20 rule, which states that 80% of inventory value will be contained in 20% of the parts. The three classes are defined as follows:

A items—The top 20% of items, representing 80% of inventory cost.
B items—The middle 30% of items, representing 15% of inventory cost.
C items—The bottom 50% of items, representing 5% of inventory cost.

In Section 5.6.2 we will see how ABC classification is used in maintaining inventory accuracy through a process known as "cycle counting." Other uses include determining appropriate types of inventory systems, prioritizing inventory, and controlling ordering policy and procedures.

Inventory Controls

Three general control mechanisms are generally acknowledged in reference to inventory: financial, operational, and physical. Financial and physical will be dealt with here; operational control deals with inventory accuracy and is the subject of the next section.

Financial controls deal with inventory accounting, which is the "function of recording and maintaining inventory status information" (APICS Inventory Management

Certification Review Course, 1987). Two general inventory accounting approaches are applied: (1) perpetual, in which all inventory transactions are recorded as they occur, providing up to the minute information; and (2) periodic, in which inventory is counted only at specified times. The latter is less expensive, but provides less current information.

General methods of determining inventory value include:

Last-in, first-out (LIFO)—Inventory value is assigned to items as they are consumed, based on the cost of the most recently received unit.

First-in, first-out (FIFO)—Inventory value is assigned to items as they are consumed, based on the cost of the oldest unit on hand.

Moving-average unit cost—Inventory value is recalculated whenever parts are added to inventory. The cost per part is calculated by taking the current average cost multiplied by the quantity on hand prior to adding the new part(s). Then the value of the new part(s) is added to this total and a new average is calculated for the combined quantity.

Standard cost—The "normal" cost of an item, including labor, material, overhead, and processing costs. Standard costs are typically used to determine anticipated costs prior to production for management control purposes. They are later compared to actual costs and revised as necessary to ensure their usefulness for future applications.

Physical controls—Ensure that inventory records are maintained accurately with regard to such things as quantity, location, and status. Controls are achieved through techniques such as part numbering conventions, lot sizing and replenishment rules, stock room security, and appropriate inventory handling practices. Each of these subjects will be dealt with in further detail in subsequent sections.

5.6.2 Inventory Accuracy

Many of the problems encountered in a manufacturing company may be attributed to (or at least compounded by) inaccurate inventory records. Parts shortages, poor schedule performance, low productivity, and late deliveries are but a few. In an MRP environment, inaccurate inventory adversely affects the system's ability to perform netting, resulting in suspect material plans, schedules, and capacity plans. Having less inventory on hand than records show leads to costly expediting, or in the extreme, work stoppage. Having more inventory than is recorded leads to the purchase of unnecessary additional inventory, thus reducing profitability. Accurate recording of inventory allows these problems to be avoided. Likewise, having accurate records of obsolete inventory allows a company to dispose of it, freeing up space for active parts and possibly generating additional cash. Two general methods of achieving inventory accuracy are "cycle counting" and "periodic physical review."

Cycle counting is the continuous review and verification of inventory record accuracy through routine counting of all items on a recurring basis. Errors encountered are immediately reconciled. Cycle counting provides a high level of record accuracy at all times, eliminating the need for an annual physical inventory, with a minimal loss of production time. Since errors are detected in a timely manner, causes of the errors are much more likely to be found and corrected in addition to simply correcting the inventory records.

Periodic inventories, by contrast, are typically done on an annual basis, requiring the plant and warehouse to shut down for the duration of the review. No correction of the causes of errors is possible, and no permanent improvement in record accuracy results. While the total time spent conducting the inventory may be less than with cycle counting, none of the benefits result.

The frequency of cycle counting is typically controlled by a part's value, often by using its ABC classification. Because A items represent the greatest value (80% of total inventory dollars), they are counted most frequently. And since they represent relatively few parts (20% of part count), the effort is generally not too great. In addition to frequency, tolerance in counting errors is also generally much tighter than with lower-valued items. Tolerances allow for some acceptable level of error, to avoid spending more time (i.e., cost) trying to reconcile an error than the parts in question are worth. A typical cycle count strategy based on ABC class codes might be as follows:

ABC Classification	Count Frequency	Tolerance
A	Weekly, monthly	0%
B	Monthly, quarterly	2%
C	Quarterly, semiannual	5%

5.6.3 Lot Sizing Rules

Lot sizing rules answer the question of how much to order. These rules are as valid for computerized replenishment systems such as MRP or automated reorder point systems as they are for manual approaches (see Section 5.6.4). This section presents some of the most common lot sizing rules.

Fixed quantity—These techniques are demand-rate-based, attempting to determine an average quantity that will satisfy all future demand over a predetermined horizon, typically 1 year. Whenever an order is triggered, the specified fixed quantity is ordered. Fixed quantities almost always produce remnants (inventory carried over from one period to the next). For example, if annual demand is anticipated to be 24,000 units, then monthly quantities of 2,000 units might be desired. However, if the vendor produces in lots of 2,500, then this quantity would likely be used instead, perhaps on a less frequent basis, or skipping several months each year.

EOQ—A specialized fixed order technique that attempts to minimize the combined cost of producing (or ordering) and carrying inventory. Section 5.3.3 (Figure 5.24) presents a graphical model of EOQ. The equivalent formulas are shown in Figure 5.30.

Lot for lot—This approach orders only the discrete quantity required to cover the period in question. Because the order quantity equals the period's

1. To determine quantity in units

$$EOQ=\sqrt{\frac{2US}{IC}}$$

U = Annual usage in units (units/yr)

S = Setup or ordering cost ($)

I = Inventory carrying cost (% /yr)

C = Unit cost ($/unit)

Example: U = 10,000 units/yr, S = $100, f = 10%, C = $50

$$EOQ=\sqrt{\frac{2x10{,}000x100}{0.10x50}} \quad \sqrt{\frac{2{,}000.000}{5}} \quad = 632 \text{ Units}$$

2. To determine quantity in dollars

$$EOQ=\sqrt{\frac{2AS}{I}}$$

Where

A = Annual usage in dollars ($/yr)

S = Setup or ordering cost ($)

I = Inventory carrying cost (%/yr)

Example: A=$500.000/yr, S=$250, I= 18% /yr

$$EOQ=\sqrt{\frac{2x500{,}000x250}{0.18}} = \sqrt{\frac{250{,}000.000}{0.18}} \quad \$37.268$$

FIGURE 5.30 Equivalent formulas for determining EOQ in number of units and dollars.

demand, no remnant results. Unlike fixed quantity techniques, lot-for-lot order quantities vary from order to order. (Refer to Section 5.5.3 for additional discussion.)

Period order quantity—This technique uses the EOQ to establish the number of periods to be covered by a single order. Remnants are only carried over within the periods covered, not between orders. Order quantity varies with each order. An example is shown in Figure 5.31. Other, less frequently used techniques include the following:

Least total cost—Attempts to minimize the total ordering cost by selecting order quantities where setup costs and carrying costs are most nearly equal.

Least unit cost—Calculates the combined ordering (or setup) and carrying costs of trial lot sizes, divides by the lot size, and selects the lot size that minimizes the unit cost.

Dynamic programming—mathematical optimization models that select an order size that minimizes the total ordering costs over the planning horizon.

5.6.4 Replenishment Systems

Replenishment systems seek to answer the "when to order" question introduced at the beginning of Subchapter 5.6. A useful model for aiding our understanding of replenishment is the "order point" system depicted in Figure 5.32. In this example the demand rate is variable, as indicated by the changing slope of the heavy "saw-tooth" line. As inventory is consumed (downward-sloping line), it eventually reaches the order-point level and an order is placed. Assuming an order size based on EOQ and instantaneous replenishment, inventory will be replenished (light vertical line) and demand will continue to consume it (perhaps at a new rate). Since instantaneous replenishment cannot occur in reality, demand will continue to consume inventory during the order lead time, further reducing it. The goal is to maintain inventory at or above the safety stock level, dipping below only in the event of unanticipated demand (such as the unusually steep demand shown in the second cycle). Upon receipt of an order, inventory is increased by the amount received, returning to a level greater than the order point and the cycle repeats.

Example: If Annual Demand = 250, and EOQ = 60, then

No. Orders/Yr = 250÷60 = 4.2 (Round to 4)

12 Months /Yr ÷4 Orders/Yr = 3 Months/order

Period	1	2	3	4	5	6	7	8	9	10	11	12	Total
Net Req't	25	20	18	18	15	12	15	10	20	27	30	40	250
Planned Order	63			45			45			97			250

FIGURE 5.31 Example of period order quantity.

The order-point level is calculated as follows (APICS Inventory Management Certification Review Course, 1987):

$$OP = DLT + SS$$

where
OP = order point
DLT = demand during lead time
SS = safety stock

For example: If demand averages 100 units/week, lead time is 3 weeks, and 2 weeks of safety stock are to be maintained, then

$$\begin{aligned} DLT &= 3 \text{ weeks} \times 100 \text{ units/week} = 300 \text{ units} \\ SS &= 2 \text{ weeks} \times 100 \text{ units/week} = 200 \text{ units} \\ OP &= 300 + 200 = 500 \text{ units} \end{aligned}$$

Other replenishment techniques include the following:

Periodic review—Inventory levels are reviewed on a periodic basis, and replenished when a predetermined order point level is reached. This review may be based on physical inspection or review of system records. Periodic

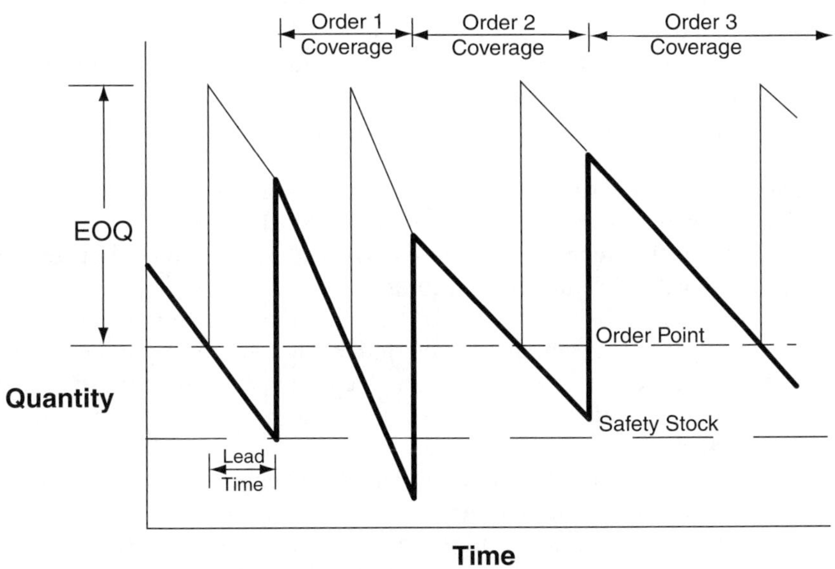

FIGURE 5.32 A model depicting the material replenishing system, assuming the demand rate is variable for this part.

review is used when recording individual withdrawals is difficult, or for items with limited shelf life.

Visual review—Inventory is reordered based on a visual review of inventory level. This may be as simple of painting a line on the inside of a barrel and reordering only if the line is visible at the time the review is conducted. Visual review is generally applicable only to low-value items (e.g., nuts and bolts).

Two-bin—Inventory is reordered when a bin of parts is emptied. The second bin covers demand during the lead time and remains in use until it empties, being replaced in turn by the first bin. Also used to control low-value items.

Material requirements planning—Time-phased material plan that determines planned orders by exploding a master production schedule against a bill of material and accounting for inventory already on hand or on order. Orders are planned in advance, providing a proactive approach to inventory replenishment as opposed to the reactive approaches described above. Refer to Section 5.2.4 and Subchapter 5.5 for further information.

5.6.5 Warehousing

Physical Control

Effective physical control of inventory is the final subject of this subchapter. Since the majority of inventory in manufacturing companies is typically maintained in controlled stockrooms or warehouses, we will focus the discussion on those areas. As companies move to reduce inventory through JIT and other initiatives, a greater percentage of inventory will be kept in flow on the factory floor. However, many of the disciplines and controls discussed here are equally valid in such a decentralized environment. A brief overview of some fundamental principles follows.

Secure storerooms ensure that only a limited number of responsible, accountable individuals have access to inventory. Stock keepers must be well educated in the importance of inventory accuracy and the practices for maintaining accurate inventory records.

Timely recording of all issues and receipts is critical to maintaining inventory records and essential for determining net space requirements for items, to ensure that inventory is available to support customer orders and production requirements, and to avoid carrying excess or obsolete inventory in stock.

Knowing inventory location is as critical as any other element of record accuracy. Various storage and location methods are presented below.

Stock keeping performance is measured to ensure that adequate control is maintained and an atmosphere of continuous improvement is fostered. As errors are found and recorded, they should be immediately corrected, along with the cause of the error, to avoid recurrence.

Storage Location Methods

Three general methods for assigning storage location for parts and materials are fixed, random, and zoned. The *APICS Dictionary* (1992) describes these as follows:

Fixed location storage—A method of storage in which a relatively permanent location is assigned for the storage of each item in a storeroom or warehouse. Although more space is needed to store parts than in a random location storage system, fixed locations become familiar and therefore a locator file may not be needed.

Random location storage—A storage technique in which parts are placed in any space that is empty when they arrive at the storeroom. Although this random method requires the use of a locator file to identify part locations, it often requires less storage space than a fixed location storage method.

Zoned location storage—A combination of fixed and random location storage in which items are assigned to random locations within a predesignated zone.

5.7 BIBLIOGRAPHY

APICS Dictionary, 7th ed., APICS Educational and Research Foundation, American Production and Inventory Control Society, Inc., Falls Church, Va., 1992.

APICS Inventory Management Certification Review Course, American Production and Inventory Control Society, Inc., Falls Church, Va., 1987.

APICS Systems and Technologies Certification Review Course, American Production and Inventory Control Society, Inc., Falls Church, Va., 1987.

Goddard, W.E., *Just-in-Time: Surviving by Breaking Tradition,* Oliver Wight Companies, Essex Junction, Vt., 1986.

Goldratt, E.M. and Cox, J., *The Goal,* North River Press, Great Barrington, Mass., 1984.

Goldratt, E.M. and Fox, R.E., *The Race,* North River Press, Croton-on-Hudson, New York, 1986.

Wight, O., *The Oliver Wight ABCD Checklist for Operational Excellence,* 4th ed., Oliver Wight Publications, Inc., Essex Junction, Vt., 1993.

Peters, T.J. and Waterman, R.H., *In Search of Excellence,* Harper & Row, New York, 1982.

Vollmann, Berry, and Whybark, *Manufacturing Planning and Control Systems,* 2nd ed., Business One Irwin, Homewood, Ill., 1988.

6 Control of Quality

Robert L. Lints

6.0 INTRODUCTION TO CONTROL OF QUALITY

Ultimate customer satisfaction with today's complex products has its roots in the quality of the design process. Quality must become an inherent characteristic of the design at its inception. The utilization of a multidisciplined review team, working in a concurrent engineering environment, has proven to be an effective means of accomplishing this goal, and the preliminary phase of design development is not too early for implementation. This melding of technical and administrative professionals, ranging from operators and inspectors on the production floor, through supporting departments, to the individual participation of executive management is the key to achieving customer satisfaction in the marketplace. This practice is commonly referred to as "total quality management" (TQM), and is being implemented in one form or another by most major manufacturing and service industries today, regardless of their product line.

Today, many businesses find that procurement costs for raw materials, detail parts, assemblies, and outside services run in excess of 50% of their sales dollar. In addition, factory downtime, schedule slips, rework and replacement costs, and lost time working with suppliers to prevent problem recurrence elevate the need for improved supplier quality for a high-priority item. This points to the need to assure that all outside procurement activity is included under the TQM umbrella and is an integral part of the up-front concurrent engineering activity. The prevention of supplier quality problems that impact just-in-time (JIT) delivery of quality materials, problems which surface during later production runs, or problems that may result in customer warranty claims and product recall actions must be identified up front to control costs and assure customer satisfaction when products reach the field. Recent reports of automotive recalls attributed to supplier problems bear mute testimony to the importance of establishing a clear understanding of specification requirements with suppliers, and the controls in place to assure they will be met.

Once the integrity and producibility of the design have been validated by the concurrent engineering team and the design has been subjected to development testing, another phase of team quality assurance comes into play. This phase of quality assurance involves the preparation and implementation of a "transition to production" plan. Development and timely completion of this plan will assure a smooth and trouble-free start of production, an area where many have faltered in the past when insufficient attention has been given to past lessons learned. The transition plan is truly a quality assurance plan and a defect prevention plan combined, as it will include a detailed review of past lessons learned, a detailed review of potential risks that may be encountered during the transition, and specific plans to progressively reduce and eliminate those risks. The Department of Defense manual titled *Transition from Development to Production* (1984) is available to the public and may be found helpful in establishing a transition plan tailored to a specific application.

With design integrity assured, procurement sources carefully selected and controlled, and an effective transition-to-production plan in place, we are ready to proceed with production and the steps necessary to assure that a quality product will be produced and delivered to the customer. Here, TQM comes into play again. The customer is defined as "the next person to receive the output of the process." In this context, the customer is not necessarily the *final* customer. In a sequence of operations where there are many individual steps, there will be *many* customers, with each one in a position to judge the quality of the "product" being delivered to him or her from the previous step. An assembler, as an example, expects to receive detail parts for assembly properly formed, free of burrs, and painted the proper color. Applying the principles of TQM to this example includes providing the proper training, work instructions, specifications, tools, work environment, clarification of responsibility, motivation, and recognition. When this applies to both the supplier *and* the assembler, we will have a supplier intent on producing a quality product and a customer who will provide feedback to the supplier if his or her expectations are not met. This activity will result in correction of the process and ultimate satisfaction of the customer. When carried on throughout the whole process, we have the process control necessary to assure delivery of a quality product satisfying the *ultimate* customer's expectations.

The aim of the quality control (QC) program should be prevention of defects by improvement and control of processes in all aspects of the business. Regular monitoring of the process will inevitably result in steady improvement through timely process adjustment and result in a corresponding reduction in cost. The following subchapters will provide greater insight into the techniques that have proven to be effective in reaching this goal.

6.0.1 Additional Reading

Feigenbaum, A. V., "Total Quality Control," *Harvard Business Review,* November–December 1958.

Juran, J. M. and Gryna, F. M., *Quality Control Handbook,* 4th ed., McGraw-Hill, New York, 1988.

Shuster, "Profound Knowledge: Source, Character, and Application," in *Proceedings of the 1st International Symposium on Productivity and Quality Improvement,* Industrial Engineering Management Press, Norcross, Ga., 1992.

Transition from Development to Production, Department of Defense Handbook on Manufacturing Management, 2nd ed., Defense Systems Managements College, Fort Belvoir, Va., 1984.

6.1 QUALITY ENGINEERING AND PLANNING

6.1.1 Introduction to Quality Engineering and Planning

The quality engineering and planning function provides the technical and administrative guidance necessary to plan the overall quality initiatives for the company, to assist in their implementation, and to oversee achievement of the desired objectives and goals. Companywide interface and collaboration with all organizational elements is required for the successful accomplishment of this task. The planning function is a key part of quality engineering and planning responsibility. This includes the necessary indoctrination and training of participants at all levels to establish a basic understanding and acceptance of the principles of total quality control (TQC).

Typical quality engineering and planning functions include the following:

Prepare overall quality plans	Establish test plans and procedures
Prepare inspection plans	Establish classification of characteristics
Establish quality metrics	Establish quality cost estimates
Prepare manuals and instructions	Establish supplier control program
Act on material review board	Establish supplier certification plans
Participate in concurrent engineering	Evaluate and assist in supplier selection
Provide statistical QC support	Conduct internal and external QC audits
Establish sampling plans	Establish metrology designs and controls
Perform process capability studies	Assure C/A on customer complaints
Prepare QC input to process specs	Assist in benchmarking quality
Plan preventative action program	Monitor handling/packaging processes
Monitor corrective action (C/A) progress	Report quality status to all levels

6.1.2 Participation in Concurrent Engineering

The successful completion of a project or production run, which will result in the ultimate customer being satisfied with the quality of the product received, may well depend on the initial concurrent engineering participation during the design process.

The following are some of the more important areas specifically addressed by quality engineering and planning during this process:

Drawing Review

Quality engineering participation in the preliminary design process provides a timely check for such things as clarity of views and dimensioning, proper call-out of specifications for materials and processes, and inclusion of required specifications for nondestructive inspection processes. Drawing review, and correction at the preliminary stage, minimizes costly drawing revisions after final release.

Review of Lessons Learned

The preparation, maintenance, and use of a file called "lessons learned" is invaluable in the prevention of repetitive discrepancies in new or updated products. The recording of problems encountered, the root cause of the problems, and the corrective action taken to eliminate them can be reviewed by the concurrent engineering team and positive steps taken to avoid the same pitfall in the new design. A new design should not be released without this important review.

Assessing Inspectability and Testability

A major responsibility of quality engineering during the initial design reviews is the evaluation of the inspectability and testability of the product. Special attention must be given to the accessibility of design characteristics for inspection with standard measuring instruments, nondestructive inspection (NDI) probes, or other special fixtures or gauges. The accessibility of test points and fixture locating pads must also be considered. If equipment currently available will not be adequate, then consideration must be given to design changes, if practical, or to the design and procurement of required special equipment. Requirements for nondestructive inspection assure that design configuration or structure will not preclude application of probes, or in the case of radiographic inspection, the alignment of the x-ray beam or the placement of film. If special inspection or test equipment will be required, adequate fabrication or procurement time should be provided for equipment to be available when needed. Where production volume or the nature of the product dictates that automated equipment is required, even greater time for design and procurement may need to be included in the manufacturing flow plans.

Classification of Defects

Quality engineering has the additional responsibility to coordinate with the designer and other specialists to establish the importance of each individual characteristic of the design to the quality and operational integrity of the product. This activity is referred to as "classification of defects" and provides the basis for determining the seriousness of any defects and the attention that must be given in subsequent inspection or testing. The characteristics are normally classified as critical, major, or minor. A *critical* defect is broadly defined as one that is likely to result in unsafe operation for personnel using or observing its use, or a defect which will preclude the product, or those used in conjunction with it, from performing its primary function. A *major* defect is one that, although not as serious as a critical defect, is likely to result in failure of the product or materially reduce its usability. It follows, then, that a *minor* defect is one that will not materially reduce the usability of the product and will have little effect on its operation. Classifications of defects are used as a basis for establishing inspection plans for either 100% or sampling inspection, and are often placed on blueprints and work instructions as aids to inspection. When placed on blueprints or other documents, they are generally placed adjacent to the blueprint note or characteristic using the letter C or M. On some blueprints, these letters may be enclosed in a symbol. The minor classification is not generally shown on documents, but may be listed in specific inspection or test instructions.

Determining Process Capability

Quality engineering can provide valuable assistance in the selection of equipment that will consistently provide products with the least variation from the specification mean. Manufacturing facilities often have more than one piece of equipment that may be used to produce a product When this situation exists, a process capability study may be conducted using data collected on quality control charts from past operations, or by gathering data on current production runs. Data thus gathered can be statistically analyzed to provide guidance to the manufacturing engineer as to which equipment will consistently produce products within specification limits and minimize the risk of costly scrap and rework.

Process capability as referred to in this section is a measure of the reproducibility of product, where the *process* is the unique combination of materials, methods, tools, human, and machine, and *capability* is the statistical ability of the process based on historical records. In other words, process capability is the inherent ability of the process to consistently turn out similar parts independent of external forces. It is expressed as a percent defective or as a statistical distribution and is most often calculated from data gathered using quality control charts. The application of quality control charts is explained further in Subchapter 6.7.

Inherent process capability is the ratio of component tolerance to 6 sigma and is expressed by the equation

$$\text{CP} = \text{tolerance} \div 6\ \text{sigma}$$

The process capability (CPK) is measured in relation to the specification mean and is expressed by the following equation:

$$\text{CPK} = \text{the lesser of:}$$

$$\frac{(\text{USL} - \text{mean})}{3\ \text{sigma}} \text{ or } \frac{(\text{mean} - \text{LSL})}{3\ \text{sigma}}$$

where

USL = upper specification limit
LSL = lower specification limit

A CPK of between 0 and 1.0 indicates that some part of the 6 sigma limits falls outside of specification limits. A CPK larger than 1.0 means that the 6 sigma limits are falling completely within specification limits.

A review of the process capability report shown in Figure 6.1 indicates that the output of machines 2, 4, and 5, with CPKs of 0.68, 0.40, and 0.45, respectively, vary outside specification limits, and the process requires adjustment. The CPKs of machine 1 (1.97) and machine 3 (3.07) are well within specification limits, and machine 3 has the smallest variation from mean (*N*). Machine 3 should be first choice for the production run.

Procurement Source Selection

Source selection is an important consideration at the beginning of any project, whether it be for raw material, detail parts, finished assemblies, or services. The concurrent engineering approach is invaluable in that it brings technical as well as administrative expertise together to evaluate all aspects of the procurement, i.e., supplier design capability, manufacturing capability, quality assurance record, and cost and schedule performance, as well as management commitment to implementation of TQM principles.

Quality engineering, working closely with procurement personnel, maintains current records of past supplier performance and has available, through industrywide evaluation programs, quality ratings on many suppliers other than those whom a particular company may have dealt with in the past. Most ratings available through

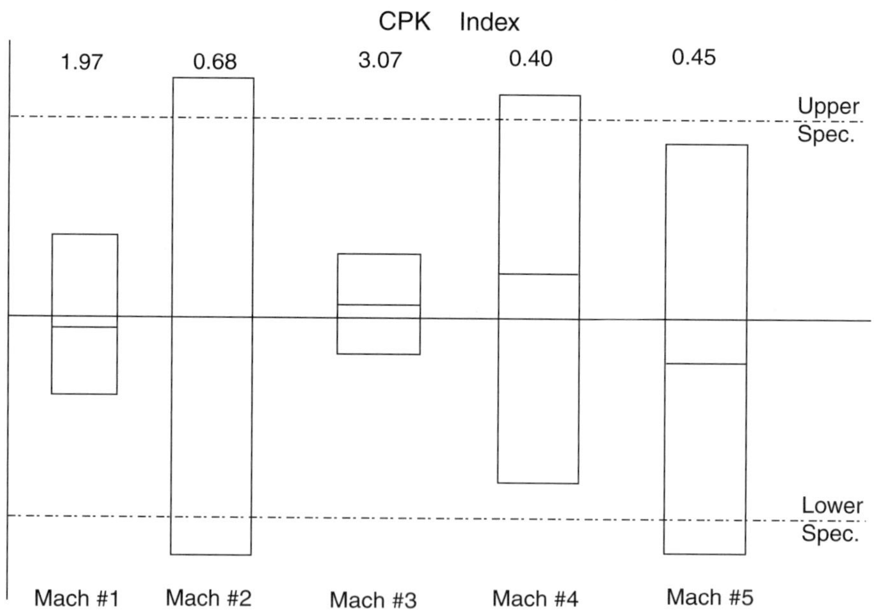

FIGURE 6.1 Process capability report.

these industry sources have been calculated using formulas based on the supplier's recent performance in the areas of quality, schedule, and cost.

In the case of large procurements, or of critical items, it is wise to supplement available ratings with source surveys and concurrent engineering conferences to assure supplier understanding of specifications and quality requirements. These conferences take on added significance when JIT deliveries are planned and the receipt of first-time quality is necessary for smooth operation within your facility. They are also an excellent time to obtain supplier input relative to any cost savings that may be obtained though modification of specification requirements.

6.1.3 Process Flow Diagrams

Figure 6.2 illustrates a typical top-level manufacturing flow diagram for a simple fabrication and assembly process. Development of similar flow diagrams for the entire manufacturing process as a part of early concurrent engineering activity is an important planning tool. Lower-level (or "subflow") diagrams should be prepared for each major element of the process in conjunction with the development of specific work instructions for the subprocesses. This planning effort provides insight for the location and type of equipment required to perform and control the process; highlights the need for special inspection and test equipment, including the need for automated equipment; and assures that proper attention has been directed to the location and type of inspection to be conducted.

Completion of the flow diagrams also aids in highlighting those operations where special training and supplementary work instructions may be required. Manufacturing operators and inspectors must be trained and understand the process they are expected to control and their authority to keep it in control. They must understand how their piece of the process fits into the total picture and what actions they may take to adjust or correct the process. Posting the flow diagrams in the work area enhances operator and inspector understanding and the importance of their contribution. Inspection points, or stations, should always be highlighted on the flow diagrams, as shown in Figure 6.2. A more in-depth discussion of inspection methods for control of processes will be found in Subchapter 6.2.

6.1.4 Quality Planning

The quality planning function is responsible for the preparation and implementation of all quality operating plans, from the top-level company plan for assuring contractual compliance and customer satisfaction to the lowest-level instruction sheets used in data gathering and other record keeping. The following elements of quality planning are among the most important.

The Quality Plan

Preparation of the company quality plan should begin with a review of contractual requirements for quality and a review of company policies, procedures, and goals that have been established by management to assure customer satisfaction with the products

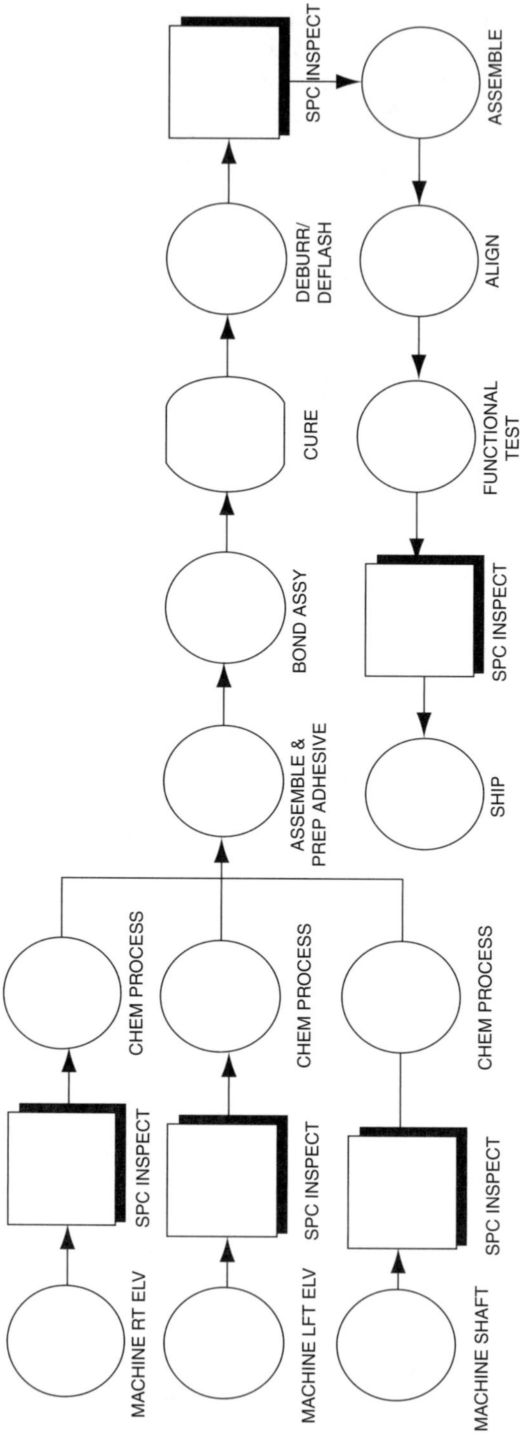

FIGURE 6.2 Typical manufacturing flow chart with inspection points highlighted.

they receive. The quality plan will describe overall policies, procedures, and interfaces required to meet company objectives, including company plans for implementing the principles of TQM. Implementing direction is generally contained in a series of lower-level manuals, directives, and specific work instructions The quality plan is often required to be submitted as a deliverable item on some government contracts.

The Inspection Plan

The inspection plan is generally a subset or lower-level plan prepared by quality planning in conjunction with the company quality plan. The inspection plan will include a detailed description of the overall approach to product inspection, including a copy of the manufacturing flow plan, which shows the location of all inspection and test points established during the concurrent engineering review of the manufacturing process. Details of the type and level of inspection (100%, sampling, etc.) that will be conducted at each point, the requirement for statistical quality control charts or other data recording, and the preparation of inspection records of acceptance or rejections will be included. Since data gathering is an important part of the inspection process, specific instructions should be included as to *what* data are to be recorded, *when* they are to be recorded, *who* is to record them, *where* and in what form they are to be recorded, *why* they are being recorded, and *when* they may be disposed of. Data must not be collected for the sake of data! There must be a valid reason for all data collection. Configuration identification of product as manufactured, as inspected in process, as tested, as inspected at final inspection, and as shipped is a good example of data that must be carefully recorded and maintained in accordance with the inspection plan.

The Process Control Plan

Process control planning should begin with a review of the manufacturing flow plan to establish the specific processes planned for use and an evaluation of historical data on the acceptance rate of each process. If data available on existing processes indicate a need for improvement, if historical data are unavailable or inconclusive, or if a new process is to be used, consideration should be given to conducting a process capability study as a first step in establishing process control. Processes to be studied should be prioritized based on their sequence in the flow plan. Manual or automated quality control charts can be used to gather the necessary data on initial runs. Several excellent software programs are currently available on the market for use with machining stations, which will gather data and provide automated computation of the machine capability in relation to part tolerance limits. The procedure for establishing these charts, their use, and the evaluation of data should be an integral part of the process control plan. Further information on the use of process control charts may be found in Subchapter 6.7.

Handling, Packaging, and Shipping

The manner in which products are handled, packaged, and shipped is often the cause for many avoidable scrap and rework actions within the production facility, and results in many customer complaints and loss of goodwill among customers who receive the products—a direct reflection on the quality of the product as seen by the customer. Quality planning, working in conjunction with production and manufacturing engineers responsible for the design and development of handling, packing, and shipping containers and procedures, can provide data on lessons learned from prior problems with similar products. In so doing, they will aid in pinpointing the need for design improvements, better handling and packing instructions, and specific training needs of personnel directly involved in these operations. In addition to the attention directed to handling and packaging concerns, quality planning receives input on all quality customer complaints related to improper count of quantities received versus quantities ordered and to discrepant product received. These complaints are entered into the corrective action system and investigated, and action is initiated to preclude recurrence. This action often involves the preparation of improved instructions and check sheets for packaging, shipping, and inspection personnel, and the addition of improved equipment, for example, digital scales to aid in counting. Quality planning also has the responsibility of assuring that proper correspondence is forwarded to the customer in response to a complaint, as an aid to improved customer relations.

6.1.5 Supplier Quality Control

Today, many industries are committing over 50% of their sales dollars to the procurement of outside raw materials, detail parts, assemblies, and special services. This, combined with the advent of JIT procurement practices, makes the assurance of supplier quality of paramount importance.

In large organizations, the quality department often has a group charged with the specific responsibility for the preparation and implementation of a supplier quality plan. In other organizations, this responsibility may be assigned to quality engineering and planning. Regardless of where the plan originates, it is important that it be prepared and put in place as a directive that has been coordinated with all major departments that may be involved with, or affected by, procurement practices and the receipt of products obtained through them.

Supplier quality control must begin with the very first supplier contact made by the purchasing department buyer. A common understanding must be established of the quality requirements included in the statement of work, the specifications to be met, authorized materials and processing sources that may be used, supplier understanding and application of statistical process control in his or her facility, type of inspections to be conducted (i.e., supplier 100% or authorized sampling, inspection by the procuring agency at the supplier's plant, or on-site witnessing of final inspections and tests), and requirements for data to be supplied at the time of shipment. In certain instances, where proprietary processes may be involved, it may also be necessary to have contractual coverage for site visits. On larger or critical procurements, it is wise to hold concurrent engineering meetings with the supplier to assure common understanding of requirements and to determine the need for any special training or other assistance the supplier may need in fulfilling his requirements. Establishing a supplier team arrangement will go a long way toward smooth procurement and receipt of quality products.

Rating supplier performance not only provides incentive to the supplier to have a better than average rating, but provides the buyer with an evaluation tool in monitoring the performance of suppliers and an aid in placing future procurements. Today, most supplier performance measuring systems compute a composite score based on quality, schedule, and contract administration data. Quality is emphasized in this rating system, and quality performance usually accounts for 50% of the total composite rating. Typically, the quality portion of the rating is based on factors such as quality costs, responsiveness to corrective action requested on quality problems, number of data delinquencies, and number and frequency of repetitive rejections. Additional factors utilized are schedule performance (late or early), number of rejected parts and/or shipments, number of shipments held up upon receipt due to missing documentation, and number of invoice reconciliation problems. These latter factors comprise the other 50% of the composite rating. The resultant ratings are quantified in dollars and reported to those with a "need to know" as the "cost of doing business" with that specific supplier. In many organizations, ratings are maintained in online computer systems that provide the buyers with real-time performance data on their suppliers. Ratings are usually published quarterly and provided to individual suppliers for review and action as necessary. Individual supplier conferences are called with suppliers who need to improve their ratings in an effort to improve the supplier base. Special recognition awards are often presented to those suppliers with outstanding ratings, and they are given special consideration in the placement of future orders.

6.1.6 Control of Nonconforming Material

Nonconforming material generated within the production facility, generated and identified at a supplier's facility, or detected during incoming inspection and test must be conspicuously identified and segregated from good production material until a disposition has been made by authorized personnel. The use of red identification tags securely fastened to the material or special banding identifying the material as nonconforming, along with appropriate identifying paperwork, are common ways of accomplishing identification. The next step is to remove the material from the normal production flow and place it in a controlled area with access restricted to authorized personnel. If the material requires special environmental storage (for example, refrigeration), special accommodations need to be made within the freezers to provide identifiable restricted storage. Similar accommodations need to be made for fluids and gases normally stored in large external tanks or drums. Material must be held in these restricted areas until its usability has been determined by authorized personnel.

Authority for disposition, or determining acceptability for use, is established with certain authorized inspection personnel for minor nonconformances. Procedural coverage, limits for approval, and specific instructions for this action must be included in the quality manual or other authorized operating procedures. Similar control procedures for the complete material review board (MRB) must also become a part of company procedures. Authority for the disposition of major defects is restricted to an established MRB consisting of a design engineer and a quality engineer. Government contracts may require that a customer or government representative also be a member of this board. It is the responsibility of the MRB to review the noted defects, evaluate the effect they might have on the ability of the product to satisfactorily fulfill the purpose for which it was designed, and mutually agree upon its acceptability "as is," prescribe rework which will make it acceptable for use, or determine that it should be scrapped. Their disposition is recorded on the rejection documentation and becomes a historical record retained as permanent record of that product. Following disposition and the completion of the required paperwork, the rejection tags are removed from the material and it is returned to the normal production flow and treated as normal acceptable material.

6.1.7 Cost Account Planning and Budgets

The estimation of costs for the performance of required tasks is required for bid preparation on all proposed work. The responsibility for the preparation of these estimates and their coordination with the project business office or others charged with preparation of the total estimate is usually assigned to quality engineering and planning. When historical-cost performance data are available on similar projects that had a comparable set of tasks, it is possible to factor these estimates up or down to arrive at the required estimate for the new work. In estimating quality tasks, including inspection and test effort, such estimates are often based on a percentage of manufacturing effort for the inspection-related tasks, and a "level-of-effort" for other quality tasks. While this approach is sometimes necessary in the interest of time, it does not provide the basis for preparation of specific work packages and cost accounts to be used later in managing and evaluating cost/schedule/budget performance relative to planned tasks. For maximum control of costs, a bottom-up estimate should be prepared starting with (1) the definition of the work to be performed, and (2) the identification of specific work packages. Individual work packages should be estimated at the level where the task is to be performed, that is, the level or classification of personnel who will be necessary to complete the task; the time period required for completion of the task; and the place in the production flow where the task will be performed, that is, the planned start and completion dates. These work packages will provide a clear picture of the job at hand and may later be integrated into cost accounts for management evaluation and budget control.

6.1.8 Quality Audits

The primary purpose of quality audits is to determine individual or organizational compliance with an established plan, procedure, process, or specification. The audits should themselves follow an audit plan, and should be conducted by professionals who are trained auditors familiar with the plan or process being audited. With few exceptions, supervisory personnel in the area being audited should be given advance notice of the impending audit in order to gain their full cooperation. Quality audits generally fall into three classifications: procedural, hardware, and system audits. Each will be discussed in the following paragraphs.

Procedural Audits

Procedural audits are reviews to determine compliance with published manuals, bulletins, work instructions, process specifications, and other contractual documents that provide regulation and direction for the manner in which work is to be performed. They may originate with the customer, the local or federal government, or internally as a matter of company policy or directives. Audits of this nature are most often conducted following a unique set of check sheets prepared in advance of the audit, which contain the elements of the procedure to be checked. These check sheets are later used to summarize audit findings.

The auditor, or audit team, should begin all audits by meeting with supervisory and management personnel responsible for the area to be audited to brief them on the purpose and scope of the audit. This meeting should be used to establish the audit ground rules and, in the case of proprietary processes, to obtain specific authorization for the audit. At the conclusion of the audit, an exit briefing should be held with the same personnel to review findings and establish definitive dates when any corrective actions required will be implemented. Depending on the nature or seriousness of certain findings, it may be advisable to plan a follow-up audit to verify implementation of required actions.

Hardware Audits

Hardware audits are designed to closely examine the product being produced and are most often conducted on the final product by an audit team. This audit team may be comprised of a design engineer, a manufacturing engineer, and a quality engineer. This audit is a more complete audit than a procedural audit in that it involves a detailed review of the production process, including handling of the product during the process flow, and may involve disassembly of a selected item for internal inspection if this can be accomplished by the producer without damaging the product. The same ground rules for entrance and exit briefings stated above for procedural audits apply to hardware audits.

System Audits

System audits are broad in scope and are conducted infrequently due to the time and cost associated with planning, coordinating, and conducting an audit of this scope. The more narrowly focused procedural and hardware audits have been found to be a more cost-effective way of monitoring performance. However, over projects that last 5 or 6 years, another type of audit, or evaluation, has found favor with many organizations as a means of self-evaluation of their "quality posture" and comparison with others in the industry. This is the activity associated with the Malcolm Baldrige National Quality Award. This award was established by Congress in 1987 in recognition of the need for improved competitiveness and improved quality across the broad range of industries in the United States. Since that time, there has been a growing interest in the competition for the National Quality Award, and a growing group of organizations utilizing the Baldrige criteria for self-evaluation and improvement regardless of their desire to enter the competition. The appeal of the criteria is their relation to "total quality" concepts. In-depth evaluations are required in seven major categories: leadership, information and analysis, strategic quality planning, human resource development and management, management of process quality, quality and operational results, and, last but not least, customer focus and satisfaction. Ground rules, information, and forms for use with the Baldrige criteria are available from the American Society for Quality Control (ASQC). Local and national Baldrige certified examiners are used in the evaluation of organizations that choose to enter the competition for the award, but any organization will find the criteria valuable for use by internal teams as a means of self improvement.

6.1.9 Continuous Improvement Programs

The aim of the continuous improvement plan should be the prevention of defects through improvement and control of processes. Once undertaken, regular monitoring of the process will inevitably result in steady improvement as a result of process adjustments, and with this improvement will come corresponding reductions in the cost of scrap and rework. It follows that since quality engineering and planning play key roles in defect prevention and data analysis, they will be instrumental in formulating and guiding the companywide effort on continuous improvement plans. The principal elements of an effective continuous improvement plan are discussed below.

Data Analysis

Plans for continuous improvement must begin with a careful analysis of historical quality records to pinpoint those processes with the greatest variation. A study of area and individual process control charts and investigations already underway will aid in selecting and prioritizing those processes with the greatest potential for gain. Special "continuous improvement charts" should be made for the processes selected, and these charts should be placed in the area that will give high visibility to all achievements.

Plan Implementation

The key element in any continuous improvement plan is the training and attention to detail of the participants. Everyone involved must have a sense of serving the customer and be ready, willing, and able to buy in. They must understand the goals to be reached, participate in their formulation, and take part in the setting of interim targets with which to measure progress. They must understand the tools of the trade—that is, the use of the statistical quality control charts that form an important part of the data gathering on the process—and they must understand how to interpret the charts relative to the need for process adjustments (see Subchapter 6.7). It has been found that reaching this point involves both formal training and on-the-job training to give the participants a thorough understanding of this phase of process improvement and their responsibility and authority in *making it happen.* It has also been found that indoctrination in the accomplishments of continuous process improvement and the actual achievement of desired results is often a slow process in itself, requiring regular coaching to preclude a slackening of effort when desired results are elusive.

Feedback Action and Recognition

Good communication is important in most endeavors, but it is of vital importance in the conduct of a continuous improvement program. Two-way communication is the mode of operation and must flow between all levels of management, customers, suppliers, and the operators on the production floor. Timely feedback to the process source, whether good or bad, is important in the evaluation of the effectiveness of past actions and the formulation of further process adjustments. Of equal importance is the coaching and recognition of accomplishments of both group and individual achievers. Attainment of targets on or ahead of schedule, a significant breakthrough, etc. should be worthy of special recognition via posters, pictures in company papers, special award ceremonies, and, as a minimum, a handshake from top management. Selection of those to be recognized has special meaning when the recommendations originate with peer groups or coworkers and awards are presented by management in the work area. Recognition of accomplishment keeps the wind in the sails of the continuous improvement program.

6.1.10 Additional Reading

Assistant Secretary of Defense, *Transition from Development to Production,* DoD 4245.7M, Department of Defense, 1985.

Crosby, P. B., *Quality Is Free,* McGraw-Hill, New York, 1979.

Crosby, P. B., *Quality without Tears,* McGraw-Hill, New York, 1984.

Crosby, P. B., *Running Things,* McGraw-Hill, New York, 1986.

Deming, E. R., *Out of Crisis,* Management Center, Cambridge, Mass., 1986.

Hall, R. W., *Attaining Manufacturing Excellence,* Dow-Jones-Irwin, 1987.

Juran, J. M. and Gryna, F. M., *Quality Planning and Analysis,* 3rd ed., McGraw-Hill, New York, 1993.

Malcolm Baldrige National Quality Award Criteria, U.S. Department of Commerce and National Institute of Standards and Technology, 1993.

Mohr-Jackson, Quality starter vs. quality advanced organizations, *Quality Management Journal,* 1994.

Shuster, *Teaming for Quality Improvement,* Prentice Hall, Englewood Cliffs, N.J., 1990.

6.2 INSPECTION

6.2.1 Introduction to Inspection

Through the 1930s, inspection was considered to be an operation that occurred at the end of the line to visually, and occasionally dimensionally, check work performed by the production operators. Since these operators worked for the production foreman, it was considered his responsibility to make sure their work was properly completed, or to have the product fixed before it left his department. This task was often time-consuming for the foreman to accomplish, so he added some inspectors to his staff to aid in the sorting. This idea of inspection continued in many companies until the late 1940s or early 1950s when the need for better control of in-process losses due to scrap and rework began to be recognized. Initially, some organizations began to add in-process inspection stations and assigned an ex-production "salvage foreman" to review findings at these stations to cause needed corrections. This action was shortly followed by the addition of trained quality engineers to replace the salvage foreman, and the establishment of an inspection department under the control of an inspection manager. This latter action was a natural outgrowth of the recognition that until that time, "the fox had been guarding the chicken house."

In today's environment, with many companies adopting some tailored version of the TQC or TQM concepts, we see a much improved approach to assignment of individual responsibility for customer satisfaction and the modification of the production flow to include and identify specific inspection points to aid in total process control. These points are addressed in the following paragraphs.

6.2.2 Inspection at Source

The decision to inspect or test purchased parts or assemblies at the supplier's facility may be based on several concerns. Among the more prevalent are the following.

1. Internal structure, wiring, components, etc., may be covered or sealed within an assembly, making inspection at a later point difficult if not impossible.
2. Specialized inspection or test equipment used by the supplier may be costly to duplicate. This is particularly true where environmental acceptance testing is required.
3. Size or sensitivity of the item being purchased may cause added risk if the item has to be returned to the supplier for any reason.
4. Transportation time to and from the supplier's facility might affect schedules adversely.

Inspection manuals or instructions should clearly prescribe the manner in which source inspection is to be conducted, and source inspection personnel should be thoroughly trained. Since source inspection may be performed by personnel resident at the supplier's facility or by transient inspectors, it is most important that ground rules for supplier interface and the completion of source inspection reports be well established. In all cases, access to the supplier's facilities for the purpose of source inspections should be a part of the buyer's initial negotiations and become a part of the resulting purchase order agreement.

6.2.3 Receiving Inspection

Incoming inspection is the norm in many industries, and is conducted upon receipt of everything from raw materials to finished subassemblies and assemblies. The extent of inspection upon receipt is usually determined by the quality history of the supplier, the complexity of the item, and the quantity or lot size received in the individual shipment. Inspection and/or acceptance testing may be conducted on 100% of the parts received, or on a random sample, and may include visual, dimensional, mechanical, chemical, electrical, and in certain situations environmental testing.

The receiving inspector should be provided with an inspection data package consisting of a copy of the purchase order, the blueprint or specification, copies of all related company specifications and procedures, and a history data card showing past receiving inspection records on the particular part and any alert notices received. In some larger companies, the inspector is also provided with a kit of required gauges or instruments needed for the inspection, to preclude the necessity of making one or more trips to the gauge crib for the gauges. As in other inspection operations, it is important that they be trained in the type of inspection to be conducted, and in the completion of all required documentation.

6.2.4 Setup Inspection

Setup inspection is a preventative action-type inspection designed to give assurance that machine surfaces are clean, holding fixtures are in good condition and properly placed, machine settings have been set or adjusted to predetermined positions, material has been properly loaded in the fixtures or will be fed into the machine at proper speeds and feeds, and any lubrication or coolant flow has been properly set. Depending on the particular assignment of responsibility in an organization, this inspection may be conducted by the line foreman, a lead man, or a floor inspector. Regardless of who performs this inspection, it is wise to have a setup check sheet prepared for each unique operation, which is used as a guide in completing the inspection and a record that it was completed. When conducted in conjunction with a "first article" or "first piece" inspection, which is usually completed by an inspector, setup inspections prove their value in precluding repetitive rework or scrap due to inadequacies or oversights in the setup procedure and establish a solid base for the process that follows.

6.2.5 Operator Self-Check

Today's production operators are being asked for their input on how to improve the quality of the products they produce and how to better satisfy their customer. And management is listening to their advice; they are rewarding the operators for their participation, giving them greater responsibility for the production of a quality part, and giving them the authority to make it happen. Along with the responsibility and authority goes additional training to give them the tools of the trade and a greater overall perspective of the big picture. In many instances, the operators complete training programs that result in their becoming "certified operators," who have authority to inspect their own work and stamp acceptance records with their "certified operator's stamp." Application of this approach is most common for in-process-type inspections and process control, with final inspection prior to shipment usually being conducted by members of the inspection department.

6.2.6 Roving Inspection

Roving inspection, otherwise referred to as patrol inspection, audit inspection, or floor inspection, is most often used where production operations are spread out over a large floor area, or where the product is large and not readily moved to a separate inspection area. It is usually conducted by experienced inspectors who have been specially trained and are familiar with the shop layout, the product, and the product flow. These inspectors perform preplanned in-process inspections and maintain surveillance over gauge, equipment, and instrument calibrations; environmental controls; and stores areas. Their records of acceptance and rejections in the area become a part of the total inspection database. Roving inspection is sometimes used to check the results of operator self-check operations on a random basis.

6.2.7 Process Inspection

Process inspection is a specialized surveillance of an identifiable element of a total production process to assure that it is being operated in accordance with specifications and variations from norm are being held to a minimum. The element might be flow soldering, chemical processing, heat treating, plating, forming, braiding, bonding, laser cutting of material patterns, and so forth. In cases where these processes are located in the same area, as in a chemical processing area, or an environmentally controlled area, a single certified operator or inspector may be assigned the task of assuring that the operation is being conducted in accordance with procedures and is maintaining control. The latter is generally monitored through the use of process quality control charts that are maintained by personnel in the area, and in some instances also posted by the floor inspector to add an independent control point to the chart. Where individual process elements are located at various parts of the operation, the use of certified operators assigned to the particular area or a roving inspector best meets the need.

6.2.8 Bench Inspection

Certain inspections are best conducted at a specific inspection station, where the product to be inspected can be spread out over a large surface or passed over or through an inspection fixture or gauge. This is also true where a product is passed down a conveyer belt and either manually or automatically inspected on a section of the conveyer or bench. At times, as with optical inspection of bearings for surface finish blemishes, it is necessary to pass the product over a special light table to readily detect defects while maintaining high productivity. Cases such as these are best suited to bench inspection.

6.2.9 Nondestructive Inspection

The inspection of materials, and parts fabricated by special processes such as forging, casting, molding, and welding, requires the use of a variety of nondestructive inspection techniques to detect and evaluate surface and subsurface defects. Most common among these techniques are radiography, ultrasonics, eddy current, magnetic particle, and several types of penetrant inspection. Description of the use and techniques utilized in these inspections is beyond the scope of this chapter, but it is important to note that the application of all of the techniques mentioned requires that personnel who have been specially trained and certified in the use of each special technique must be available to plan, establish techniques, and conduct these inspections. In determining inspection personnel needs, it is important to keep this requirement in mind. It is also necessary to consider the availability of the special equipment required to conduct these inspections. If equipment and certified personnel are not available, there are commercially available companies that specialize in nondestructive inspections that may be called upon to meet this need.

6.2.10 Automated Inspection

Increasing use is being made of automated inspection to reduce hands-on labor costs, reduce cycle time, assure uniform inspection results (eliminate personal judgment in evaluation), and reduce monotony, stress, or eye strain. Automated inspection should be considered when you are faced with inspection of repetitive characteristics on a large volume of parts. Automated inspection and testing is also of particular value in the process control of electronic equipment, where designs accommodate ready access to test points. However, careful consideration must be given to the advantages to be realized before investing in special-purpose automated equipment. Some flexibility may be realized through the application of the latest developments in microcomputers, artificial intelligence (AI), and computer-aided manufacturing and inspection setups, as well as through the use of robotics that may be fitted with universal probes.

6.2.11 Software Inspection

The identification, proofing, and control of computer software in manufacturing, inspection, and testing is of paramount importance, since the operational performance will be no better than the software that controls it. Software code must be carefully checked and verified through proofing trials. Configuration must be maintained from the outset and all design changes must be reflected in appropriate configuration updates. Software libraries must be maintained in secure areas, and only authorized personnel should be permitted to remove software from the library for installation in production or inspection equipment.

6.2.12 Sampling Inspection

Experience has proven that 100% inspection, or even 200% or 300% inspection, will not yield risk-free assurance that defects present within a large lot of material will be detected. The principal reason for this is that monotony and physical fatigue set in and affect the effectiveness of personnel performing the inspection. The use of automated inspection and computer-assisted inspection equipment has helped materially to reduce this problem in instances where cost trade studies have justified the procurement. In cases where this is not true, and in cases where inspection or testing may be destructive, statistical sampling provides a more economical approach, and may even prove to be more accurate than 100% inspection.

There are several approaches that may be used in sampling, and many variations to each. There are *single sampling* plans, where the decision to accept or reject a lot of material is based on the evidence obtained from one sample; there are *double sampling* plans, where it is possible to put off the decision until a second sample has been evaluated; and there are *multiple* or *sequential* sampling plans that defer the decision to accept or reject until several samples have been taken from the lot. There are plans for lot-by-lot inspection by attributes and variables. When an attribute plan is used, the decision to accept or reject the lot is based on a "go" or "no go" basis—that is, whether the number of defective parts in the sample exceeds the acceptance number for the plan. With variable plans, acceptance is based on the acceptability of individual characteristics on the sample parts, or on selected specific characteristics. This type of inspection is more expensive than attribute inspection, but in certain instances it is desirable to gain more information on a particular characteristic for guidance in product improvement, and variable-type plans meet this need. There are also plans for continuous sampling for use with large production runs.

The majority of sampling plans in use today are based on desired achievement of a predetermined acceptable quality level (AQL). The AQL is the maximum percent defective that, for the purpose of sampling inspection, can be considered satisfactory as a process average. It is the designated value of percent defective that the consumer has indicated will be accepted most of the time by his or her incoming inspection plans. In other words, an AQL plan favors the producer, as it gives assurance of probable acceptance by the consumer. The AQL value may be specified in contracts, or may be established internally by quality engineering and planning for guidance in the application of sampling plans.

Another quality value, referred to as the average outgoing quality limit (AOQL), covers plans that may be utilized in conjunction with sampling when rejected lots of product can be 100% inspected and have rejects replaced with good product. AOQL plans assume that the average quality level of multiple lots will not exceed the AOQL because all rejected product detected during sampling and the 100% screen have been replaced with good product.

Sampling plans prepared by Harold Dodge and Harry Romig utilize still another classification index, the lot tolerance percent defective (LTPD). The LTPD represents an allowable percent defective which favors the consumer due to a decrease in the risk of accepting a lot at or below the lower quality limits.

Sampling tables are based on operating characteristic curves. The tables specify the size of the sample to be randomly selected based on the lot size, the inspection level desired, and the selected AQL. When these parameters have been selected, reference to the table will establish the acceptance and rejections numbers governing the acceptability of the sample, and thus the lot, for various sample sizes, lot sizes, and given AQLs.

Due to the number of sampling plans available, it is not possible to adequately address selection of plans in this handbook. However, many texts have been published on this subject that include descriptions of the various plans available. A partial list of texts and military specifications for sampling is included at the end of this subchapter under "Additional Reading."

6.2.13 Final Acceptance Inspection

This inspection is the last visual, dimensional, and operational evaluation of the product before it is delivered to the purchaser or ultimate customer. At final inspection, the routing traveler or other documentation should be reviewed to assure that all previous operations have been completed and accepted by authorized personnel. The inspection is then completed in accordance with final inspection check sheets and instructions prepared by inspection supervision or quality planning The inspection kit and data package used by the inspector should include the top-level drawing and any lower-level drawings necessary for the final inspection, as well as copies of all related specifications and process specifications. It should also include any inspection aids and necessary charts to be used in the inspection. If a log book, instruction book, manual, or other documentation is to be delivered with the product, these items should be available for acceptance at this time.

Final inspection also serves as a final process control point, since it provides a positive evaluation of the effectiveness of process controls built into the production process flow. Feedback to preceding control points provides a closed-loop corrective action system for continuous process improvement.

6.2.14 Shipping Inspection

There are many horror stories told by customers who waited patiently for the product needed to keep the line moving, only to open the carton upon receipt and find the product inside damaged beyond use. The receipt of a damaged product has never been known to result in customer satisfaction with the quality of the product purchased. Rather, the memory of the damaged product lingers long after it has been replaced and due apologies sent. The message is clear: careful attention must be given to the design of packages or containers to be used for shipping, and these designs must be submitted to testing in all environments that the package may reasonably be expected to encounter in service. In addition, packing and handling instructions must be written, and personnel must be trained in proper packing procedures to assure the product is not damaged during the packing operation or in transit. Specific tests plans must be prepared and both the product and the empty container inspected for conformance to design requirements before the testing is conducted. Inspection of both items is again necessary following the tests to assure that the container adequately protects the product and is capable of reuse if it has been designed as a reusable container.

6.2.15 Data and Records

One of the most important functions of the inspection department is the generation of acceptance and rejection records, which become the basis for future action throughout the company.

Standard forms, usually in multiple copies, are used for this purpose, and inspection personnel are trained to describe clearly any nonconformances found in order to facilitate further disposition of the material. Inspection is also responsible for the collection and review of all production work orders to assure that all planned work has been properly completed and accepted prior to final inspection. These documents are combined with all other certifications, log records, and check sheets and forwarded to a central data center. A more detailed discussion of data and records collection and management will be found in Subchapter 6.6.

6.2.16 Training and Certification

With the increased use of computerized inspection equipment capable of extreme precision in multiple planes and computer-assisted gauging, specific training of inspection personnel becomes a necessity. The more complicated coordinate measuring equipment, and special-purpose electronic harness and board inspection equipment, often requires training at the equipment manufacturer's facility, or special in-house training programs. Once trained, it is important that personnel regularly use the equipment on which they were trained, and have periodic refresher training to practice techniques that are not utilized frequently. Similar training needs exist for inspection personnel responsible for environmental control of fluids, atmospheres, furnace and refrigeration equipment, and processing equipment. In the case of certain nondestructive inspection solutions, regular inspections must be conducted for contamination, and tank additions must be properly controlled. Personnel who are required to establish operating techniques, perform setups, and operate radiographic or ultrasonic inspection equipment and other similar equipment require special training for certification and must have qualifying experience for many applications.

Personnel handling and inspecting electronic components must receive specific training to avoid damage to the hardware as a result of static electricity. Proper handling procedures, proper clothing, and special workstations that include provision for personal grounding are required, and personnel must be trained and understand the importance of their use.

Personnel working in areas that must be maintained at specific cleanliness levels, as in space systems and electronics clean rooms, must also be trained and provided special clothing for work in these controlled areas. Training, monitoring, and control of these special areas is critical to the production of reliable, quality hardware.

6.2.17 Additional Reading

Bowker, A. H. and Goode, H. P., *Sampling Inspection by Variables,* McGraw-Hill, New York, 1952.

Dodge, H. F., Sampling plans for continuous production, *Industrial Quality Control Magazine,* November 1947.

Dodge, H. F. and Romig, H. G., *Sampling Inspection Tables: Single and Double Sampling,* John Wiley, New York, 1944.

Foster, R. A., *Introduction to Software Quality Assurance,* 3rd ed., R. A. Foster, 1975.

Juran, J. M., *Quality Control Handbook,* McGraw-Hill, New York, 1951.

Shewhart, W. A., *Economic Control of Quality of Manufactured Product,* D. Van Nostrand Co., Princeton, N.J., 1931.

6.3 TESTING

6.3.1 Introduction to Testing

Testing is that part of the inspection acceptance task that determines the functional acceptability of the product. It may be accomplished on-site at a supplier's facility if the product is large and difficult to move or if test equipment is unique, expensive, or otherwise unavailable. Product testing covers the broad spectrum of items from minute electronic components to major space-age systems, from the smallest mechanical fasteners to major structures, and from the simplest fluids and fabrics to raw materials rolled, cast, forged, extruded, and formed, both before and after special processing. Today, most testing is performed in-house by the purchaser or by a special laboratory, contractor, or agency under contract to the purchaser. The latter is particularly true when specialized equipment and expertise are required, as in environmental and vibration testing and product safety testing. There are a number of companies today that specialize in this type of testing. Whatever the scope of testing, a specific test plan should be prepared and carefully followed to assure the quality of the product to be delivered.

6.3.2 Test Planning

Test planning starts with the development of a written test plan prepared by a test engineer. It is the responsibility of the test engineer to develop the concepts and procedures to be used for the test, based on the specific product specifications that must be tested for conformance to requirements. Plans may be prepared for development testing, for simple in-process testing prior to close-up of a component, or for complex testing of a major assembly.

In preparing development test plans, the test engineer may consider the use of fractional factorial concepts in the design of the test in order to obtain the greatest amount of information in the shortest period of time. The fractional factorial approach provides for testing individual characteristics in combination with one or more other characteristics during the same test. Statistical analysis of test data thus obtained not only provides valuable information to the design engineer, but provides the test engineer with additional information that he or she may use in the cost-effective design of production tests.

Plans for testing production items should specify the following elements:

1. Test location
2. Environments to be used in test
3. Specific equipment to be used
4. Test sequence and procedures to be followed
5. Check sheets for recording test results
6. Pass/fail criteria for each step of test
7. Summary record of complete test results
8. Signature or identification stamp of tester and date

It is important that the product to be tested undergo an inspection prior to the start of the test to establish and record both its configuration and conformance to specifications. This is especially true during development of the product, when configuration revisions may be the order of the day, but holds true for all production testing. Following completion of the tests, it is also important that the product be inspected to assure that defects have not been induced during the testing. This inspection may be a simple visual inspection, or may require a more extensive check, but it is important to assure the quality being passed on to the next "customer."

6.3.3 Test Equipment

Test equipment must be included in the metrology calibration and control system of the production facility and be subject to established calibration cycles. All gauges, instruments, meters, and recording devices must be calibrated against working standards traceable to the National Bureau of Standards' masters. They must be identified with current calibration labels, and conformance to the required calibration cycles must be confirmed prior to the start of the test. In addition, all control cabinets for test equipment should be sealed with tamper-proof seals following calibration, to preserve the integrity of the equipment.

Design of test equipment and test-equipment fixtures must be carefully reviewed to assure that the product will not be unintentionally stressed by forces induced through the moment arms of the fixture, or by clamps used to secure the product to the test fixture. This is particularly important in vibration testing, where improper fixture design may subject the product to destructive amplitudes.

6.3.4 Test Software Control

Test software must be inspected and controlled in the same manner as other production or deliverable operational software. All software media, such as punched cards, magnetic tape, punched paper or Mylar tape, cassettes, disks, and computer programs must be under the control of a software library accessible only to authorized personnel.

Test software must be designed and prepared by software test engineers and subjected to design review by software technical specialists to identify risk areas. This design review will assure that the software will test all significant parameters, and establish traceability to specification requirements. Following design review, all software must be subjected to tests witnessed by a quality assurance software specialist. Any anomalies or discrepancies noted must be documented and the documentation held open until the anomaly or discrepancy has been investigated and any needed corrections made. Following acceptance, the software may be released and placed under configuration control. It must be prominently identified with the current change letter and placed in the software library with other working and master media. The software now becomes subject to change control and must be submitted to a software change control board for technical review of any desired change. If approved by the change board, the software media may be changed and subjected to the same acceptance and validation process as the original software for approval.

6.3.5 Conduct of Tests

All testing should be accomplished in accordance with the written and approved test plan and in the sequence dictated by the work order. (Some contracts may require customer review and approval of the test plan.) Development testing may be conducted by authorized test engineers or development technicians with monitoring by roving inspection personnel. Personnel assigned to conduct production acceptance tests should be trained production operators or inspectors, rather than test engineers, and acceptance tests should be witnessed by inspection when not specifically conducted by inspection. In production facilities where training and certification of production operators has been incorporated, many tests are conducted by certified operators without benefit of inspection witness. However, final acceptance tests are normally performed or witnessed by inspection. On some government and commercial contracts, the customer may also request to be present to witness acceptance tests. Records of all acceptance tests become a part of the product record and are retained in the central record center.

6.3.6 Additional Reading

Juran, J. M. and Gryna, F. M., *Quality Planning and Analysis,* 3rd ed., McGraw-Hill, New York, 1993.

6.4 RELIABILITY

6.4.1 Introduction to Reliability

The standard definition for reliability established years ago by those in the electronics industry states: "Reliability is the *probability* of a device performing its purpose *adequately* for the *period of time intended* under the *operating conditions encountered.*"

This definition has stood the test of time, especially when considering design reliability, but it is currently being reexamined in the eyes of "the customer," who often equates reliability with dependability, and in the eyes of those producers who are alert to consumer demands. If this definition were to be written by the customer, it would read: "Reliability is the *assurance* that a device will *satisfy my expectations* for *as long as I expect to keep the product* and *under the operating conditions that I may need to rely upon it.*" In other words, "I can depend on it!"

It is generally recognized that inherent product reliability is dependent on the excellence and maturity of the design process, and that reliability cannot be *inspected* into the product. At the same time, the customer is showing an increased awareness of the importance that manufacturing workmanship plays in the quality and reliability of the end product. The competition in the automotive field between American and foreign-made automobiles and between manufacturers of electronic components here and offshore provides positive evidence of this trend.

6.4.2 Reliability in Manufacturing

The key to assuring that designed-in reliability is not degraded during manufacturing is indoctrination and training. Everyone who is associated with the manufacturing process must be indoctrinated with the concept of "first-time quality" and *accept* this concept as a personal commitment. Achievement of this goal at each step must be the result of training, coaching, and personal attention to detail in all aspects of planning the work flow, tooling, procedures, and work instructions that will be used in production, with special attention paid to past lessons learned. This must be followed by the establishment of a positive system of manufacturing process control. It is here that the greatest assurance against any degradation of the designed-in reliability will be gained. The use of statistical process control charts to monitor critical processes by either "items produced" or "time samples" provides the most efficient means of accomplishing this task. Additional details on the training and use of statistical control charts can be found in Subchapter 6.7.

6.4.3 Reliability Testing

One of the lessons learned from the past is that bench testing and even 100% inspection leave an element of risk that defects undetected in the design, and latent defects introduced during production, will find their way to the field. Reliability testing of the final product was introduced to aid in minimizing this risk. The precise form of reliability testing chosen will depend on the product. Most plans include subjecting the product to the extremes of the environmental elements expected to be encountered during use for a period of time that statistical evaluation has proven to be effective as a screening measure. It is important that products be operationally tested both before and after exposure to these environments in order to assess the effectiveness of the testing as a reliability enhancement tool. The complexity of electronics in the manned space program, and the need for maximum assurance of their reliability for the mission, focused attention on the need for reliability testing. It was on this program that the value of temperature cycling and random vibration were proven to be cost-effective means of reducing the risk of latent failures occurring in the field. Reliability failure investigations established that foremost among the causes for these failures were all types of poor solder joints, shorts and opens in circuitry, damaged insulation, inadequate mounting of components, and damage induced in multilayer boards and harnesses due to installation procedures or mishandling.[1]

Temperature cycling of completed circuit boards and/or black boxes has been found to provide an effective method of exposing latent defects or weaknesses in electronic assemblies. The author's experience has shown that three to ten thermal cycles provide the best screen, with the number of cycles depending on the complexity of the unit under test. Over 70% of latent defects are normally detected during the first three cycles when hardware is subjected to cycling between −40°C and +75°C with a temperature rate of change of 10°C per minute. Soaking at temperature extremes is required only to permit temperature stabilization of internal parts within ±5°C of the extreme. The use of "power on" during the thermal rise provides some additional screening power, but the added cost of test setup makes this addition of marginal value. Likewise, monitoring during the test should be used only if additional test data are needed for evaluation. For some complex equipment, operational testing at temperature extremes (as a minimum, on–off cycles) may be considered, but use of these added tests needs to be evaluated against the complexity of the equipment and its intended use. Operational tests at ambient temperature should always follow thermal cycling, with results monitored and investigated by reliability engineering to assure that needed corrections are implemented as part of the continuous improvement program.

Subjecting completed units or assemblies to random vibration has also been found to be an effective way of reducing the risk of delivering hardware with latent defects. Vibration testing may be accomplished at ambient temperatures, or in combination with temperature cycling if equipment is available and ultimate use indicates the advisability of such testing. However, under most circumstances, testing at ambient temperatures provides an adequate screen. Random vibration has been proven to be more effective than sinusoidal excitation (±2 *g* at fixed, nonresonant frequencies between 20 and 60 Hz), which was used for years for vibration testing.

Vibration equipment and fixtures used for mounting hardware to the vibration table must be carefully designed to assure that moments will not be induced that will overstress the item under test. In addition, care must be exercised to assure that hardware is securely mounted to the shaker table and that the axis of vibration is perpendicular to the circuit boards in the unit. When this has been accomplished, the unit should be subjected to the power spectral density (PSD) characteristics shown in Figure 6.3 for 10 min. If the circuit boards or major components of the item are located in more than one plane, the duration of vibration should be 5 min. in each

1. A visit to the IBM Laptop computer plant in Austin, Texas, revealed that each completed computer must pass a continuous power-on, 24-hr. functional test. This operation is fully automated, including the insertion of the test floppy disk and a printout at the end showing that the computer is accepted—or, if there is an anomaly, which test sequence or component failed, and what specific rework steps are to be taken.

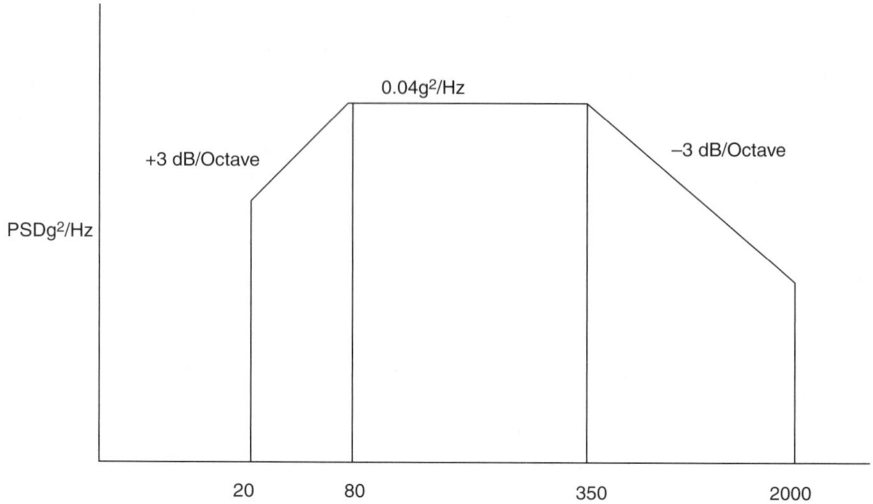

FIGURE 6.3 Typical power spectral density characteristics versus frequency used for vibration testing of electronics assemblies.

plane. No significant difference in effectiveness has been detected whether temperature cycling is performed before or after the vibration, but as in temperature cycling, a complete operational test should be performed after completion of the environmental exposure to assure the effectiveness of the test and the delivery of reliable hardware.[2]

6.4.4 Additional Reading

Calabro, S. R., *Reliability Principles and Practices,* McGraw-Hill, New York, 1962.

6.5 COST OF QUALITY

6.5.1 Introduction to Cost of Quality

For years, the cost of quality was equated to the labor cost of inspectors on the payroll. Then, some enlightened pioneers began to suggest that the cost of repairing or reworking product that was improperly made should also be considered when

2. A survey of a factory in England revealed normal (low-cost) precautions during fabrication and assembly of electrical and mechanical components. This was followed by a very sophisticated environmental test, including high- and low-temperature cycling and vibration in a "white" spectrum for 24 hrs. This was while undergoing an electronic functional test that reproduced the takeoff, maneuvering, and landing of a French fighter aircraft—the final customer. Although good products were delivered, a better manufacturing process control would have been more cost-effective.

assessing the cost of quality. Today, we find a much broader focus on what *really* constitutes the cost of quality, and an across-the-board commitment to do something about it. This change of attitude was brought about by many things, but one of the more important was an appreciation of just how much, in dollars and cents, the cost of quality amounted to, and its relation to other business expenses. Industry surveys made in recent years have established that quality costs, as a percentage of sales, range from a low of 1% for small, uncomplicated products to a high of 25% for complex, man-rated space hardware. Typically, products produced using basic mechanical processes run in the range of 3–5%, with costs for precision industries running in the range of 10–12%. It is recognition of these facts that has resulted in recent attention to the collection and monitoring of costs and the establishment of specific programs and goals for the reduction of quality costs.

6.5.2 Elements of Quality Costs

The approach most commonly used to identify quality costs for collection, evaluation, and reduction allocates costs into four major cost categories: (1) prevention, (2) appraisal, (3) internal failure, and (4) external failure. It is emphasized that these costs exist in all areas of the company and must be collected, monitored, and controlled on that basis. Cost collections should be the responsibility of the accounting department to assure proper allocation and reporting.

Cost categories, and typical activities that should be considered in them, are listed here for guidance. They may be tailored to suit your unique organizational structure:

Prevention

Prevention costs are those of all activities specifically designed to prevent poor quality in products and services. Some examples are the costs of new product review, design review, quality planning, review of specifications and specific work instructions, supplier capability surveys, supplier performance measurement, process capability evaluations, equipment calibration, tool control, preventive maintenance, environmental controls, audits, education, and training.

Appraisal

Appraisal costs are those associated with measuring or evaluating products or services to assure conformance to quality standards or performance requirements. Costs in these categories might include receiving and/or source inspection and test of purchased equipment and parts, environmental testing, supplier surveillance, process controls, roving inspection, bench inspection, final inspection, test inspection, and shipping inspection.

Internal Failure

Failure costs are those resulting from products or services not conforming to requirements or customer needs; these are divided into internal and external failure costs.

Internal failure costs are those occurring prior to delivery or shipment of the product, or the furnishing of a service to the customer. Examples are the costs of material and labor expended on material that is scrapped; rework, reinspection, retesting, and dispositioning of discrepant material for possible use; engineering changes or redesign; purchase order changes; and corrective action activities.

External Failure

External failure costs are those occurring after delivery or shipment of the product, and during or after furnishing a service to the customer. Examples of these costs are the costs of processing customer complaints, customer returns, warranty claims, and product recalls.

6.5.3 Quality Cost Metrics

A key element in the process of measuring, documenting, and improving processes is the selection of parameters to be measured, the baseline from which to measure progress, and the metrics of measurement. Of equal importance is the establishment of these parameters, baselines, and metrics in conjunction with those responsible for the element being measured. In order for metrics to be an effective tool for control of costs and process improvement, those being evaluated must have input in establishing the measurement criteria and agree that the selection of parameters is representative and fair.

Metrics are the measurable characteristics of products or processes used to assess performance and track progress in attaining improvement. They facilitate measurement of *all* processes, both product *and* administrative. For proper control of the cost of quality, both should be utilized. When this is done, the relative importance of costs in various areas is highlighted, and provides the information necessary for sharpening focus on areas needing improvement.

The first step in establishing a metrics chart or graph for any area is the selection of the parameter to be measured and establishment of the baseline from which to measure progress. This is usually done based on past performance records, even though these records may not have been used for this purpose in the past. This is followed by selection of the measurement technique to be used and the frequency of measurement. Now comes the most difficult task: selection of an improvement goal that will be both realistic and attainable while still requiring the "process owners" to *stretch* to attain the goal. The tendency in human nature is to set the goal *low* in order to assure attainment and look good on the next performance review.

A good deal of coaching is required here to assure that the improvement goal is, in fact, a goal that will require considerable *stretching,* and yet be a goal *that the process owners are willing to accept!* An understanding of *why* the goal is important to them personally, as well as why it is important to the company, is critical to goal attainment and the success of any improvement program, especially to a goal that requires a good deal of effort on the part of the process owners.

Once agreement has been reached on the metrics to be used, a decision on the manner and frequency of distribution of the information must be made. The frequency should be tailored to the format, content, and type of presentation of data: that is, detailed process charts and graphs, area charts, departmental summary charts, and top-level management reports. The important thing is that the charts, graphs, or reports must be very visible to the process owners, i.e., posted prominently in their work area, and equally visible to upper-level management. Management review and recognition of progress is vital to the continuation of process owner interest.

The selection of the metrics to be used in various areas must be based on the unique size and organizational structure of the company, but it is important that all areas be included if a true picture of the contribution of quality costs to the bottom line is to be obtained.

6.5.4 Additional Reading

Crosby, P. B., *Quality Is Free,* McGraw-Hill, New York, 1979.

Juran, J. M. and Gryna, F. M., *Quality Planning and Analysis,* McGraw-Hill, New York, 1970.

Juran, J. M. and Gryna, F. M., *Quality Control Handbook,* 4th ed., McGraw-Hill, New York, 1988.

6.6 DATA MANAGEMENT

6.6.1 Introduction to Data Management

One of the most important functions in any organization is the maintenance of historical performance records that become the basis for future action. Company procedures should provide specific direction as to who is responsible and who has the authority to generate these records, what methods and forms are to be used in the collection process, why the records are being generated, who will use them and for what purpose, where they are to be maintained, and for how long they are to be retained.

6.6.2 Types of Data in the System

There are many types of quality data generated during most development and production programs to aid in both day-to-day and trend evaluation of product quality and progress in quality improvement. The specific data maintained will vary by industry, but some of the more common types are summarized below.

Acceptance and rejection records. Records of the number of items in a production lot as well as the number accepted and the number rejected are maintained for every inspection performed, whether at the source, at receiving, in process, at final, at test, or at shipping. If the use of sampling procedures is authorized and utilized in the inspection, the sample size selected is also recorded. These records provide the basis for performance evaluation and process control of all product, whether purchased or produced within the facility, and are key inputs to the computation of the cost of quality.

Research and development (R&D) test results. These records should include identification of the exact product configuration being tested, as well as inspection data taken both before and after tests, in order to provide adequate information for test evaluation.

Data on supplier tests and inspections. Some purchase orders may require the supplier to maintain all data and provide it upon request, while others may require certain data to be provided as a data item at time of delivery to the purchaser. If the tests or inspections are witnessed by representatives of the purchaser's organization, all records should be so noted.

Environmental test data. Environmental tests may be conducted at the component, subassembly, or assembly level. Data collected is essentially the same, regardless of level, and should include identification of the product being tested, the environment utilized, the specific test equipment used, calibration date of equipment, date of the test, and the identification of the person performing the test. Date and test personnel identification are common requirements on all inspections and tests.

Incoming (receiving) inspection data. In addition to the normal visual and dimensional inspection results, incoming data should include material certifications from the supplier, laboratory test results, review of supplier data requirements specified in the purchase order, and the results of any functional or nondestructive tests and inspections performed.

Process inspection data. Included in this category are records associated with processes such as heat treating, plating, other surface treatments, and all other special processes used in the production operation. Records of equipment uniformity surveys, temperature cycles, periodic solution checks, and safety checks should form a part of these data files.

Process control data. These files should include statistical quality control charts used for process control, all automated data collection charts used for temperature and humidity control, contamination control records on both air and fluids, and charts or automated data printouts from computer-controlled equipment.

Final inspection and test data. In addition to acceptance inspection results, this data will include log books or other deliverable data along with a complete accounting of the configuration of the product at time of shipment.

Data on scrap and rework. Special reports on investigations and corrective measures taken as part of continuous improvement efforts, both in-house and at supplier facilities, will be required.

Data on product returns, recalls, and warranty replacements.

6.6.3 Data Collection

Procedures and forms to be used for the collection of inspection and test data should be specified in company manuals. For years, information has been recorded on multicopy forms that are forwarded to an information center for distribution to other organizations within the company that must take action. Typically, these forms are distributed to purchasing, production, production control, quality engineering and planning, design engineering, production engineering, reliability, accounting, and any other department affected by the acceptance or rejection of material. Following individual department action, the documents are updated to reflect status and are forwarded to a central control area, usually in quality control, for final action and/or filing. This and similar collection methods are still in use in many industries, but the availability of portable computers, workstation data entry terminals, and bar-code readers, and the marked reduction in the cost of this equipment, is bringing real-time data collection and retrieval to the forefront.

Today, many companies are transmitting test data directly from automated test equipment to mainframe or desktop computers, which are networked to provide real-time information on quality status to all personnel with a need to know. Portable handheld computers and digital measuring devices enable certified operators and mobile inspectors to enter data as they are processing or inspecting material. This permits timely investigation of any problems and expedites incorporation of solutions. In addition, computerized data entry prompts individuals to enter needed information and will not let them unconsciously omit data entry. This has been a major problem with paper systems: key information is omitted and often permanently lost because it was not entered at the time of rejection, even when space was provided on the form for the data. Computerized data entry has the added advantage that data is entered into the systems only once (there is no transcription of data from paper to computer) and may be analyzed and summarized in reports in a cost-effective and timely manner.

6.6.4 Operations Data Center

"Central files" is another name frequently associated with the area within a company where all data comes together and is cataloged and filed for either current or future reference. These files often become large, requiring a staff of trained personnel to properly maintain them and adequately service company reference needs. In some instances, the data being retained or used in specific processes may be company-proprietary, as in the case of competition-sensitive processes or formulas; or the data may be classified, as is the case of certain work for the government. Special procedures for handling such records must be maintained, and limited-access storage facilities are required to protect this data adequately. Likewise, access to computerized data files must be protected by an authorizing password system to limit access. Computer disks and tapes require regular backup and limited-access storage in an area free from magnetic interference.

6.6.5 Records Retention

Many contracts contain clauses specifying the period of time that inspection records must be retained, either in active files or in permanent files. Since record storage is costly, procedures governing records retention must be carefully prepared. Records should not normally be maintained in active files any longer than necessary to permit adequate trend analysis, and to provide information in response to field problems and prompt servicing of customer complaints. This will vary by industry, but 2 years is probably a good norm. Permanent files will vary depending on the product and company desires to maintain historical files on certain items. However, records should not be maintained for the sake of records, and in most cases should be disposed of within 7 years.

6.7 STATISTICAL PROCESS CONTROL

6.7.1 Introduction to Statistical Process Control

Statistical process control is, as the name implies, the control of a specific process through the use of numbers. In this instance, *control* means to operate within prescribed boundaries with a minimum of variation from an established goal. The *process* selected may be as simple as a single element in a production flow—like applying a label to a product—or it may be a complex combination of operations consisting of many human/machine interfaces. In any case, it is necessary to gather numerical data on the process fluctuations, establish operating boundaries based on statistical evaluation of the data, and evaluate the results to determine the process adjustments necessary to minimize the fluctuations and maintain control within the boundaries.

Successful process control requires evaluation of data gathered in an effort to understand the causes of variation. Experience has shown that process variations are either caused by system management action (or lack of it) or by factors within the control of the operator. Typical causes of process variation are machine wear, inadequate maintenance, fixtures that are not foolproof, power surges, variations in material, lack of adequate training, and inattention by the operator. Historically, the ratio of causal factors has been found to be about 80% within management's ability to correct and 20% within the operator's ability to correct. Process improvement, then, is dependent on identification and removal of those detrimental factors that may be corrected by management before the process starts, and continuation of the process monitoring/correction cycle while the process is running, through use of process control charts.

6.7.2 Selection of Charts for Process Control

Two types of charts are commonly used for process control in production operations: charts for measuring variable data, and charts for measuring attribute data. Each type serves a unique purpose in process control. Their applications are described below.

Variable Charts

Variable charts are commonly referred to as X-bar and R charts, and are particularly useful for several reasons:

1. Most processes have characteristics that are capable of being recorded numerically.

2. A measured value, such as “2 pounds, 7 ounces” or “7 mm,” has more value than a simple evaluation of compliance.
3. Fewer pieces need to be checked when exact measurements are taken in order to determine variation.
4. Since decisions can be reached sooner when exact data are available, it is possible to get near-real-time feedback to the operator for needed correction—a key point in effective process control.
5. Variable charts can explain process data in terms of two important categories: short-term variation (that is, piece to piece) and typical performance (at a point in time). These two characteristics provide for very sensitive process control tracking.
6. Because smaller samples need to be taken than with other systems, flow time can be improved and costs lowered even though additional time is needed to take exact measurements.

X-Bar Charts

X-bar charts plot the averages of each subgroup (collection of a given number of consecutive pieces) over a specific period of time. This produces an estimate of the process average over time. It also provides a picture of long-term variability that could include tool wear, machine drift, or adjustments.

The *R* Chart

The *R* chart is used in conjunction with the *X*-bar chart and plots the ranges (smallest value to largest value) of each subgroup of measurements taken over a period of time. This measures and quantifies an estimate of short-term variability. Short-term variability is defined as the inherent process variability within each subgroup that is independent of tool wear, machine drift, or adjustments.

The best applications for *X*-bar and *R* charts are:

First-run jobs, when specific data will aid in design tolerance evaluation
When specific data is needed to determine the cause of variation in a troublesome operation
When obtaining data is expensive or destructive
When operations are consistently running out of control and causes seem varied

To establish *X*-bar and *R* charts:

Calculate the average and range of each subgroup of 4 or 5 items Plot *X*-bar and *R* where:

X = average measurement of each subgroup
R = range of measurements in each subgroup
n = number of measurements in each subgroup

Calculate the control limits using the following equations:

$$\text{UCL for } X = \overline{X} + A_2\overline{R} \qquad \text{LCL for } X = \overline{X} - A_2\overline{R}$$

$$\text{UCL for } R = D_4\overline{R} \qquad \text{LCL for } R = D_3\overline{R}$$

where

UCL = upper control limit
LCL = lower control limit
$\overline{X}$ = process average
$\overline{R}$ = range average for k subgroups

The factors to use in the above equations are given below:

n	4	5	6	7	8
D_4	2.28	2.11	2.04	1.92	1.86
D_3	0	0	0	0.076	0.136
A_2	0.73	0.58	0.48	0.42	0.37

Attribute Charts

Charts based on attribute data are often referred to as "go–no-go" charts because they accept or reject an entire item on the basis of measurement of a single characteristic. These charts are less complicated to use than variable charts and therefore find frequent use in process control during production. They are particularly useful when it is desired to obtain a picture of operating trends without the cost of variable data, or when variations are known to be operator controllable.

The three most common attribute charts are the p chart, which plots percent defective; the c chart, which plots number of defects; and the u chart, which plots defects per unit. The p chart plots p as the number of defectives in each subgroup divided by the number inspected in each subgroup (n). UCLs and LCLs are calculated with the following equations:

$$UCL_p = \overline{p} + \frac{3\sqrt{\overline{p}(1-\overline{p})}}{\sqrt{n}}$$

Index

D

E

F

J

K

L

N

T

U